D1544029

Interpreting Construction Contracts

Other Titles of Interest

Alternative Project Delivery, Procurement, and Contracting Methods for Highways, **edited by Keith R. Molenaar and Gerald Yakowenko** (ASCE Committee Report, 2007). Comprehensive and objective presentation of alternative methods for highway construction issues. (ISBN 978-0-7844-0886-5)

Construction Contract Claims, Changes & Dispute Resolutions, Second Edition, **by Paul Levin** (ASCE Press, 1998). Merges the principles of construction law with practical advice to aid those involved in the construction claims process. (ISBN 978-0-7844-0276-4)

Geotechnical Baseline Reports for Construction: Suggested Guidelines, **edited by Randall J. Essex** (ASCE Committee Report, 2007). Explains the use of geotechnical baseline reports to manage risks associated with subsurface construction and emphasizes compatibility among the report and other contract documents. (ISBN 978-0-7844-0930-5)

Managing Innovation in Construction, **edited by Martyn Jones and Mohammed Saad** (Thomas Telford, Ltd., 2004). Applies innovative best-practices to the adversarial relationships and fragmented processes that characterize much of the construction industry. (ISBN 978-0-7277-3002-2)

Preparing for Design-Build Projects: A Primer for Owners, Engineers, and Contractors, **by Douglas D. Gransberg, James E. Koch, and Keith R. Molenaar** (ASCE Press, 2006). A professional reference that covers the basics of developing a design-build project. (ISBN 978-0-7844-0828-5)

Principles of Applied Civil Engineering Design, **by Ying-Kit Choi** (ASCE Press, 2004). Details the guidelines, principles, and philosophy needed to produce design documents for heavy civil engineering projects. (ISBN 978-0-7844-0712-7)

Quality in the Construction Project, Second Edition, **by the Committee on Professional Practice** (ASCE Manuals and Reports on Engineering Practice, 2000). Comprehensive guide to principles and procedures that effectively enhance the quality of constructed projects. (ISBN 978-0-7844-9596-2)

Surety Bonds for Construction Contracts, **by Jeffrey S. Russell** (ASCE Press, 2000). Describes and analyzes the surety bond process in basic terms. (ISBN 978-0-7844-0426-3)

Ten Commandments of Better Contracting, **by Francis T. Hartman** (ASCE Press, 2003). Outlines ten basic rules for improving contract performance. (ISBN 978-0-7844-0653-3)

The Engineers Joint Contract Documents Committee (EJCDC) Contract Documents. Fair and objective standard documents that represent the latest and best thinking in contractual relations between all parties involved in engineering design and construction projects. For more information, please visit http://pubs.asce.org/contracts.

Visit us on the Web at http://pubs.asce.org

Interpreting Construction Contracts

Fundamental Principles for Contractors, Project Managers, and Contract Administrators

H. RANDOLPH THOMAS, PH.D, P.E.
RALPH D. ELLIS JR., PH.D., P.E.

Library of Congress Cataloging-in-Publication Data

Thomas, H. Randolph, 1945-
   Interpreting construction contracts : fundamental principles for contractors, project managers, and contract administrators / H. Randolph Thomas, Ralph D. Ellis Jr.
      p. cm.
   Includes bibliographical references and index.
   ISBN-13: 978-0-7844-0921-3
   ISBN-10: 0-7844-0921-8

   1. Construction contracts—United States. 2. Construction contracts—United States—Interpretation and construction. 3. Contractors—United States—Handbooks, manuals, etc. 4. Project managers—United States—Handbooks, manuals, etc. I. Ellis, Ralph D., Ph.D. II. Title.

   KF902.T48 2007
   343.73'078624—dc22

                                                                    2007042067

Published by American Society of Civil Engineers
1801 Alexander Bell Drive
Reston, Virginia 20191
www.pubs.asce.org

Any statements expressed in these materials are those of the individual authors and do not necessarily represent the views of ASCE, which takes no responsibility for any statement made herein. No reference made in this publication to any specific method, product, process, or service constitutes or implies an endorsement, recommendation, or warranty thereof by ASCE. The materials are for general information only and do not represent a standard of ASCE, nor are they intended as a reference in purchase specifications, contracts, regulations, statutes, or any other legal document.

ASCE makes no representation or warranty of any kind, whether express or implied, concerning the accuracy, completeness, suitability, or utility of any information, apparatus, product, or process discussed in this publication, and assumes no liability therefor. This information should not be used without first securing competent advice with respect to its suitability for any general or specific application. Anyone utilizing this information assumes all liability arising from such use, including but not limited to infringement of any patent or patents.

ASCE and American Society of Civil Engineers—Registered in U.S. Patent and Trademark Office.

Photocopies and reprints. You can obtain instant permission to photocopy ASCE publications by using ASCE's online permission service (www.pubs.asce.org/authors/RightslinkWelcomePage.htm). Requests for 100 copies or more should be submitted to the Reprints Department, Publications Division, ASCE (address above); email: permissions@asce.org. A reprint order form can be found at www.pubs.asce.org/authors/reprints.html.

Contents

Chapter 10.   Misrepresentations 200

Preface

Little work is done on construction projects without a contract. Even if there is no written document, verbal commitments are usually recognized judicially as valid contracts. Therefore, all parties involved in a construction project must read, understand, and correctly interpret the language of key provisions of the contract. We wrote this book because we have observed a widespread lack of understanding about construction contracts.

We are engineers, and we wrote this text for engineers, contractors, and administrators. Although the text is based on case law research, we focus on understanding and application. By reading the text and working some of the problems at the end of each chapter, readers build confidence that they can render judicially correct interpretations of construction contracts. Thus, we designed this book to build confidence.

Four key elements make our text unique. First, the chapters are easy to read and comprehend. Second, they contain numerous examples. Third, we limit the use of legal terms and concepts. And fourth, readers do not need to master legal terminology; where we use legal concepts, we give many examples so the concepts can be readily understood.

We present the resolution of conflicts in flowcharts or decision trees. The flowcharts are based on the review of hundreds of appellate court cases. From our review, it is evident that judges in all jurisdictions, public or private, repeatedly ask the same questions, and these questions have been incorporated into the flowcharts. Depending how each question is answered, the outcome is reasonably predictable and the law stays fairly consistent.

In each chapter, we include case studies and exercises based on actual court cases. The case studies may involve several issues, but they are strategically placed to reinforce a key point in the text. The exercises may be used for workshop discussion or as classroom problems. Solutions to the exercises are presented in Appendix B. We derive the solutions using the flowcharts and then present the resolution of the dispute. In all but a few instances, the flowchart solution is consistent with the judicial outcome.

Some disputes can be quite complicated and are beyond the scope of this book. This book is not a substitute for sound legal advice, which should be sought in the drafting of contracts and for complicated disputes.

Acknowledgments

I would like to acknowledge my graduate students, without whose legal research this book would not be possible. They are:

Dennis Wright
Robert Mellott
Martin Ponderlick
Steve Wirsching
Michael Clark
Michael Cummings

—Randolph Thomas

First, I thank my friend and partner in this project, Randolph Thomas, for his dedication and untiring work. I also must thank my wife and daughter for their unselfish support of this and every other professional activity I have undertaken.

—Ralph Ellis Jr.

Chapter 1

Introduction

Technical competence, common sense, and experience are not enough to ensure that one is a good construction manager or contract administrator. Today, one may be called on to use communication or negotiation skills, interpret the OSHA regulations, or help develop an environmental impact statement. Managers are frequently called on to interpret construction contracts, a task for which they usually have had little or no formal training, and so they often resort to intuition, common sense, or hearsay. Each may yield a different interpretation.

Objective

This book defines rules for construction contract interpretation for the most troublesome contract clauses. The rules were developed for contractors and contract administrators to improve the quality of contract administration. The view is that by following these simple rules, the parties to a dispute can reach a contractually correct decision without resorting to the courts or other resolution forums. In this book, a contractually correct decision is assumed to be the same as a judicial decision.

Scope

In general, a correct judicial contract interpretation is determined by (1) the application of well-defined rules of law applied to the facts or (2) results from complex legal determinations and procedures applied to the facts. These choices are illustrated in Fig. 1-1, which also includes the gray area between the two. The application of well-defined rules applied to the facts is the focus of this book. Knowing the rules is the key to successful contract interpretation. Most of the problems arising from construction disputes can be correctly resolved by knowing the facts and applying well-defined and established rules in an orderly fashion. Complex disagreements involve more than the contract clauses discussed in this book. Such disputes are the arena for attorneys and are not the focus of this book. In complex disputes, this book is not a substitute for sound legal advice.

The analytical approach described in this book is conveyed in the form of flowcharts. The flowcharts provide a consistently accurate analysis of contract clauses. They may not be suitable where there are allegations of a breach of contract because these disputes tend to be complicated and can often involve multiple issues. Engineers and administrators tend to leave the resolution of these complicated disputes to their attorneys.

The rules described in this book are based on the careful examination of more than 500 judicial decisions related to the most troublesome and frequently disputed contract clauses. Little or no difference has been observed for different jurisdictions or based on whether the contract is

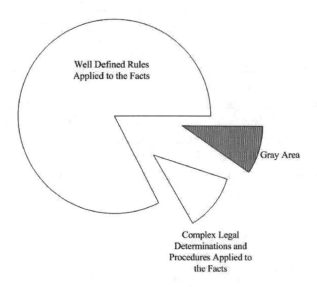

Well Defined Rules
Applied to the Facts

Gray Area

Complex Legal
Determinations and
Procedures Applied to
the Facts

Figure 1-1. Scope of Judicial Determinations.

public or private. This is not to say that there are no differences, only that the differences are likely based on legal rules, not factual determinations. Where legal interpretations or statutes are involved, legal counsel should be sought.

The inquiries made by the court as part of their decision-making process have been determined. These are factual determinations and do not require that the reader understand complex legal theories or ambiguous terms. Rules were validated using numerous other cases and were found to be correct in all but the most complex circumstances. Thus, the authors believe that these principles can be applied in almost all situations.

Many legal terms are routinely used in most legal writings about contract interpretation. Fortunately, most of these terms are not needed in this book, and where used, they are defined in understandable terms.

How to Use This Book

The rules or questions (inquiries) have been organized in easy-to-follow flowcharts. The user should answer all relevant questions. Sometimes, to resolve a dispute, all questions must be answered. However, in most situations, certain rules are more important than others. When a definitive answer is obtained for one of these questions, there may be no need to proceed further.

The flowchart figures should be treated as only a guide for decision making. They should be used along with the text, which is coordinated with the figures. Varying degrees of experience need to be applied in some situations to arrive at a judicially correct decision. Nevertheless, the flowcharts yield correct decisions in many situations or, at a minimum, highlight inquiries where more facts are needed or where inquiries are most problematic.

Throughout the text, there are case studies. These are cases intended to reinforce a single point of the analysis process. Although most cases involve multiple issues, the case studies are not intended to address multiple issues.

In many places in the text, quotes from judicial decisions are given. The reader should pay careful attention to these statements because they represent what is referred to as common law. Common law is the standard by which contractor claims are evaluated.

There are approximately 60 problems at the end of the chapters and 7 more in Appendix A. Solutions from a flowchart analysis are included at the end of the text in Appendix B. Because the problems are based on actual disputes that were litigated, the judicial decision can also be cited. The problems detail a wide variety of site-related situations. The administrator can expect that most ordinary site problems are similar to one or more of the cases. At a minimum, the same rules as outlined in Chapter 4 will be applicable.

Outline

This book is divided into multiple parts. The first part, Chapters 2 through 3, covers basic concepts. Chapter 2 addresses the reasons why there are apparent inconsistencies in the legal interpretations, and Chapter 3 covers contract formation principles.

Chapters 4 to 6 cover other preliminary information. Chapter 4 outlines some of the fundamental principles used by courts in rendering determinations. Chapter 5 points to the need for identifying the nature of the dispute. It highlights the types of changes and discusses how different rules are applied to different types of disputes. Chapter 6 covers notice requirements that are a prerequisite to any assertion by a contractor of a claim.

The next part is the first section where the basic rules of law are presented. It includes Chapters 7 through 9. These three chapters cover the rules for extra work, oral changes, contract interpretation, and differing site conditions (DSCs). The rules for oral changes are applied to disputes where the work is outside the scope of the contract. Where the disputed work is considered outside the scope of the contract by one party and within the scope of the contract by the other party, the rules in Chapter 8 on contract interpretation are used. Chapter 9 shows how to resolve disputes over DSCs or concealed conditions.

The next part consists of Chapters 10 and 11. Both chapters cover disagreements that are resolved outside the bounds of the contract, which are breach-of-contract issues.

Chapters 12 through 14 address issues related to the no-damages-for-delay clause, substantial completion, and liquidated damages.

Chapter 2

Basic Concepts

In discussing the reliance of the American judicial system on the concept of *precedence*, Sweet says that one rationale for precedence is that it provides reasonable certainty of the outcome. That is, when faced with a dispute, the outcome is reasonably predictable (Sweet 1989). If this idea is true, then why does the judicial system seem so unpredictable? Three primary reasons are explained below. The three reasons relate to the (1) appellate process, (2) situation that the application of fundamental principles is very fact-sensitive, and (3) existence of a hierarchy of documents.

The Appellate Process

When a lawsuit is brought to court, it is done so in a court of limited or general jurisdiction. This is where testimony is given by factual, and sometimes expert, witnesses before a judge or jury. This court is called "the finder of fact" because so much energy is expended in defining who did what to whom and when. The court applies certain common-law principles originating from other similar cases (precedence) to the facts of the case to reach a decision. The trial court does not make new law; it only applies existing law.

Each party to a trial has one free appeal to an *appellate court*. The rationale behind an appeal is that the trial court erred when it applied common law to the facts. The attorneys argue the appeal before the court, and there are no witnesses. It is too late to argue that the facts as determined by the trial court are incorrect; those stand as they are. The arguments are

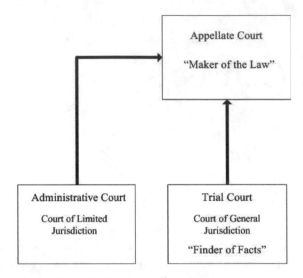

Figure 2-1. Relationship between Trial Courts and Appellate Courts.

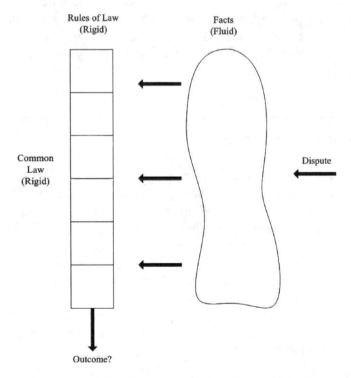

Figure 2-2. Relationship between Rules of Law and the Facts.

all legal and are over points of law. The appellate court is often referred to as "the maker of the law" because it is here that common law is made or reaffirmed. The judicial hierarchy is shown in Fig. 2-1.

When reading judicial decisions, one must always rely on appellate decisions to understand judicial reasoning. Relying on trial court verdicts can lead to erroneous conclusions. In this book, only appellate decisions have been reviewed.

Fact-Sensitive

The application of judicial rules is sensitive to the facts in the dispute. Slightly changing the facts, even small ones, and applying the same rules can lead to a different outcome. The relationship between the rules and the facts is shown in Fig. 2-2. When reading a brief synopsis of a decision, the complete statement of facts is seldom given. This, too, can lead to erroneous conclusions. To understand a decision fully, one needs to study the complete decision in detail.

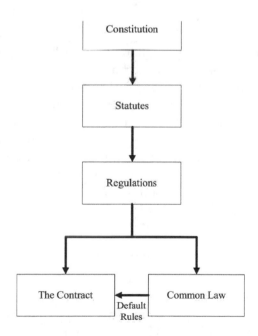

Figure 2-3. Hierarchy of Documents.

Hierarchy of Documents and Rules

Generally speaking, there is a hierarchy of documents and rules. These are shown in Fig. 2-3. Statutes cannot overrule the U.S. Constitution; regulations cannot overrule statutes, and so forth. Although this hierarchy may not apply in all but a few instances, it is a useful tool for the contractor and the contract administrator. Many contracts contain an order-of-precedence clause, which lists the hierarchy of documents should there be a conflict among various contract documents.

A contract establishes the "private law" between contracting parties. This contractual relationship cannot supersede applicable statutes, regulations, or the U.S. Constitution, but contract terms can and do routinely modify otherwise applicable common-law rules. In this sense, common law provides "default" rules if the parties have not spelled out different rules in their contract. In some instances, common-law rules cannot be modified by an agreement between the parties.

Consistency

The authors have studied more than 500 complete appellate decisions for this book. This research showed that there is remarkable consistency in the inquiries made by the courts. These inquiries or questions are based on the facts of the dispute and not on questions of legality. Although not all questions are asked in every case, an extensive review yields a clear and definite pattern. In some types of disputes, there is also a clear indication that some rules are more important than others. In other types, there are indications that any negative response to one inquiry or a series of inquiries is sufficient to defeat the position. Both situations are recognized in this book.

Equity

Sometimes contractors base a claim on theory of equity, unjust enrichment, or fairness. These are complex legal theories that have application in relatively few cases. For the topics covered in this book, equity, unjust enrichment, and fairness receive little discussion in the literature, and the case-law review performed for this book indicates that courts seldom render decisions in favor of the contractor based solely on equitable considerations.

Reference

Sweet, J. (1989). *Legal Aspects of the Architecture, Engineering, and Construction Process*, 4th Ed., West Publishing Co., St. Paul, MN.

Chapter 3

What Is a Contract?

Any discussion of contract interpretation must begin with the question "What is a contract?" The attitude of the judicial system toward contract formation provides the basis for understanding how construction contracts are interpreted when a dispute arises. According to Sweet (1989, p. 40),

> Generally, American law gives autonomy to contracting parties to choose the substantive content of their contracts. Since most contracts are economic exchanges, giving parties autonomy allows each to value the other's performance. To a large degree autonomy assumes and supports a marketplace where participants are free to pick the parties with whom they deal and the terms upon which they will deal.

The importance of this judicial attitude is that courts seek to enforce the provisions of the construction contract as a whole. The terms of the contract will be enforced even if certain provisions prove to be harsh.

A contract is a legally binding agreement between two or more parties to exchange something of value. In construction, it is usually money in exchange for construction services to build a facility. A contract imposes both contractual and legal obligations on both parties that are difficult or impossible to change.

According to Sweet, the principal function of enforcing a contract is to encourage economic exchanges that lead to economic efficiency and greater productivity (1989). Contracts are enforced no matter how harsh the terms, provided the contract was freely agreed on. As stated by one court, parties cannot ignore provisions of the agreement to suit their own convenience or profit (Stamato v. Agamie).

Formation Principles

To be a valid contract, the following elements are essential:

- competent parties,
- offer and acceptance,
- reasonable certainty of terms,
- proper subject matter, and
- considerations.

Additionally, contracts for certain types of transactions and where statutes or regulations apply need to be in writing.

Competent Parties

To enter into a contract, a party must be of proper age and have sufficient mental awareness. This is rarely a problem in construction.

Offer and Acceptance

When parties agree, it is said that there is a *meeting of the minds* or mutual assent. The agreement must be unequivocal and not propose new terms. Even small, seemingly insignificant issues are sufficient to preclude a determination of mutual assent. The case of Western Contracting Corp. v. Sooner Construction Co. is a classic illustration of this issue. The only point of disagreement was the manner of payment, but that was enough to preclude a meeting of the minds. Other legal issues that may be intertwined with the assent issue are economic duress, fraud, mutual mistake, and apparent authority. Therefore, legal counsel is often advised.

With respect to changes, all construction contracts expressly give the owner the unilateral right to direct changes, and the contractor is obligated to perform the work. Through the changes clause, the contractor can object to such things as changes, price, scope, or duration. Even when they are directed to perform the work, contractors in some instances may refuse, leaving the owner to hire another contractor to do the work. Legal advice is recommended before refusing to do the work.

Case Study

Meeting of the Minds/Economic Duress

Enid Corp. v. H. L. Mills Construction Co.
101 So.2d 906 (1958)

Enid Corp. and H. L. Mills Construction Co. entered into an oral contract whereby Mills was to build certain roads in Biscayne Key Estates. The roads were to be built to grade stakes that were set by Enid's engineer. Enid continually inspected the work from day to day and at all times had control of the elevation to which the finished roads were to be built. The stakes were set at 5-ft elevation intervals. Both parties knew the character of the subsoil, both at the time of entering into the work and at the time the work was being done; each was fully aware of the possibility that roads built on such soil might settle. The roads were inspected during the progress of the work by a Dade County inspector who was principally concerned with the depth of rock and with the surfacing. The roads were completed according to the Dade County specifications and at the elevation set and required by Enid.

During the course of the work, Mills was fearful that the roads might settle and wanted to build them higher than the elevation set by Enid. Mills advised Enid that the roads might settle, but Enid would not permit Mills to build the roads higher because a higher road would have required additional fill on each of the lots.

Soon after the work was complete, there was a slight settlement of some portions of the roads so that the roads were below the 5-ft elevation as required by Enid. Enid insisted that Mills raise the elevation of the roads at Mills' expense. Mills refused.

Simultaneous with the Biscayne Key Estates project, Mills had another contract with Enid to perform identical work on the Tropical Isle Homes, First Addition project. Enid insisted that Mills raise the elevation of the roads on this project at Mills' expense, just as Enid had insisted on the Biscayne Key Estates project. Mills again refused. When Enid said it would not accept the work (and would not pay) unless Mills altered its performance in accordance with this new directive, Mills stopped work on the project.

Are Enid's actions a form of economic duress?

The Third District Court of Appeals of Florida did not directly address the issue of economic duress, but it would appear that Enid's behavior is consistent with other situations where duress was found. Instead, the court focused its attention on the terms of the oral contracts. Enid insisted that the terms of the second contract were the same as the terms and conditions of the first. The court reasoned that because each party was contending a different interpretation of the Biscayne Key Estates contract, there was no agreement on the second contract, and therefore, no meeting of the minds: "A 'meeting of the minds' is essential for the existence of any contract."

Reasonable Certainty of Terms

Terms should be reasonably clear so that an independent third party can determine if the parties substantially performed as promised. If performance cannot be determined, there may not be a valid contract. This issue is rarely a problem in construction but is potentially a greater problem in private-sector design contracts.

Proper Subject Matter

Parties cannot contract to do something that is illegal.

Considerations

Contracts are economic exchanges; therefore, something of value must be exchanged. Valid contracts require considerations or an exchange of something of value; this rule is sometimes called the *preexisting duty* rule. Courts typically validate a transaction where an exchange took place, even if the exchange is unequal, because one party is providing something of little value. Referring to changes, Sweet (1989, p. 431) states,

> This preexisting rule is criticized. It limits the autonomy of the parties by denying enforceability of agreements voluntarily

made. Implicit in the rule is an assumption that an increased price for the same amount of work is likely to be the result of expressed or implied coercion on the part of the contractor, as if the contractor is saying, "Pay me more money or I will quit and you will have to whistle for the damages." However, suppose the parties have arrived at a modification of this type voluntarily. There is no reason for not giving effect to their agreement...In the construction contract, minor changes in the contractor's obligation have been held sufficient to avoid the rule even where the increase in price was not commensurate with the change in obligation on the part of the contractor.

Form

Sometimes, statutes preclude oral contracts. This is particularly true for public agencies and local governmental authorities. Thus, in the public sector, the writing creates the contract. Where there are no governing statutes or regulations prohibiting them, the validity of oral contracts and modifications is determined by common law and the contract language.

In dealing with private contracts, Trachtman (1987) states that the writing does not create the contract, but rather the writing comes into being after a *meeting of the minds* occurs. Private contracts are created by what people say and do, not by what papers they sign. Two illustrations reinforce this point.

Illustration No. 1: Texaco, Inc.

In a 1985 nonconstruction case, Pennzoil brought suit against Texaco, alleging that they, Pennzoil, had an "agreement in principle" with Getty Oil to permit Pennzoil to purchase the controlling interest in Getty stock. Pennzoil argued that Texaco interfered with the deal by offering Getty a higher price, thereby causing Getty to back out of the deal with Pennzoil. Texaco countered by saying that the talks between Getty and Pennzoil were informal and that there was never any real contract. At best, it was only a *handshake deal*. There was no formal, written agreement that would be binding.

The court said Texaco was wrong. A jury found that the handshake deal was a legally binding contract. Texaco was required to pay an $11 billion verdict, the largest damage award in U.S. history.

Illustration No. 2: The Philadelphia Eagles

Mr. Leonard Tose, owner of the Philadelphia Eagles, entered into complex negotiations with several investors who wanted to buy the team. Eventually an agreement was reached, but before the papers were signed, Mr. Tose changed his mind and decided not to sell.

The investors sued, saying that the deal was beyond the backing-out stage; it was a "done deal." Mr. Tose faced the media and assured Philadelphia fans that no deal had been finalized. There might have been an "understanding" or the outline of an "arrangement," but there was no signed contract.

Mr. Tose believed that a contract had to be in written form. It was an expensive lesson, and it cost Mr. Tose a seven-figure settlement.

Exercise 3-1: Rosen & Morelli and Kreisler Borg

Rosen and Morelli was the low bidder for certain masonry work let by Kreisler Borg, the general contractor. Rosen and Morelli never commenced work and did not execute a written subcontract with Kreisler Borg. However, over a three-month period, Rosen and Morelli did attend job meetings, submit a trade payment breakdown and a certificate of insurance and, according to Kreisler Borg, take other actions that demonstrated the existence of a binding subcontract between them.

When several disputes arose between Rosen and Morelli and Kreisler Borg over the contract specifications and disagreements over the requirements of the proposed contract documents were not resolved, Rosen and Morelli withdrew from the project, contending that it never finalized an agreement. Kreisler Borg took a contrary position and commenced legal action against Rosen and Morelli for breach of contract.

Rosen and Morelli argued that the steps it took on the project were in furtherance of negotiations and were undertaken at Kreisler Borg's request as the normal accommodations extended by a potential subcontractor to a general contractor. Kreisler Borg argued that the steps taken by the subcontractor were in furtherance of an enforceable agreement that the parties had reached. The court sided with Rosen and Morelli, saying that there was no offer and agreement and thus there was no contract.

References

Stamato v. Agamie 131 A.2d. 745,748 (1957).
Sweet, J. (1989). *Legal Aspects of the Architecture, Engineering, and Construction Process,* 4th Ed., West Publishing Co., St. Paul, MN.
Trachtman, M. G. (1987). *What Every Executive Better Know about the Law,* Simon and Schuster Publishing Co., New York.
Western Contracting Corp. v. Sooner Construction Co. 256 F.Supp. 163 (1966).

Additional Cases

The following are additional cases related to issues associated with contract formation, such as meeting of the minds, duress, and unjust enrichment. The reader is invited to review the facts of the case and determine the rationale behind the decision.

Blake Construction v. Upper Occoquan Sewage Authority, Supreme Court of Virginia, 10/31/03 (legality of contract provisions).

Cox & Floyd Grading v. Kajima, Court of Appeals of South Carolina, 11/24/03 (duress).

G. R. Osterland Co. v. Cleveland, Court of Appeals of Ohio, 11/20/00 (unjust enrichment).

KW Construction v. Stephens & Sons Concrete, Court of Appeals of Texas, 6/8/05 (oral contracts).

Lichtenberg Construction & Development v. Paul W. Wilson, Court of Appeals of Ohio, 9/28/01 (revocation of subcontractor's bid).

Roger's Backhoe Service, Inc., v. Nichols, Supreme Court of Iowa, 6/16/04 (implied contract).

Ry-Tan Construction v. Washington Elementary School District No. 6, Supreme Court of Arizona, 5/25/05. (meeting of the minds).

Chapter 4

Fundamental Principles

Construction disputes involving contractual obligations are not uncommon and arise for many other reasons. In general, disputes are the result of an ambiguity or error in the contract documents, or a contract that is silent on a particular issue, or conditions that are not covered or anticipated by the contract. This chapter shows that there are fundamental principles and an organized structure to determine what the parties intended. Principles allow the outcome of disputes to be reasonably predictable or, at a minimum, to tell how the outcome is determined.

Common-Law Rules

Common law establishes rules for interpreting construction contracts. As shown in Fig. 4-1, the rules are organized into two major divisions: procedural and operational. Procedural rules of interpretation are guidelines within which the court must operate. Operational rules are ones that are applied to the facts of a dispute. The fundamental principles are summarized in Table 4-1.

Procedural Interpretation Rules

Procedural rules support the interpretation process. They furnish guidance on how to control and evaluate an interpretation. The procedural rules that are shown in Fig. 4-2 establish (1) the objective of interpretation,

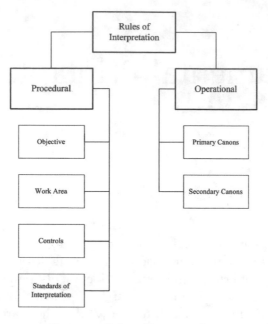

Figure 4-1. Rules of Interpretation.

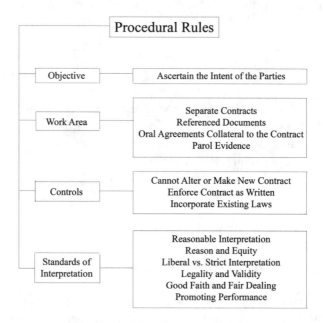

Figure 4-2. Procedural Rules of Interpretation.

Table 4-1. Fundamental Principles of Contract Interpretation.

RULE #	FUNDAMENTAL PRINCIPLE
1.	The fundamental rule of contract interpretation is to ascertain the intention of the parties.
2.	Secret or undisclosed intentions will not control.
3.	To ascertain the intent of the parties, one must examine what is often termed the four corners of the agreement.
4.	A court cannot change or rewrite an existing contract. Contract reformation is not allowed.
5.	One must generally enforce a contract as made or written. One cannot ignore or alter harsh or unreasonable terms that are clearly written into the contract. The contract says what it says.
6.	Existing laws are a part of the contract.
7.	Words used in a contract will be assigned their ordinary meaning unless it is shown that the parties used them in a different sense.
8.	The actions of the parties during execution of the contract may speak volumes about the parties' mutual intentions. (Actions speak louder than words.)
9.	A contract must be construed as a whole and, whenever possible, effect will be given to all its parts. All provisions of the contract should be read in such a way as to be in harmony with each other.
10.	Read the contract as a whole.
11.	An interpretation will not be given to one part of a contract that will annul another part of it. Proper interpretation means that no part can be treated as useless.
12.	Specific terms should govern over general terms.
13.	Where words and figures are inconsistent, the words govern.

(2) the work area, or measures for the admissibility of evidence, (3) controls on what interpretation can be adopted, and (4) standards of interpretation.

Objective of Interpretation

To determine the contract meaning, a court must determine what the contracting parties wanted the contract to mean. Therefore, the intent of the parties at the time the contract was made must be determined.

Common law provides a rule that establishes the objective of interpretation. The principle states,

In the construction or interpretation of contracts, the primary purpose and guideline, and indeed the very foundation of all the rules for such construction or interpretation, is the intention of the parties. Generally speaking, therefore, the fundamental and cardinal rule in the construction or interpretation of contracts is that the intention of the parties is to be ascertained and, if it can be done consistently with legal principles, is to be given legal effect. (*Corpus Juris Secundum* 1963, Sec. 295, p. 66)

Clear intentions are relevant to interpretation, not undisclosed or secret intentions. Normally, a court does not uphold hidden intentions, particularly when a hidden intention differs from that expressed in the contract.

A secret or undisclosed intention of the parties that is at variance with the expressed or inferable intention will not control (Williston 1961, Sec. 620, p. 750).

A *manifested intent* is that which has been outwardly expressed by the written agreement or by acts and deeds during performance. It is the true or actual intent that a party has given to the agreement. In the view of one court,

Greater regard is to be had for the clear intent of the parties than to any particular words which they may have used in the expression of their intent. (*American Jurisprudence* 1990, Sec. 264, p. 688)

Work Area

The work area contains the evidence that a court is permitted to consider. The scope of the work area is established by applying principles in four basic areas. Generally, these principles define what documents and evidence make up the *entire* agreement. To ascertain the intent of the parties, a court must determine what is often termed the four corners of the agreement, that is, "What is the whole contract?"

Separate Contracts

Courts examine other contracts separate from the disputed contract, provided that they are part of the same transaction. A Construction

Management delivery system is particularly subject to this principle. The general rule of law states,

> In the absence of anything to indicate a contrary intention, instruments executed at the same time, by the same contracting parties, for the same purpose, and in the course of the same transaction will be construed together, since they are in the eyes of the law, one contract. (*American Jurisprudence* 1990, Sec. 263, p. 666)

Referenced Documents

Construction contracts usually refer to other documents and standards that are necessary for the proper execution of the work. Common law holds that these documents are to be used in determining the meaning of the entire agreement. As a rule,

> Where a written contract refers to another instrument and makes the terms and conditions of such other instrument a part of it, the two will be construed together as the agreement of the parties. (*American Jurisprudence* 1990, Sec. 263, p. 667)

To illustrate, in McCarthy Brothers Construction Co. v. Pierce, a contractor was awarded a HUD-financed contract. The contract form was a HUD standardized agreement, and it expressly incorporated AIA Document A201 (1970) as the general conditions. A dispute subsequently arose over conflicting definitions of substantial completion between the HUD and AIA documents. When read as a whole, the U.S. District Court noted that the HUD document contained an order-of-precedence clause that expressly stated that the HUD agreement took precedence over all inconsistent provisions within AIA A201, so the HUD language was applied.

Special attention is given to project memoranda and other correspondence because courts may view these documents as evidence of the intent of the parties (*Corpus Juris Secundum* 1963, Sec. 298, p. 135). If the parties agree to a change via correspondence, the meaning of the agreement is gathered from both the written contract and the correspondence (*Corpus Juris Secundum* 1963, Sec. 299, p. 136).

An important element in determining if a document should be incorporated into an agreement is the requirement for explicit referral. The reference to another document must be clear, and both parties must have known of its incorporation (*Corpus Juris Secundum* 1963, Sec. 299, p.

136). However, the incorporation of reference documents may be limited in their use as an aid to interpretation. Where the reference is made to another document for a specific purpose, then the other writing becomes a part of the contract for that specified purpose only (*American Jurisprudence* 1990, Sec. 262, p. 666).

Oral Agreements as Collateral of a Contract

Oral agreements are construed together with the written agreement. Where permitted by statutes, a written contract may properly be varied by oral agreements only where it is collateral and is not inconsistent with the express or implied conditions of a written contract (Williston 1961, Sec. 631, p. 949). Parties seeking to include oral collateral agreements face formidable hurdles, and the outcome may be based on legal rules rather than purely factual ones.

Parol Evidence

Ordinarily, preliminary agreements and negotiations are not considered when interpreting a written contract (Williston 1961, Sec. 630, p. 947). When contracting parties have reduced their agreement to writing, then all previous agreements and negotiations are believed to have been incorporated into that agreement. However, in limited cases involving misunderstandings or ambiguities, there may be an inquiry into the meaning attached to the words of a contract. To facilitate this inquiry, common law allows previous agreements and testimony to be used.

> Oral evidence may be admitted if used to establish the meaning of a word or term, or to establish grounds for granting or denying reformation, specific performance, or other remedies (*Corpus Juris Secundum* 1963, Sec. 296(3), p. 88).

The use of parol evidence is limited and is more likely to occur with private design contracts. As with oral collateral agreements, contractors face formidable legal hurdles, so legal advice is recommended.

Controls

Before providing legal effect to an agreement, a court must ensure that the legal consequences of its interpretation do not violate the contract.

Therefore, before adopting a meaning, the meaning must be evaluated to ensure that it is lawful and does not infringe on the contractual rights of the parties.

Contract Alteration

Competent parties are entitled to make their own contractual arrangements, and courts have limited rights to rewrite or make a new contract. The rule states,

> A court cannot alter an existing contract or make a new contract for the parties, but can only construe the contract which they have made for themselves. (*American Jurisprudence* 1990, Sec. 242, p. 628)

Courts have no right to revise or modify contracts parties made for themselves, including contracts that appear unfair to one side (*Corpus Juris Secundum* 1963, Sec. 296(3), pp. 97–98). This rule protects the right of freedom to contract. Disputing parties cannot ask the courts to rewrite the contract (*Corpus Juris Secundum* 1963, Sec. 296(4), p. 98).

Enforce Contract as Written

The function of a court is to interpret and enforce a contract as written. The general rule states,

> In the absence of any ground for denying enforcement, a court must generally enforce a contract as made or written. (*Corpus Juris Secundum* 1963, Sec. 318, pp. 185, 187)

However, courts cannot enforce contracts or contractual provisions that are illegal or against public policy, or where there is evidence of fraud (*American Jurisprudence* 1990, Sec. 257, p. 656).

Incorporate Existing Laws

Existing laws are a part of the contract. The laws encompassing the validity, operation, and effect of the contract that exist at the time and place of its making form a part of the contract (Dunham and Young 1958). In construction contracting, the place where the contract is made is seldom the place of performance, and this sometimes results in conflicting laws. The general rule is that

> Contracts are to be governed by the law of the place where made, unless the parties clearly appeared to have had some other law in view. (Dunham and Young 1958)

In contrast, performance of a contract is governed by the law of the place of performance (Williston 1961, Sec. 630, p. 948).

Standards of Interpretation

Common sense and sound judgment are essential when applying the standards of interpretation. Common law has adopted the following standards of interpretation in choosing among possible meanings:

- A reasonable interpretation is favored over an unreasonable one.
- An equitable interpretation is favored over an inequitable one.
- A liberal interpretation is favored over a strict one.
- An interpretation that promotes the legality of a contract is favored.
- An interpretation that upholds the validity of a contract is favored.
- An interpretation that promotes good faith and fair dealing is favored.
- An interpretation that promotes performance is favored over one that would hinder performance.

There can be other standards. An explanation of each standard listed above is given below. The fourth and fifth standards are combined.

Reasonable Interpretation

A reasonable interpretation is preferred to one that is unreasonable. An interpretation should be given a meaning that would be adopted by a reasonably intelligent person acquainted with all operative usages and knowing all of the circumstances before and at the time of the making of the agreement (*Corpus Juris Secundum* 1963, Sec. 319, p. 191).

Equity of Interpretation

An interpretation that is equitable to both parties is preferred (*American Jurisprudence* 1990, Sec. 252, p. 646). Courts avoid harsh terms if possible, i.e., when one party is at the mercy of the other. However, the court does not ignore or alter harsh or even seemingly unreasonable terms that are clearly written into the contract because ignoring the foolishness of a contractual undertaking is not within the function of the court (Corbin 1960, Sec. 546, p. 169).

Liberal Construction

Interpretations can either be strict or liberal. In practice, however, courts do not narrowly or loosely interpret a contract to violate its obvious purpose or to relieve a party of an obligation. Courts look at the language used and determine its realistic limitations (*Corpus Juris Secundum* 1963, Sec. 318, p. 187).

Legality and Validity

An interpretation that promotes the legality and validity of a contract or contractual provisions is preferred to an interpretation resulting in the contrary. If two interpretations render two reasonable meanings, with one offering legal effect and the other not, the former interpretation is adopted; and, if a provision has two meanings and one validates the contract, whereas the other voids the contract, the interpretation that validates the agreement is selected (*Corpus Juris Secundum* 1963, Sec. 318, p. 187).

Good Faith and Fair Dealing

Contracts imply good faith and fair dealing between the parties. The standard of preference states that if an interpretation implies bad faith or fraud against one of the parties, and the other interpretation does not, the court should choose the interpretation that promotes good faith and fair dealing (*Corpus Juris Secundum* 1963, Sec. 301, p. 142).

Contractual Performance Favored

An interpretation that renders performance possible is preferred to an interpretation that makes performance impossible. No matter how clear the words may appear, courts should not adopt a meaning that will render performance impossible (*Restatement of the Law* 1981, Sec. 202, p. 89).

Operational Interpretation Rules

Fig. 4-3 summarizes the operational rules for interpretation that are applied to determine the meaning of a contract. The primary rules are discussed in more detail in Chapter 8. Operational rules are primary or secondary according to the nature of their application. Primary rules are applied to the entire agreement. Secondary rules, called *canons*, are ap-

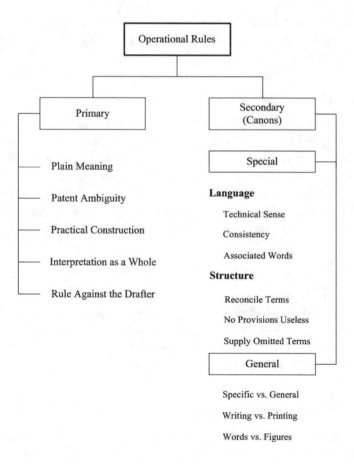

Figure 4-3. Operational Rules of Interpretation.

plied only when required by a specific fact or situation. The primary rules are used for interpretation disputes, whereas the canons are used in several types of disputes when they are needed.

Primary Rules of Interpretation

Relative to Fig. 4-3, the primary rules of interpretation address broad questions to ascertain the meaning of the contract. Each rule views the contract from a somewhat different vantage point. For instance, the plain meaning rule focuses on establishing the meaning of single words or phrases, whereas interpretation of the contract as a whole focuses on the interaction of the relevant documents. The patent ambiguity rule looks at prebid duties of the contractor, and the practical construction rule examines the actions of the parties.

The Plain Meaning Rule

Where words, terms, or phrases appear to have a clear and unambiguous meaning, then that meaning should be adopted. The plain meaning rule states that

> Words employed in a contract will be assigned their ordinary meaning unless it is shown that the parties used them in a different sense. (Calamari and Perillo 1990)

This rule is invoked where the evidence clearly suggests that the word was to be used in an ordinary sense. However, the rule is not to be used where it allows evidence of a contrary meaning to be excluded (*Corpus Juris Secundum* 1963, Sec. 294, p. 37).

The determination of whether the language is to be given its ordinary meaning or some other meaning, or if it is ambiguous, is made by the court. Recent cases indicate that courts consider trade custom and industry practice only where it is conclusive (*Restatement of the Law* 1981, Sec. 201, p. 83).

Patent Ambiguity

A *patent ambiguity* is an obvious or drastic conflict within the contract. A *latent ambiguity* arises when apparently clear language, coupled with an extrinsic fact, creates an ambiguity. A latent ambiguity is not obvious.

Without regard to the type of ambiguity, a court will seek to determine the parties' intentions from the entire agreement and its surrounding circumstances (*American Jurisprudence* 1990, Sec. 258, p. 658). As a rule,

> In case of ambiguity, a contract is generally given that meaning which a party knew, or had reason to know, was in accordance with the other party's understanding. (*Restatement of the Law* 1981, Sec. 202, p. 86)

Good faith and fair dealing form the basis of this rule. The knowledge of, or reason to know, that the other party's understanding is different, and not bringing it to the other party's attention, implies bad faith and unfair dealing of the party with the knowledge, particularly when that party may profit. Where a party recognizes or should have recognized an ambiguity that is patent, there is a duty to inquire imposed on that party. Courts have held that a failure to inquire often results in the denial of a claim.

Practical Construction

The parties' intentions are best demonstrated by their actions during performance. Because the parties know best what they meant, their actions are often the strongest evidence of the contract's meaning (*Corpus Juris Secundum* 1963, Sec. 297, p. 113). The rule of practical construction states,

> A reasonable construction of an ambiguous contract by the parties thereto, although not conclusive, will be considered and accorded great weight, and usually will be adopted by the courts. (*Restatement of the Law* 1981, Sec. 206, p. 105)

The interpretations of the parties must be mutual. Knowledge of the other party's interpretation and acquiescence or silence is a mutual interpretation (*American Jurisprudence* 1990, Sec. 240, p. 625).

Interpretation as a Whole

It is common practice in a dispute for each party to focus on specific clauses or provisions in the contract supporting their position. However, a court may not approach an interpretation so narrowly. It is a universal principle that the agreement in its entirety must be examined (*Corpus Juris Secundum* 1963, Sec. 302(1), p. 147). When interpreting the contract as a whole, the rule states,

> A contract must be construed as a whole and, whenever possible, effect will be given to all its parts. (*American Jurisprudence* 1990, Sec. 251, p. 643)

All provisions of the contract should be read in such a way as to be in harmony with each other. The agreement should not be interpreted so as to render part of the contract inoperable (*Corpus Juris Secundum* 1963, Sec. 303, p. 150). Common sense suggests that each provision in a contract is included for a reason and should therefore be given equal consideration. The isolation and interpretation of particular clauses or provisions within the contract may not provide the true intent of the parties (*Corpus Juris Secundum* 1963, Sec. 302, p. 149). Courts read and consider the "four corners" of the agreement before arriving at a conclusion of the meaning of any particular provision. In the view of one court,

> It is necessary to consider all parts and provisions of a contract in order to determine the meaning of any particular part, or of particular language, as well as of the whole. (Monroe M. Tapper v. U.S.)

Rule against the Drafter

An interpretation may lead to more than one reasonable interpretation. In this situation, a court must apply a tiebreaker rule. Common law provides the following rule:

> In choosing among reasonable meanings of a promise or agreement or a term thereof, that meaning is generally preferred which operates against the party who supplies the words from whom a writing otherwise proceeds. (Patterson 1964)

The rationale is that the drafting party failed to express the desired intent clearly and that the drafting party is more likely to provide more carefully for the protection of its own interests. This rule should only be applied as a rule of last resort, after all other rules have been exhausted, and only where both parties possess reasonable meanings (Patterson 1964).

Canons of Interpretation

In some cases, using the primary rules of interpretation may not be sufficient to interpret a contract properly. When assistance is required, the canons of interpretation are applied (*American Jurisprudence* 1990, Sec. 267, p. 673). There are two categories of canons: special and general. The special canons support the primary rules. General canons do not support a particular primary rule.

Special Canons

The special canons deal with specific ambiguous situations found within the language or physical structure of the written instrument. Special canons dealing with ambiguities surrounding the meaning of a word or phrase are applied, when necessary, with the plain meaning rule. Special canons associated with conflicts, errors, and omissions within the contract documents are applied with the rule of interpreting the contract as a whole.

Ambiguous Language
Three special canons deal with interpreting the ambiguous language in a contract, and these are

- Technical words are given their technical meaning.
- Words are given consistent meaning throughout the agreement.
- The meaning of a word may be indicated by the words associated with it.

Contracting parties often use the language of a particular trade or geographical region. New words may be developed, or older, more common words may be given new meaning. To support this situation, common law provides the rule that

Technical words are to be ordinarily taken in a technical sense, unless they are clearly used in a different sense. (*Corpus Juris Secundum* 1963, Sec. 309, p. 165)

Courts interpret technical language from the viewpoint of how a person in the profession or business with which it is associated would normally understand it (*Restatement of the Law* 1981, Sec. 204, p. 98). Courts also look at local custom and usage, and that meaning, if conclusive, is adopted (*American Jurisprudence* 1990, Sec. 270, p. 677).

The words in a contract are given consistent meaning. The rule states

Where there is nothing in the context to indicate otherwise, words used in one sense in one part of the contract are deemed to have been used in the same sense in another part of the instrument. (*Corpus Juris Secundum* 1963, Sec. 313, p. 176)

The context in which the word is found may play a key role in determining its meaning. A word can possess more than one meaning in a contract if each provision clearly defines the word (*Corpus Juris Secundum* 1963, Sec. 313, p. 177).

The consistency rule is supported and further defined by the rule of associated words. As commonly stated, "words are known by the company they keep," and common law provides that

The meaning of words in a contract may be indicated or controlled by those words with which they are associated. (*Corpus Juris Secundum* 1963, Sec. 310, p. 168)

Also, where several subjects of a class or group are enumerated and there are no general words to show that other subjects are included, the subjects or items not listed were intended to be excluded (*Corpus Juris Secundum* 1963, Sec. 311, p. 171). The rule of associated words is further defined by two Latin maxims: *Ejusdem generis* and *Expressio unius*. The rule *Ejusdem generis* states that

A general term joined with a specific one will be deemed to include only things that are like (of the same genus as) the specific one. (Patterson 1964)

The rule of *Expressio unius* states that

If one or more specific items are listed, without any more general or inclusive terms, other items although similar in kind are excluded. (Patterson 1964)

Ambiguous Structure

Three special canons deal with ambiguities occurring within the physical structure of contract documents. These rules are

- The court reconciles conflicting provisions.
- No provision is treated as useless.
- The court may supply an omitted term.

Common sense, founded on good faith and fair dealing, dictates that conflicting provisions would not have been deliberately inserted into a contract. Thus, a court takes the viewpoint that

> Where there is an apparent repugnancy between two clauses or provisions of a contract, it is the province and duty of the court to find harmony between them and to reconcile them if possible. (*American Jurisprudence* 1990, Sec. 267, p. 673)

To accomplish this, the court looks at the entire agreement (the "four corners") and the surrounding circumstances associated with it.

It is a common consideration in interpreting disputes that no provision in a contract should be treated as useless. It must have been supplied to serve a purpose. Common law states,

> If possible, an interpretation will not be given to one part of a contract which will annul another part of it. (*Corpus Juris Secundum* 1963, Sec, 309, p. 165)

A necessary term of the contract, whether omitted inadvertently, or where the parties have not agreed to a term when signing the agreement, may be supplied by the court to aid in determining the intent of the agreement. The rule states,

> When the parties to a bargain, sufficiently defined to be a contract, have not agreed to a term which is essential to a determination of their rights and duties, a term which is reasonable in

the circumstances is supplied by the court. (*Restatement of the Law* 1981, Sec. 204, p. 98)

Case Study

Enforce Contract as Written/No Provision Is Useless/One Logical Interpretation

Florida Department of Transportation v. MacAsphalt, Inc.
429 So.2d 1281 (1983)

On April 23, 1979, the Florida Department of Transportation entered into a contract with Jasper Construction Co. to construct a portion of Interstate 75. The standard specifications for road and bridge construction were a part of the contract. In accordance with Sec. 9-5.6 of the specifications, Jasper elected to subject the freight costs for coarse aggregate to the rate adjustment provisions found in Sec. 9-5.5(c). The applicable contract language read as follows:

> 9-5.1 General: Except as provided herein for certain railroad freight rates, no allowance or deduction will be made for any increase or decrease in common carrier rates or transportation costs on materials . . .

> 9-5.2 Material on Which Adjustment is Allowable: Allowance or deduction for any changes in railroad freight rates may be made under the provisions of this Article, only for the construction materials described and limited below.

> . . .

> (6) Fine and coarse aggregate for asphaltic mixtures of bituminous surface treatments (mineral filler excluded).

> 9-5.5 . . . (c) The amount of the contract adjustment shall be further limited to 90 percent of the excess base freight cost increase or decrease over the $1,000 deductible amount.

> 9-5.9 Changes in Shipping Methods or Points; Errors in Rate Quotation:

> . . .

> The Department will not make any allowances for increased freight cost due to errors in the affidavit or quotations, or change in type of transportation.

MacAsphalt, Inc., was the subcontractor and filed a form with the department listing the current freight rate for coarse aggregate as $5.65 per ton. Although MacAsphalt intended to ship the aggregate from Miami to its asphalt plant in Port Charlotte by rail, at the time of shipping, there was an embargo on that route, making rail transportation impossible. Instead, MacAsphalt shipped the aggregate by truck at an excess cost of $26,000. MacAsphalt now seeks reimbursement for this amount.

Should MacAsphalt be allowed to recover the $26,000 in excess cost?

First District Court of Appeals of Florida reversed an award by the Board of Arbitration saying that

> If the purpose of §9-5 were to remove freight cost considerations from the bidding process when the contractor initially intends to ship by rail, the last quoted provision (§9-5.9) would not be included in the contract.
>
> We therefore find that a proper interpretation of the foregoing contract provisions can only result in a denial of MacAsphalt's claim for an allowance.

General Canons

The general canons of interpretation are applied to all primary rules when additional assistance is needed. Three general canons are discussed below:

- Specific terms should govern over general terms.
- Written words should prevail over printed words.
- Written words should prevail over figures.

When interpreting an agreement, the specific provisions control over a conflicting general provision. Common law states

> Where there are, in a contract, both special and general provisions relating to the same thing, the special provisions prevail. (*American Jurisprudence* 1990, Sec. 270, p. 677)

The rationale is that special provisions more thoroughly and accurately define the matter in question, compared to the more general, "boilerplate" provisions of the contract. Where, however, both provisions are deemed reasonable, both are to be retained if possible (*Corpus Juris Secundum* 1963, Sec. 313, p. 176). This rule is to be applied sparingly, cautiously, and only when absolutely necessary (*Corpus Juris Secundum* 1963, Sec. 313, p. 177).

When a contract consists of both printed and typed, or printed and handwritten matter, and a conflict exists between them, the typed or handwritten portion normally prevails. The rule provides

> . . . all parts of a contract, whether written, typed, or printed, must be considered in interpreting the contract. Where there is inconsistency, matter deliberately added by the parties to a contract form must prevail. Thus, written or typewritten matter, or even stamped matter, will ordinarily prevail over printing, and handwriting will prevail over typewriting. (*Corpus Juris Secundum* 1963, Sec. 310, p. 168)

This rule is based on the proposition that the language added by handwriting or typing is a more immediate and reliable expression of intent.

Where there is an ambiguity between the words and drawings, the words normally govern. The rule states,

> Where words and figures in a contract are inconsistent, the words govern. (*Corpus Juris Secundum* 1963, Sec. 311, p. 171)

Caution must be exercised in using these rules, however, because more specific information may be included in the figures.

References

American Institute of Architects (AIA). (1970). *General Conditions of the Construction Contract*, Document A201, American Institute of Architects, Washington, DC.

American Jurisprudence. (1990). 2d, Vol. 17, Lawyers Cooperative, Rochester, NY.

Calamari, J., and Perillo, J. (1990). *Contracts*, 2nd Ed., Black Letter Series. West Publishing Company, St. Paul, MN.

Corbin, A. (1960). *Corbin on Contracts*, Vol. 3, West Publishing Co., St. Paul, MN.

Corpus Juris Secundum. (1963). *Contracts,* Vol. 17, American Law Book Co., Brooklyn, NY.

Dunham, C., and Young, R. (1958). *Contracts, Specifications, and Law for Engineers,* McGraw-Hill Book Company, NY.

McCarthy Brothers Construction Co. v. Pierce. 626 F. Supp. 981 (1975).

Monroe M. Tapper & Associates v. United States, 602 F.2d 311 (1979).

Patterson, E. (1964). "The Interpretation and Construction of Contracts," *Coll. Rev.,* 64, 833, 853.

Restatement of the Law. (1981). 2d., Contracts, American Law Institute, St. Paul, MN.

Williston, S. (1961). *Williston on Contracts,* 3rd Ed., Vol. 4. Baker, Voorhis & Co., Inc., Mt. Kisko, NY.

Chapter 5

Types of Changes and Disputes

When confronted with a dispute, the first thing one must do is identify the type or nature of the dispute. This identification is usually done by applying some judgment to a careful reading of the contract to determine if the contract provides any relief for the alleged problem. The contract wording is critical. Fortunately, for the topics and clauses addressed by this book, all standard contract forms say essentially the same thing. For instance, all contracts require that notice be given in writing before the work is disturbed. The main difference is to whom the notice is given and how much elapsed time is allowed.

Identifying the Type of Dispute

Knowing the type of the dispute is of paramount importance because the primary rules are different for different types of disputes. To illustrate, if the dispute involves an orally directed change, the owner's role is important to the outcome. If the allegation is over work that one party claims was part of the contract work, then owner knowledge is seldom an issue. Likewise, if the specification was supposedly defective, then the intention of the parties is not an issue. Instead, the first question to address is what caused the contractor not to be able to perform as required. To aid in understanding the different types of changes, Fig. 5-1 was developed.

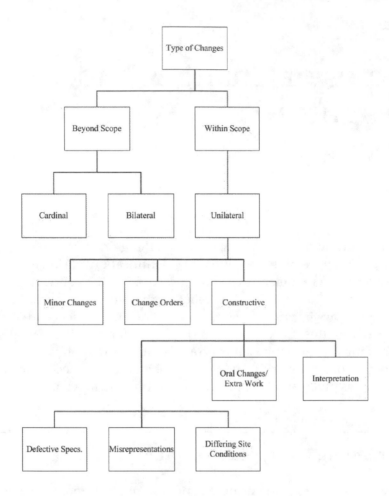

Figure 5-1. Types of Changes.

Bilateral Change vs. Unilateral Change

All contracts require contractors to perform changes to the work that are within the general scope of the work. The contractor's only recourse is to protest and file a claim; however, the contractor must still do the work. These are called *unilateral* changes.

A contract does not allow the owner unilaterally to require the contractor to perform work that is beyond the scope of the contract work, even if the contractor is compensated. The contractor can agree or disagree, and this is called a *bilateral* change. AIA A201 (AIA 1987) calls this a *contract modification.* Contractor consent is required, and the contractor may rightly refuse to do the work. Suppose that a subcontractor has a contract to paint the interior of a building. The prime contractor may not require the subcontractor to paint a building down the street as well. Obviously, there are many gray areas, and legal advice is recommended before one refuses to perform the work.

Cardinal Changes

A *cardinal* change is a term that is sometimes used to describe a single change or accumulation of changes that are beyond the general scope of the contract. Usually, cardinal changes are an accumulation of many small changes that make the project different from the project on which the contractor bid. Therefore, cardinal changes are a breach of contract and are beyond the scope of this book.

Minor Changes

Minor changes are changes that can be ordered by the owner or owner's representative and do not affect the time or cost of the project. If a dispute arises over a minor change, it is usually because the contractor feels that the change will cost money or the contractor deserves more time. An example of a minor change is when the owner changes the color of paint before the contractor purchases the paint.

Change Orders

This term refers to any change that is done in accordance with the terms of the changes clause. There are many common elements about changes clauses, but there are differences, too. Contractors must follow the provisions verbatim to assert a claim. The change order proposal only covers the direct cost of the change. Contractors should always add a written statement reserving the right to claim impact costs at a later time.

Case Study

Meeting of the Minds and Unilateral v. Bilateral Change

Florida State Road Department v. Houdaille Industries, Inc.
237 So.2d 270 (1970)

In 1964, Houdaille Industries, Inc., entered into a contract with the Florida State Road Department to construct project 3502, which was part of the Everglades Parkway (Alligator Alley). The plans showed 408 yard^3 of muck and 622,968 yard^3 of embankment material to be excavated from a borrow canal to be dug adjacent to the roadway. There were numerous problems, including a gross underestimation of muck removal. In fact, Houdaille removed 278,392 yard^3 of muck. Partway through the project, there was a meeting in Tallahassee between the road department and the contractor.

At this meeting, the plans were officially changed in several important respects, and provisions were made regarding payment for overhaul and progress payments. The district engineer was directed to prepare a document showing the changes and to include an estimate of anticipated units of overhaul, increased excavation, and increased embankment material. This document reflected the contract unit prices and was signed by a representative of the contractor. This document represented a change order request (COR). This document was never executed as a change order or supplemental agreement. Regarding this document, the road department representative led the fight for reduced unit prices at the meeting, arguing economies of scale, but the contractor's representative insisted that they should be increased. The road department representative concluded that in the absence of an agreed-on price, the contract unit prices should govern. The contractor's position was that the COR authorized progress payments for the extra work required, but that final settlement of the unit prices could not be determined until after the work was completed. The road department later asserted that the document was a change order, and payment should be made at the contract unit prices contained in the document. Is this COR a valid agreement to apply the contract unit prices? Is this a change order (unilateral) or a contract modification (bilateral), and why?

The First District Court of Appeals agreed with Houdaille. Because the entire method of construction was changed and there were enormous quantity overruns, the court apparently viewed the COR as an attempt by the Florida State Road Department to consummate a new agreement. The court concluded that there was no "meeting of the minds" because of the following facts:

1. The road department did not execute or process the document in the manner required for a valid supplemental agreement.
2. The document did not dispose of other important parts of Houdaille's claim.
3. The road department admittedly refused to negotiate on that phase of the dispute.
4. The road department acknowledged that the parties never reached mutual agreement on final unit prices.

Thus, the COR was not considered a new agreement because the parties failed to reach a meeting of the minds.

Constructive Changes

A constructive change is any change that is not in accordance with the terms of the changes clause. As can be seen in Fig. 5-1, there are several

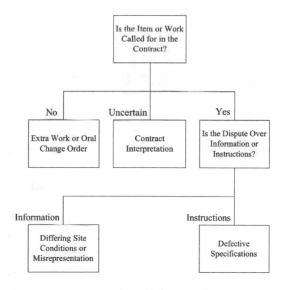

Figure 5-2. Guide to Deciding What Rules Apply.

kinds of constructive changes, and different rules apply to each. Fig. 5-2 demonstrates which set of rules to apply.

Examples

A contractor is ordered to remove existing conduit, which he believes is additional work not called for in the contract. The owner's position is that it is required. Is this extra work? Probably not. The dispute is most likely concerning the interpretation of one or more of the clauses.

Another contractor installs an extra door that is not shown in the contract documents and now seeks payment for the installation materials. The additional work can be evaluated using the rules for extra work or oral change orders given in Chapter 7.

A contractor installs a footer for a structural steel building. Excessive loads and unforeseen soil conditions later cause the footer to fail. Is this a differing site condition? It is more likely an issue of defective specifications. Only if the cause of the failure is faulty information provided is it likely to be a differing site condition. If the problem arose over instructions, in this case, the size of footer, strength of concrete specified, or the amount and location of steel reinforcement, then it could be considered a defective specification issue.

Exercise 5-1: Diamond and the U.S. Assay Office

On March 31, 1939, Arnold M. Diamond entered into a contract with the U.S. Treasury Department for $26,400 to furnish all labor and materials required for improvements to the U.S. Assay Office, 32 Old Slip, New York City.

The U.S. Assay Office was a stone structure in which smelting processes had been carried on. Gold was electrolytically separated from base metals, and for this process, nitric and hydrochloric acids were used as solvents. The windows of the building were protected by steel bars, which, over time, had been corroded by the acids. The building itself had also been affected by the acid because rain had washed the acid down the side of the facade, causing the acid to penetrate the stone beneath the windows and produce peculiar yellow stains.

Diamond's work consisted, in part, of cleaning and pointing the exterior of the building. This work involved two operations: exterior cleaning of stonework by a wet sandblast process and removing stains from the exterior stone surface. Sandblasting cleans stones but does not remove the stains. The sandblast cleaning was satisfactory in all respects and was accepted by the government.

To remove the stains, paragraph 9 of the specifications required that

Stains shall be removed as follows:
a. Thoroughly wet the stone to be treated before applying the acid.
b. Apply with a fiber brush a solution of 85% phosphoric acid composed of one part acid to four parts water.
c. Keep the stone wetted with this solution until the stain disappears. This time will vary from 15 to 25 minutes, depending upon the depth of the stain.
d. Wash thoroughly with a hose to remove all acid.

Diamond proceeded with the work in strict accordance with the procedures outlined in paragraph 9, and at the direction of the government inspector, prepared a sample of stone cleaned according to the specified requirements. However, the phosphoric acid solution had almost no effect in removing stains from the stone. Subsequently, the inspector directed Diamond to use another method. The inspector suggested experiments to determine which acid solution would do the job best. Experiments were made accordingly, and hydrofluoric acid (instead of phosphoric acid) was found to be most suitable for removing stains.

On receiving this information, the inspector suggested that Diamond submit a letter to the Treasury Department requesting permission to use hydrofluoric acid. On May 15, 1939, Diamond wrote a letter addressed to the Procurement Division, Public Buildings Branch, Washington, D.C., as follows:

I propose, in lieu of the method specified, to clean the exterior stonework of the building, by means of steam cleaning.

This will provide the use of a mild Hydrofluoric acid % about 4 followed by a blast of steam.

This work done by the above method will be more satisfactory and I propose to do it at no additional cost to the Government.

Please advise if this is satisfactory.

On May 22, 1939, Diamond received a reply that stated, in part, that "no change of method was desired, except that dilute hydrofluoric acid may be used on the stains in lieu of the phosphoric acid which is specified."

Diamond wet all of the stained stone surfaces thoroughly and applied with a fiber brush a solution of hydrofluoric acid of the strength outlined in the May 15, 1939, letter, keeping the stone wetted by repeated applications with this solution from 15 to 25 min. Diamond then washed the stone surfaces thoroughly to remove all traces of the acid, thus fully complying with specification requirements, except that hydrofluoric acid was used in lieu of phosphoric acid. This procedure resulted in cleaning the building more satisfactorily than had been accomplished through the use of phosphoric acid, but the outcome was still unsatisfactory. When the building was finally inspected, the government construction engineer suggested that Diamond experiment further and submit another proposal for the removal of the stains still appearing under the windows.

Diamond experimented further and devised a method for cleaning the stains and submitted a proposal to the construction engineer in a letter dated June 15, 1939, which read in part,

> Reference is made to the areas under the windows from the fourth floor down, on the east end of the Old Slip side of the building, and under the windows on the South Street side of the building, which although cleaned and treated as required by the contract still show the staining caused by the action of acids on the old windows.
>
> For the sum of $565 I propose to do the following:
> a. Cover the windows as required to protect the glass.
> b. Treat with 8% hydrofluoric acid and wash off after 1 minute.
> c. Treat with 85% phosphoric acid, diluted one part acid to four parts water, and wash off after 20 minutes.
> d. Repeat treatment (b).
>
> Special care will be taken to use sufficient water to remove the acid.
>
> Kindly advise if this meets with your approval.

Diamond's proposal to do the work in accordance with the method outlined in the letter was rejected by the construction engineer in a letter dated July 18, 1939, the last paragraph of which reads as follows:

> You are hereby directed to proceed with this work in accordance
> with the specification covering your contract.

On July 31, 1939, Arnold M. Diamond wrote a letter to the government's supervising engineer in Washington, D.C., in which he took issue with the decision of July 18, 1939, and in closing, stated,

> I wish to protest that I consider this additional work, and believe
> that I should be compensated for doing it.

On August 11, 1939, the government's supervising engineer wrote a letter to Diamond, the third paragraph of which reads in part as follows:

> Since the specifications call for...treating of the exterior
> stonework by the specified method until stains disappear, additional compensation cannot be allowed for this work which is a
> part of the contract requirements.

On August 22, 1939, Arnold M. Diamond wrote a letter to the supervising engineer asking that the matter be reviewed. On November 22, 1939, the government's supervising engineer wrote a letter to Diamond refusing to reconsider the matter.

The work as proposed in Diamond's June 15, 1939, letter was performed by Diamond with the government's knowledge at no additional compensation. Diamond now seeks an equitable adjustment for delay damages.

What set of interpretation rules are applied to this dispute?

Exercise 5-2: Grand Forks and Moorhead Construction

The city of Grand Forks, North Dakota, entered into a contract with Moorhead Construction Co., Inc., to build a sewage treatment facility. The city had divided the construction project into two phases, with each phase to be performed by a different contractor.

Phase I was designed by the city's engineering department and covered primarily the earthwork and site preparation for four aerated anaerobic treatment ponds, including the installation of piping and appurtenances, such as foundations for the compressor and meter buildings. The four ponds or earthen cells were to be formed by building earthen embankments in a square pattern divided into four large, square, watertight sections.

The phase I contractor, Valley-Mayo, was scheduled to complete its work in September 1969, before the phase II work (to be done by Moorhead) was to commence. However, the phase I contractor did not substantially complete its contract until November 1970, some 14 months late. The final acceptance by the city of the phase I work was not given until October 1971, when the contractor was paid in full and discharged.

Phase II of the project was designed by Richmond Engineering, Inc., of Grand Forks, the city's agent and supervisor for the project. Phase II consisted of completing the buildings, constructing manhole installations and access bridges into and over the ponds, and installing all electrical and mechanical equipment. When completed, the aeration equipment would treat the city's sewage primarily in the aerated cells, with secondary treatment occurring in lagoons. A separate phase II contract was awarded to Moorhead in July 1969, with completion scheduled for October 30, 1970. The contract stated in part,

> F20. CHANGED CONDITIONS. Should the contractor encounter or the Owner discover during the progress of the work subsurface or latent physical conditions at the site differing materially from those indicated in this contract, or unknown physical conditions at the site of an unusual nature, differing materially from those ordinarily encountered and generally recognized as inherent in work of the character provided for in the contract, the Engineer shall be promptly notified in writing of such conditions before they are disturbed. The Engineer will thereupon promptly investigate the conditions and if he finds they do so materially differ and cause an increase or decrease in the cost of, or the time required for performance of the contract, an equitable adjustment will be made and the contract modified in writing accordingly.

At the time Moorhead bid on the phase II contract, the phase I earthwork had just commenced. An inspection of the site by Moorhead would not then have disclosed the difficult site conditions that it would later face due to excess moisture and lack of compaction. Moorhead, in estimating its bid, relied on the city to provide a construction site prepared in accord with the specifications of phase I. Those specifications called for 90% compaction of the soil embankments and cell bottoms.

As early as December 2, 1969, Moorhead expressed concern about increased costs related to the delay. On December 19, 1969, Moorhead wrote to the city,

> I am very deeply concerned and perturbed in regards to the contract we hold with the City. We bid the project under certain stipulations. We had to follow scope and time limits to be adhered to. As of this date we have not even been given access to the project.

Before January 1970, when Moorhead was notified to proceed, Moorhead's president inspected the site and refused to take responsibility for it.

> ...I went to inspect the project site and I definitely would not accept accessibility to the site and be responsible for it in its present condition.... We are going to incur additional costs, as to increased labor, material, sales tax, warehousing and scheduling material shipments and placement of our crews to complete this project.

When Moorhead began work, the bottom surfaces of the lagoon cells were extremely soft. As a result of the unstable soil condition actually encountered in the cell bottoms and on the embankments, Moorhead was forced to work by different and more expensive methods, without heavy equipment. Most of the foundation footings for the mechanical installations and access bridges had to be redesigned and spread apart for greater support.

On February 17, 1970, Moorhead wrote to Richmond,

> It is very difficult to construct a job under existing information and complete the same when the conditions and time of availability are not the same as stated under the bidding plans and specifications.
>
> According to the specifications of Phase I, the bottom and slopes shall have 90% compaction. This definitely is not there.

On February 18, 1970, Richmond, relaying Moorhead's letter to the city, stated,

> As we all know, the soil conditions at the site are treacherous. It is entirely possible that 90% compaction by the Phase I Contractor

was a physical impossibility considering the time of year the work was done.

On March 23, 1970, Richmond wrote to the city,

> In discussing the unstable soil matter with Moorhead, they again noted the soil conditions were beyond their control and that they bid the job under the premise that the Phase I Contractor would obtain 90% compaction in the cell bottoms and dikes. Since the dikes are not completely finished and are still frozen it is probably too early to comment on their density, but *the bottoms are definitely a changed condition.*

Due to adverse weather, soil conditions, and delays, phase II was not completed until November 1971. Moorhead claims that the phase II job was entirely changed and greatly increased its construction cost.

What set of contract interpretation rules should be applied to this dispute?

Reference

American Institute of Architects. (AIA). (1987). *General Conditions of the Contract for Construction*, AIA Document A201. American Institute of Architects, Washington, DC.

Chapter 6

Notice Requirements

All contracts require that a contractor give written notice within a specified time as a prerequisite to filing a claim.

Contract Language

Most construction contracts contain procedural requirements regarding how and when knowledge is communicated about situations that may affect the project costs and schedule. Typical contract language can be found in Article 12.3.1 of AIA A201 (1987). It states,

> If the contractor wishes to make a claim for an increase in the Contract Sum he shall give the Architect written notice thereof within twenty days after the occurrence of the event giving rise to such claim.

and in AIA A201 (1987), Article 4.3.3 states,

> Claims by either party must be made within 21 days after occurrence of the event giving rise to such Claim or within 21 days after the claimant first recognizes the condition giving rise to the Claim whichever is later. Claims must be made by written notice.

and in AIA A201 (1987), Article 4.3.7 states,

If the Contractor wishes to make Claim for an increase in the Contract Sum, written notice as provided herein shall be given before proceeding to execute the Work.

Other standard contract forms have language similar to the 1987 version, differing only with respect to who receives the notice and its timing.

Background

Need for Clarification

Two opposing viewpoints have been advanced relative to notice requirements. The decision of the U.S. Supreme Court in Plumley v. United States is generally regarded as a landmark case providing for strict interpretation of the notice requirement; that is, where the contract requires written notice, formal writing is the only form of communication that will suffice. However, there is a substantial body of case law in which compliance on this point has not been enforced. The case of Hoel-Steffen Construction Co. v. United States is often cited to support the proposition that written notice is a mere technicality, and courts have sought to avoid strict enforcement of the written notice. In the cases cited in this chapter, the courts often found that the owner committed certain acts or that the course of conduct between the parties was such that the court could not allow the written notice to be the basis for denying the contractor an equitable adjustment.

Caution should be exercised. As stated by Anderson, there are many instances where the Plumley doctrine has been applied. She also stated that

> Some states will seize upon any fact or circumstance growing out of the conduct of the parties to show waiver of strict compliance with the contract requirements. Reliance on a court to reasonably construe your contract to avoid forfeiture or analyze your conduct to determine if strict compliance with the contract has been waived is ill advised. (Anderson 2005)

Thus, the outcome of a notice dispute may depend somewhat on jurisdiction. It seems clear that if the owner consistently insists on strict compliance with the contract provisions (requiring written notice), there is a stronger likelihood that the Plumley doctrine will be applied. A simple letter may be sufficient to avoid costly legal action. The remainder of this chapter is focused on situations where the Plumley doctrine is not applied (there was not strict compliance with the contract).

Notice requirements have received some discussion in the published literature (Loulakis 1985, Transportation Research Board 1986). However, much of this material is vague and confusing. Rules of application are imprecise and incomplete and are not logically arranged. In some instances, the notice requirements are superficially or simplistically presented, often misleading the reader. More often, the requirements for notice are fragmented and hidden in discussions of differing site condition claims, delays, and changes.

The need for clarification with respect to notice issues is shown by the treatment of the case of State of Indiana v. Omega Painting, Inc., in the "Legal Trends" column of the January 1985 issue of *Civil Engineering*. The column argued that the decision marked a return to the strict application of the contract notice requirements, the Plumley doctrine. The case involved the sandblasting and painting of a bridge structure. Problems arose when the state changed inspectors, and the new inspector required more sandblasting than the contractor felt was necessary to attain the specified finish. The contract contained the provision: "If the Contractor deems that additional compensation will be due him for work or material not clearly covered in the contract or not ordered as extra work, he shall notify the Engineer in writing..." The case was complex and confusing and contained limited factual information. The issues of oral change order and notice were inseparably intertwined. Neither the owner nor the contractor asserted that the change was classified as extra work; rather, the contractor stated that the owner somehow modified the contract. However, the contractor failed to produce evidence showing how the owner waived the notice provision. Furthermore, the state argued that the additional sandblasting had been warranted because of poor-quality workmanship. The principal issues in the case did not relate to notice, and the record seems insufficient to conclude a return to the Plumley doctrine.

The State of Indiana v. Omega Painting case is used at this point to demonstrate that not all judicial decisions follow the flowcharts. At times, courts may revert to some other theory for resolution. The case also points out the danger of reading brief synopses of judicial decisions. One must read the entire case to understand the rationale behind the decision.

Before such a conclusion can be drawn, a study of numerous significant cases must be made and the body of case law evaluated. By isolating a single case, authors can portray courts as capricious and incongruous in their decisions, while missing the remarkable overall consistency that exists relative to certain construction contract issues.

Purpose of Written Notice

The owner has the right to know the liabilities that accompany the bargained-for item. Contractually, the owner preserves this right by requiring that the contractor notify the owner in writing if situations arise that may increase project costs to the owner or may delay completion. As stated by an Illinois court,

> In a building and construction situation, both the owner and the contractor have interests that must be kept in mind and protected .... [T]he owner has a right to know the nature and extent of his promise, and a right to know the extent of his liabilities before they are incurred. Thus, he has a right to be protected against the contractor voluntarily going ahead with extra work at his expense. He [the owner] also has a right to control his own liabilities. Therefore, the law required his consent be evidenced before he can be charged for an extra. (Watson Lumber v. Guennewig)

Additionally, the Court of Appeals of North Carolina stated,

> We are not blind to the possibility that the Contractor in this case encountered considerably changed conditions and extra work. But the position of the Contractor must be balanced against the Commission's compelling need to be notified of "changed conditions" or "extra work" problems and oversee the cost records for the work in question. The notice and record-keeping procedures of these provisions are not oppressive or unreasonable; to the contrary, they are dictated by considerations of accountability and sound fiscal policy. The State should not be obligated to pay a claim for additional compensation unless it is given a reasonable opportunity to ensure that the claim is based on accurate determinations of work and cost. The notice and record-keeping requirements constitute reasonable protective measures, and the Contractor's failure to adhere to these

requirements is necessarily a bar to recovery for additional compensation. (Blankenship Construction v. North Carolina State Highway Commission)

The various courts handling notice-related cases seem to agree that the notice should allow the owner to

- investigate the situation to determine the character and scope of the problem,
- develop appropriate strategies to resolve the problem,
- monitor the effort and document the contractor resources used to perform the work, and
- remove interferences that may be limiting the contractor's ability to perform the work.

Rules of Application

Primary Issues Governing Notice

In deciding disputes involving notice where the Plumley doctrine is not applied, three general questions must be answered:

- Does the notice clause apply?
- Did the owner have the opportunity to limit liabilities?
- Was the requirement waived?

The case law review indicates that these questions are hierarchical and can be organized as a flowchart, as shown in Fig. 6-1.

Does the Notice Clause Apply?

Generally, the question of applicability only occurs with nonstandard contract forms where the notice clause is inadvertently limited to certain situations. For example, in the case of the State of Indiana v. Omega Painting, Inc., the notice clause applied only to "extra work." The additional sandblasting was not extra work, so theoretically, the clause does not apply. As written, the notice clause did not apply to situations

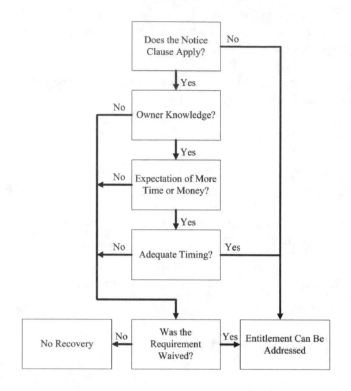

Figure 6-1. Decision Diagram for Disputes Involving Notice.

involving other work unless it could be argued that the sandblasting was truly "extra," which could not be successfully argued.

Did the Owner Have the Opportunity to Limit Liabilities?

Written notice implies that a formal letter has been delivered to the owner, or the owner's authorized agent or representative, clearly identifying the problem and applicable contract provisions and stating that the contractor expects to be compensated or have the contract time extended. If written notice is delivered in a timely manner, the requirements have been satisfied. However, notice can be communicated in ways other than by formal letter. Use of alternative forms, such as verbal statements, often constitute notice, and some courts may elect to set aside the formalities if it is determined that the purpose of the notice requirement has been satisfied. However, reliance on verbal statements is ill-advised. The principal issues considered in making this determination are

- owner knowledge of the events and circumstances,
- owner knowledge that the contractor expects compensation or a time extension under some provision of the contract, and
- timing of the communication.

The owner can protect himself or herself by always following the contract verbatim.

Owner Knowledge

Owner knowledge can be in two forms: actual knowledge and constructive knowledge. Constructive knowledge can be further subdivided into implied and imputed knowledge. Each of these forms of knowledge is defined and illustrated with an actual case.

Actual Knowledge

Actual knowledge refers to knowledge that is clear, definite, and unmistakable. If the facts of a situation have been conveyed orally or in writing so that there is no doubt that the party who requires the knowledge has received it, then actual knowledge has been substantiated. The case of New Ulm Building Center, Inc., v. Studtmann demonstrates the essential elements of actual knowledge. The case involved a couple who negotiated for the construction of a house. The builder refused to sign a written contract for a lump sum, but the parties orally agreed to proceed with construction based on an estimated price. According to the court record,

> The Studtmanns took the plans and material list to New Ulm Building Center, Inc., who agreed in writing to furnish all of the material for the sum of $11,385, plus 3-percent sales tax. That agreement contained the following postscript: "If job runs less Owner will receive credit, but not any extra unless owner is notified." It is undisputed that as the work progressed, there were extensive changes and "extras" and that although the Building Center furnished the Studtmanns with monthly statements of the cost of materials, no specific notice was given to them that these (costs) included extras.

The Studtmanns visited the site daily and were fully aware of the progress of the work. The monthly materials listing from the building center and the daily site visits provided them with the information necessary to compare actual construction with the negotiated quantity and quality of construction. The court found evidence showing that the Studtmanns were fully aware that extras were being included as the work progressed. At the trial, the Studtmanns acknowledged that they knew of most of the extras and had talked with the contractor about them at the time. The Studtmanns knew the contractor expected additional compensation but mainly objected to the price of the extras. Because they did not object to the extras, they were responsible for payment.

Implied Knowledge

Implied knowledge is communicated by implication or necessary deduction from the circumstances, job site correspondence, or conduct of the parties. Although this type of knowledge may not be complete in and of itself, it is usually sufficient to alert the owner of the need for action or further investigation. Implied knowledge is illustrated by the case of Hoel-Steffen Construction Co. v. United States. The case concerned the construction of the Gateway Arch of the Jefferson National Expansion Memorial in St. Louis, Missouri. Several contractors were simultaneously involved with the construction. Hoel-Steffen contracted with the U.S. Interior Department to construct various interior features of the arch, including the ductwork. Working space inside the arch was limited, which resulted in substantial interferences between contractors. According to Hoel-Steffen, some contractors received preferential access to the construction site in a way that was not specified in any of the contract documents. The court stated,

> Where duct work contractor... brought dispute between the prime contractor, transportation system subcontractor and duct work contractor to the government's attention, it was the contracting officer's duty to take action to remedy the difficulty; it was not necessary that duct work contractor specifically accuse the government of "unreasonable or unfair measures in attempting to resolve the problem," it was enough... that the government knew or should have known that it was called upon to act.

In another case, Weeshoff Construction Co. v. Los Angeles County Flood Control District, the contractor recovered because the awarding agency knew that a site inspector was directing unauthorized changes. Although no written notice was ever given, Weeshoff was able to show that the district knew that work outside the contract requirements was being ordered and that the contractor expected additional compensation. Furthermore, evidence of owner knowledge is more convincing if the problem is the owner's fault or involves something within the owner's control (Chaney and James Construction Co. v. U.S.).

Imputed Knowledge

Imputed knowledge can be established when a person in an organization is given actual notice of a fact or circumstance and that individual has the duty to report it to the person affected. The case of Powers Regulator Co. provides insight into the circumstances under which courts recognize imputed knowledge. The contract provided for Powers to install emergency control centers in three Social Security Administration Program Centers constructed for the U.S. General Services Administration. The specifications were highly technical, and the installation was complex. The General Services Board of Contract Appeals found that

> Notice of a specification dispute to a supervising architect employed by the government constituted notice to the contracting officer within the meaning of the changes clause of the contract. The regional architect on the project had the authority to approve or reject the contractor's submittals. Under the circumstances, the actual notice of the architect who had authority to issue changes could be imputed to the contracting officer because the architect was the technical expert to the contracting officer and this was a highly technical claim. The law is settled that a directive need not come from the contracting officer personally, and that he need not necessarily even be aware of it. (Powers Regulator Co.)

The board apparently felt that the circumstances were of such importance that it was the duty of the supervising architect to communicate the problem to the contracting officer. Had the dispute not been of a highly technical nature, the court may have ruled otherwise. However, if the person who makes a decision or knows of a contractor's predicament is properly acting within that individual's scope of authority, the owner may be committed by the agent's actions without personally being aware of the situation. The board further stated in the Powers case that

We thus hold that the contracting officer cannot insulate himself from the operating level by layers of construction managers, architects, and consultants, then disclaim responsibility for the actions of one of his agents because the contractor failed to give him notice. (Powers Regulator Co.)

Implied and imputed knowledge are not always found to exist. Obviously, the outcome of any case depends on the facts. Certainly, courts carefully examine the technical nature of the problem, the authority of those involved, and the project management structure before deciding if the knowledge requirement has been satisfied.

Additional Compensation Expected

Mere knowledge that additional expenses may be incurred is not sufficient to make the owner liable for cost increases. As stated by Anderson,

While owners often have actual knowledge of a contractor's delay, a contractor is ill advised to rely on this knowledge as excusing the need for factual written notice. (Anderson 2005)

And quoting the decision in DOT v. Fru-Con Construction Corp.,

... if DOT's mere knowledge were sufficient, the provision requiring timely written request for an extension of time would be meaningless and superfluous. (Anderson 2005)

The extra cost and time needed may be due to contractor error, and courts have held that if the owner is unaware that the contractor expects additional compensation, the owner is not liable.

Schnip Building Co. v. United States presents this important consideration in a unique setting. Schnip was awarded a contract to build a hobby shop at the navy submarine base at Groton, Connecticut. The contract documents for the foundation showed the presence of high-quality rock and provided for precision blasting and excavation. The smooth rock face was to serve as the exterior concrete formwork as a way to reduce the expense of manufactured wooden forms. During construction, the contractor had to remove excessive excavation spoil caused by overbreak from the blasting operation because, according to the contractor, the rock was not of the quality shown in the contract documents. Schnip sought and received approval from the government representative to alter the formwork requirements so that he could use exterior wooden forms. Whereas the dispute concerned subsurface conditions, which the contractor later alleged differed materially from those represented in the contract drawings and specifications, the court found that

> ... (the government representatives) were personally unaware of existence of such conditions (a DSC [differing site condition] that would entitle the contractor to additional compensation), and that their observations at the job site did not alert them to such condition.... The plaintiff infers that the government should have known of the changed conditions as they were obvious. Whether the government representatives reasonably should have known from the circumstances that subsurface conditions differing from those described in the contract documents were being encountered was a question of fact. The (contractor's) extensive backfill and grade fill requirements could have been caused either by a subsurface condition or by improper blasting technique. The Board considered the evidence and said that it was "unable to charge the Government with constructive knowledge (of a DSC) under these circumstances." The burden was on the appellant (Schnip) to prove to the government why such extensive fill needs existed. The government had no obligations to ferret out the reason. (Schnip Building Co. v. U.S.)

The Schnip case is particularly instructive because it establishes that owner knowledge alone is insufficient to satisfy the notice requirement. The owner representative was present daily and was fully aware that additional costs were being incurred but was unaware of the alleged differing site condition. A reasonable person could have inferred that the extra excavation and formwork costs were caused by an inexperienced blasting subcontractor or other contractor-caused problems rather than from a DSC. The court clearly assigned to the contractor the duty to ensure that the government was aware of the conditions and that the contractor expected additional compensation. In this case, the contractor failed to be specific regarding the incurrence of additional costs and did not assign the cause of the added costs to the alleged DSC. The government did not have the opportunity to limit its liabilities because it did not know that the contractor expected more monies.

The requirement for positive representations relative to additional compensation cannot be overemphasized. For example, in another case, the contractor's claim was rejected because notice was characterized as a complaint rather than formal notice, and no intent to file a claim was ever asserted (Blankenship Construction v. North Carolina State Highway Commission).

Timing of Notice

The timeliness of the notice is the final consideration used to determine if the intent of the notice provision has been met. Occasionally, an owner can have knowledge that there is a situation outside the contract for which the contractor expects additional compensation, but that knowledge becomes available so late as to prejudice the owner; that is, the owner cannot control his or her liabilities. The Schnip Building Co. case also illustrates this point. Notice was not given by Schnip until the contractor filed his claim, which was long after the work was completed. The court found that

> The lack of a timely notice was prejudicial to the Government because it effectively prevented any verification of appellant's claim and also the employment of alternate remedial procedures. (Schnip Building Co. v. U.S.)

In the Powers case, the board stated,

> Regardless of terminology, the issue is whether the government has been unnecessarily put at risk—either the risk of additional liability to the contractor or the risk of being unable to prepare and present its defense against the contractor's claim—by the contractor's delay in notifying the government of pertinent facts. (Powers Regulator Co.)

Clearly, notice must be provided in time for the owner to make an independent assessment of the situation, decide what action to take, and monitor the additional work, if desired. This right to control one's liabilities is the key consideration in determining timeliness. Generally, if the contract specifies a time limit and the contractor is several days late in filing the notice, the contractor is not precluded from recovery so long as the owner has been given an opportunity to control his or her liabilities other than for minor inconveniences (C. H. Leavell & Co.; Bramble and Callahan 1987). Waiting until the end of the project to file notice clearly does not afford the owner the opportunity to control liabilities.

Owner prejudice occurs when a delay in providing the owner with timely notice prevents the owner from mitigating damages. The owner may need to demonstrate that extra expenses could have been minimized or avoided or that the passage of time obscured information relevant to verification of the claim.

Was the Requirement Waived?

If constructive notice has not been given, the only recourse available to the contractor is to show that the notice requirement was waived. According to *Anson's Law of Contract*, an owner's actions can waive the owner's right to insist that the contractor perform in accordance with the contract requirements (Guest 1984). The owner is prevented from insisting on strict compliance with the contract requirements where the owner's actions have clearly been in conflict with the same requirements. (See Table 4-1 in Chapter 4, Rule No. 8.)

Sweet views waiver in terms of three subissues (Sweet 1989, p. 468):

- Is the requirement waivable?
- Who has the authority to waive the requirement?
- Did the facts claimed to create waiver lead the contractor to reasonably believe that the owner had eliminated the requirements?

Requirement for Waivability

If statutes or ordinances exist that require written notice, the requirement cannot be waived. This situation would most likely occur when the owner is a municipality, township, school board, or some other local public entity.

Authority to Waive

The contract provision requiring written notice can only be waived by the owner or the owner's designated representative. In the case of Crane Construction Co., Inc., v. Commonwealth of Massachusetts, the court determined that the architect had no authority to waive the notice requirements, and the contractor claim was denied.

Conditions of Waiver

As defined by Black's Law Dictionary,

> A waiver is implied where one party has pursued such a course of conduct with reference to the other party as to evidence an intention to waive his rights or the advantage to which he may be entitled . . . provided that the other party concerned has been induced by such conduct to act upon the belief that there has been a waiver, and has incurred trouble or expense thereby. (Black 1979)

Waiver is the first step in a two-step process, the second step being estoppel. Waiver leads to estoppel when a party relies on the waiver and acts on it accordingly. Thus, waiver is a two-step process of the owner waiving a right and the contractor acting on the waiver in such a way that prevents the owner from reasserting that right.

The principles of waiver and estoppel are illustrated in the case of E. C. Ernst, Inc., v. Koppers Co., Inc. Koppers was the turnkey prime contractor for the design and construction of an A-5 coke oven battery and related facilities at Aliquippa, Pennsylvania. The oven was to be used to produce coke as part of the steel-making process. Ernst was the electrical subcontractor. Koppers was nearing completion of a similar facility in the Midwest and was using that design as a basis for the Aliquippa project. The technology of the project was state of the art, and throughout construction, Koppers was altering the design to incorporate lessons learned from the Midwest facility, which was experiencing numerous start-up problems. All drawing revisions delivered to Ernst by Koppers were marked "Approved for Construction." The design changes, coupled with requests from the owner and engineering difficulties created by Koppers, increased Ernst's actual work hours to more than double the original estimate. A provision of the electrical subcontract required that any requests for additional compensation, including the amount requested, be submitted to Koppers within 30 days after receiving revised drawings. Because of the volume and magnitude of the changes received from Koppers, Ernst was unable to comply with the 30-day requirement. Ernst wrote Koppers and asked Koppers to waive the 30-day requirement. Koppers did not respond to Ernst's letter. Ernst wrote again, stating that because there had been no response to the earlier letter, Ernst assumed that Koppers was waiving the 30-day requirement. Again, Koppers did not respond, even though the second letter was circulated internally among several departments. In ruling in favor of Ernst, the court stated,

> We find that the conduct of Koppers in failing to insist on the 30-day notice provision in light of their "approved for construction" orders to proceed and their failure to reply to Shannon's [Ernst superintendent's] letters, prevents Koppers from now using this clause as bar to Ernst's actions. (E. C. Ernst v. Koppers Co.)

By failing to respond, Koppers waived its right to insist on notice. Furthermore, by Ernst's continued performance under the contract in reliance on Koppers' silence, Ernst gave up its ability to comply with the 30-day requirement on a major portion of its work. Koppers was aware that the "extras" were not being priced and that written authorization had not been given as required by the contract, yet Koppers willingly accepted Ernst's performance. Koppers was thus prevented (estopped) from using the notice clause as a defense to avoid payment to Ernst for delay damages and compensation for extra work.

Additional circumstances also exist under which an owner can waive the notice requirements. For example, if the owner pays for a change order or extra work for which notice was not provided, the owner may be precluded from insisting on notice for other similar work performed thereafter unless the contractor has been advised accordingly.

Other Issues

Several issues, although not directly included in the primary rules governing notice, are nevertheless important considerations in certain instances.

Would the Owner Have Acted Differently?

Situations can arise when the owner may not have been prejudiced despite the lack of notice. In such a situation, the contractor must show that the owner would not have acted differently had notice been properly communicated. A case illustrating this consideration is Sante Fe, Inc. The contract called for the construction of a 520-bed hospital at the Veterans Administration Medical Center in Bay Pines, Florida. The dispute involved the proper installation of lighting fixtures. The supervising architect, who was also the government's technical expert, was aware of problems but failed to communicate these to the authorized government representative. The Board of Contract Appeals stated,

> Boards of contract appeals, in practice, will not enforce this technical clause [notice provision] absent a showing of prejudice by the Government. The Government has the burden of proving that prejudice resulted from its lack of written notice. To meet its burden, the Government must demonstrate affirmatively "how the passage of time in fact obscured the elements of proof" or "how the Contracting Officer might have minimized or avoided possible extra expenses" ... There is no indication that the Government would have acted differently, with respect to its rejection, regardless of a notice of claim. That is, the lack of written notice did not prejudice the Government. (Sante Fe, Inc.)

The board felt that the government was not prejudiced because it would have merely referred the matter back to the supervising architect. However, the Sante Fe case may not reflect the prevailing judicial attitude because even though the government had no other alternative, it nevertheless was not afforded the opportunity to document the actual costs to the government.

Was the Contract Breached?

Some courts have been willing to set aside the requirements for notice if there has been a breach of contract. Nat Harrison Associates, Inc., v. Gulf States Utilities Co. illustrates this concept. The contract called for the construction of approximately 158 mi of 500 KVA single-circuit, three-phase transmission lines in Louisiana. After the contract was awarded, a revision to the contract was negotiated for the construction of an extra tower arm to a portion of the transmission line on which an additional line was to be strung. The pertinent facts are as follows:

Gulf States Utilities Co. was late in providing both right of way and materials and did not extend the contract completion time as provided for in the contract. In addition, Gulf States either ignored or refused time extensions requested by Harrison resulting from delays due to inclement weather, strikes, and other conditions. Harrison sought damages resulting from acceleration of the contract and claimed damages for breach of Gulf States' duties under the contract to furnish the right of way and materials so that Harrison could perform its work in a timely manner and in sequence. One important issue centered on whether the costs allegedly incurred by Harrison were "extra costs" within the meaning of the notice provision. The court stated,

> There is a point, however, at which changes in the contract are to be considered beyond the scope of the contract and inconsistent with the "changes" section (clause). Damages can be recovered without fulfillment of the written notice requirement where the changes are outside the scope of the contract and amount to a breach. Since the evidence supports the jury's finding that there was a breach of contract, we are unable to hold, as a matter of law, that Harrison was required to give prior notice of the additional costs it claims here or that it is not entitled to damages for fundamental alteration of the contract. (Nat Harrison Associates v. Gulf States Utilities Co.)

Thus, the Louisiana court refused to enforce the notice requirement because there was a contract breach. Importantly, the court decision dealt little with other relevant issues, such as owner knowledge, prejudice, and waiver. The implication of the Harrison decision is that the question of breach is supreme in the decision hierarchy (in some jurisdictions) and that where a breach occurs, the remaining questions need not be addressed.

The judicial attitude toward breach and notice is not always unanimous. For instance, in the case of Buchman Plumbing Co., Inc., v. Regents of the University of Minnesota, the court refused to set aside the requirement even though a breach occurred. The court stated that "compliance with provision in construction contract requiring written notice... for damage by way of extra cost was condition precedent to contractor's maintenance of action for breach of contract."

Form of Communication

A formal letter is normally the anticipated form of communication; however, other forms may suffice. Notice can occur in job site correspondence, letters, memos, and other documents. Minutes of project meetings that summarize discussions about problems requiring notice may also be sufficient, provided that they are accurately drafted. One court held that the form of the contractor's statements and objections made at meetings and requests for reconsideration of the government's rejection of submittals was sufficient notice (Chaney and James Construction Co. v. U.S.). In another case, a document drafted by the government agent that clearly indicated the agent's knowledge constituted notice (Crane Construction Co. v. Commonwealth).

Critical path method (CPM) schedule updates have been found to be adequate to alert the owner of a delay. In Vanderlinde Electric v. City of Rochester, it was determined that monthly updates kept the owner "fully and continuously aware" of delays. However, updates that contain errors, are inaccurate representations of job progress, or fail to assign responsibility for the problem are probably insufficient to constitute notice. Also, nonperiodic updates or mere submissions of schedule adjustments are probably inadequate (Lane-Verdugo).

Verbal notice can also suffice (New Ulm Building Center v. Studtmann). However, in one instance, an extended phone conversation with the chief engineer was not considered sufficient for notice (Blankenship Construction v. North Carolina State Highway Commission). An analysis of the various decisions clearly illustrates that the content of the communication is more important than the form, and the Plumley doctrine is not always applied.

Requirements for Additional Detail

Some contracts require that the notice be accompanied or followed immediately by submission of detailed information regarding cost or delay effects. However, compliance with such provisions is sometimes difficult on an active construction site. In some cases, courts have found such requirements too onerous to enforce, considering the brief time allowed for submission of the notice (E. C. Ernst v. General Motors Corp.).

Apparent Authority

As a general rule, communication of delays and problems affecting costs must be made to the person having the authority to initiate or issue changes. Communications to others may result in the claim being denied. The law is difficult to analyze in this area. However, there is some legal basis for stating that notice conveyed to an agent or a person with apparent authority is the same as conveying notice to the person with authority (Powers Regulator Co.). Caution suggests that a contractor deliver notice to a corporate officer (in a private case) or an executive officer of a public agency (in a public case).

Repetition of Events

Once notice is given, no further notice is required when the same conditions recur throughout the job (Ginsburg and Bannon 1987). This statement may not be true when the conditions lead to separate costs. The wise thing to do is to continue to provide notice.

Notice to Others

In some situations, it is necessary to provide notice to others, such as the performance bond surety. The bond may or may not waive the requirement for notification of claims or changes. Failure to notify the surety of the change may invalidate the performance bond.

Illustrative Example

The following example is based on a 1989 case involving McDevitt & Street Co. and the Marriott Corp.

Statement of Facts

McDevitt & Street Co. entered into an agreement with the Marriott Corp. for the construction of a new motel. Part of the work included the installation of elevators.

The contract specifications required the contractor to install a "pre-engineered" elevator (PX-2, Specs., Section 14240, Part 1.02). The specifications also listed three pre-engineered elevator manufacturers, the name and telephone number of a contact person, and the names or model numbers of acceptable models (PX-2, Specs., Section 14240, Part 2.02). The "Fastrack III" model manufactured by the Westinghouse Elevator Co. was one of the models listed. On April 25, 1986, Westinghouse informed McDevitt by letter that its Fastrack III jamb design could not accommodate the wall thicknesses called for by Marriott's architect in revised elevator shop drawings (PX-56, at 4; Peters, Tr. at 296, 298–00, 423). Westinghouse ultimately agreed, at McDevitt's urging, to modify its standard, pre-engineered elevators to fit the architect's drawings for an additional $6,364.

On or about August 29, 1986, McDevitt issued a change order to Westinghouse, authorizing the necessary modifications. The modified elevators were subsequently installed in the project. McDevitt submitted Proposal No. 21, dated October 8, 1986, to Marriott requesting an additional $6,364 as compensation for the costs of providing and installing wider, nonstandard door jambs at six elevator doors.

The relevant contract language came from Articles 17 and 19, which stated, in part,

> Art. 17(a) ... Contractor shall not perform any change in the Work without prior written authorization.

> Art. 18 ... If the Contractor claims that any instructions by Drawings (including Shop Drawings) or written clarification or otherwise involve extra costs under his Contract, he shall give the Owner's Representative written notice within seven (7) days, complete with detailed labor and material costs. Except in an emergency which endangers life or property, the Contractor shall not proceed without written acceptance in the form of a Change of Contract or written notice to proceed from the Owner. The failure of the Contractor to provide such written notice shall deprive the Contractor of its right to claim such extra costs.

McDevitt failed to give Marriott the seven days' written notice of the additional costs required to accommodate the Marriott architect's drawings for the elevators as expressly provided for in Article 18. Until its submission of Proposal No. 21, McDevitt did not formally notify Marriott of this problem or of the additional costs to modify the elevators.

Analysis

Does the Notice Clause Apply?

The provisions of Articles 17 and 18 clearly include shop drawings and cover all types of changes or instructions. So the notice clause does apply.

Did the Owner Have Knowledge?

McDevitt & Street had no formal conversations with Marriott before issuing a change order directing Westinghouse to modify the elevator
doors. Another six weeks elapsed before Proposal No. 21 was presented
to Marriott. This was the first time Marriott knew of the cost increase.
There is no indication in the facts that Marriott should have known or
that knowledge was implied or imputed. This is true even though the architect was perhaps somewhat aware that there was a problem. On this
basis, McDevitt & Street should not be allowed to recover the added
costs because there was no knowledge.

Was McDevitt & Street's Expectation of Additional Compensation Known?

This question is irrelevant because Marriott did not know of the change,
but it is important to note that even if Marriott had known of the need
for the elevator revision, it was not automatic that there might be an increased cost. Marriott may have chosen to modify the walls (possibly at
no additional cost) rather than the elevator door.

Was the Timing Adequate?

The question does not apply in this situation.

Was the Requirement Waived?

There is no indication that sloppy contract administration created a
waiver to the requirement.

Discussion

In this disagreement, McDevitt & Street must incur the cost of modify-
ing the elevator. This was the outcome of flowchart analysis (Fig. 6-1)
and the court decision. Failure to provide notice is a common problem
for construction contractors, and potentially a fatal one. Both parties
can easily protect their rights by following the contract. In this case,
Marriott followed the contract (no waiver was found); McDevitt
& Street did not. Some jurisdictions, such as Virginia and Maryland,

require strict adherence to the notice provisions of the contract, whereas other jurisdictions may take a more liberal approach in evaluating whether a waiver has occurred.

Exercise 6-1: Granger Contracting and Chiappisi Brothers

Granger Contracting Co., Inc., entered into a contract to build a high school. Granger made a subcontract with the Chiappisi Brothers, under which Chiappisi agreed to furnish all labor and materials for completing the lathing and plastering work covered in paragraph 19 of the specifications. By the subcontract, Chiappisi agreed to be bound to Granger by the plans and specifications, including all general conditions. The general conditions of the prime contract included (Article 16),

> If the contractor claims that any instructions by drawing or otherwise involve extra cost under this contract, he shall give the Architect written notice thereof within a reasonable time of receipt of instructions or otherwise, and in any event before proceeding to execute the work, except in emergency endangering life or property ... No such claim shall be valid unless so made ...

and Art. 37,

> The contractor agrees to bind every sub-contractor and every subcontractor agrees to be bound by the terms of the ... General Conditions, the Drawings, and Specifications as far as applicable to his work including the following provisions of this article ... The subcontractor agrees ... to make all claims for extras ... to the Contractor in the manner provided in the General Conditions of the Contract ... for like claims by the Contractor upon the Owner, except that the time for making claim for extra cost is one week.

The work to be done by Chiappisi included the installation of Zonolite spray insulation on the metal flute deck over the science area of the building. The flutes were to be filled with insulation, and then a 1¼-in. coat of Zonolite was to be applied over the entire surface. Chiappisi, from a drawing that dealt with the science area of the building, measured the size of the flute openings as shown on that plan. From these measurements and the area of the deck, an estimate was made of the materials and labor necessary to fill the flutes and cover the surface. Chiappisi used these computations in bidding the job. The drawing that was used by Chiappisi in making the calculations, if measured to scale, shows the flute openings in the metal deck to be 1¼ in. wide at the mouth. This drawing was, however, intended by the architect who prepared it to be schematic or symbolic, rather than an actual diagram of the roof deck. The drawing did not disclose this intention clearly. The drawing used by Chiappisi may have been inconsistent in some respects with other drawings and paragraph 19 of the specifications for the metal deck.

The metal deck was installed in accordance with the manufacturer's instructions referred to in paragraph 19 and the shop drawings (approved by the architect). The shop drawings were never shown to Chiappisi until the time of installation. As installed as per the manufacturer's instruction and the approved shop drawing, the roof deck had flute openings of a width of 4¼ in. If installed as per the diagram in the drawings, the metal flute spaces to be filled would be 1¼ in. wide and ½ in. deep. If installed as required by the shop drawings, the flute spaces to be filled would be 4¼ in. wide and of the same depth. The architect's drawing relied on by Chiappisi showed the narrow flute spaces on the underside of the flute deck. They were installed with the narrow flute spaces pointing upward and the broad flute spaces pointing downward, as the manufacturer intended.

While the work was in progress, more material was needed than Chiappisi had estimated. Alphonse Chiappisi went to the school and found that the flute openings that were being filled were wider than had been estimated. The orientation of the flutes was opposite to the plan sheet on which Chiappisi had relied. Before Chiappisi left the area, he saw Mr. Gillette, the job superintendent for Granger. He told Gillette that the work on the flutes was extra. Gillette replied that the roof and flutes were installed as per the manufacturer's instructions. Chiappisi then said that he had to use more material than estimated and he wanted to be paid for it. Gillette replied that Chiappisi would have to take it up with "the office." Chiappisi ordered more Zonolite, and his men continued with the science area roof deck and completed filling the flutes. In a bill sent to Granger on April 29, 1963, after the work in the science area was completed, Chiappisi made no reference to any extra work in the science area roof deck.

On May 23, 1963, Chiappisi wrote to Granger asking for $2,927 for added work involved in connection with the roof deck. This was the first written notice that Granger received from Chiappisi, and it was the first notice given to anyone at Granger other than to Gillette.

Should Chiappisi be paid? Was constructive notice given? Was the notice requirement waived? Did Granger have an opportunity to control its liabilities?

Exercise 6-2: Linneman Construction and Montana-Dakota Utilities Co.

Linneman Construction, Inc., is a Colorado corporation engaged in the construction of gas, water, telephone, and electric lines. Montana-Dakota Utilities Co., Inc., is a regulated public utility engaged in supplying gas and electric service in North Dakota, South Dakota, Montana, and Wyoming, and gas service only in Minnesota. As part of Montana-Dakota's "Progress 70" plan to extend its gas service in North Dakota to an additional 12 towns, Montana-Dakota issued invitations to bidders in February 1970 for the construction of gas distribution and service lines in the 12 towns. Bidders were allowed to bid on all 12 towns or on the six steel-gas pipeline towns or the six plastic-gas pipeline towns.

Provided to the bidders was a list of estimated total pipe footage to be laid and an estimate of the number of service lines to be installed, a copy of the proposed contract and specifications, and maps of each town showing where the distribution mains were to be located within each town. The specifications and the maps were expressly made a part of the contract to an extent consistent with the contract provisions. Linneman's bid for the six steel-gas pipeline towns was accepted by Montana-Dakota, and the contract was signed March 16, 1970.

The contract was a unit-price contract. The pertinent unit-price provision provided for a single unit price based on the diameter of the main and covered

> For unloading and hauling millcoated steel pipe, fittings, wrappers and other materials, clearing right-of-way, ditching, coating, laying, construction welding, connecting, cleaning, testing, repairing, backfilling, roll packing, and grading...

The contract further provided

Distribution mains in general shall be approximately 5–10 feet from the property line in the street and 5 feet from the property line in the alley. Where more specific information is required it will be furnished by the Company.

2. Location of Lines

The lines shall follow the locations as shown on maps furnished by the Company or staked by the Company engineers. The Company reserves the right to make any changes in location which it deems necessary and such changes shall in no manner alter the items of this contract except as to linear measurements.

The "extras" clause of the contract states,

Contractor shall be allowed no additional compensation for any extras or any work performed by the Contractor not contemplated by the agreement or by said plans and specifications, except under written order signed by the Company's representative, which order shall specify the amount payable to the Contractor on account of such extra work. Where no specific amount is agreed upon for extra work and for which no unit price is set forth in this proposal, the Contractor agrees to execute such work on the basis of cost plus 15% for overhead and profit. Contractor to furnish hourly cost rates to be charged for any extra work. In no event will bills or claims for extras or extra work so ordered be allowed unless submitted to the Company within thirty (30) days from the date of furnishing or completion of extra work.

Linneman began construction of the gas distribution systems on May 5, 1970, in Jamestown, North Dakota. During the course of construction, completed in November 1970, more than 766,000 ft of gas mains were laid in the six towns. All of the footage was paid for at the unit price specified in the contract. Linneman demanded payment for an additional $460,000 for the laying of gas distribution mains outside the paved portions of the streets. The computation was based on the cost of laying some 160,145 ft of mains behind the curb, plus 15% for profit and overhead as called for by the "extras" clause of the contract.

Linneman contends that all the lines were to be located in the paved portion of the streets, where it was faster and cheaper for Linneman to lay the mains. After the Linneman contract was signed, Montana-

Dakota received another bid for repaving the streets that was much greater than Montana-Dakota had estimated. Linneman alleges that Montana-Dakota began moving as much of the construction as possible out of the paved portion of the street and behind the curbs to avoid substantial repaving costs. Moving the gas lines away from the paved streets saved Montana-Dakota the (Linneman) contract prices for cutting and removing the old pavement in addition to the cost of repaving, while increasing threefold the cost to Linneman of laying the mains behind the curbs. To further buttress its contentions that the location of mains behind the curbs was not contemplated by the contract, Linneman contends that the maps attached to the contract show that the mains were to be located within the paved portion of the streets.

No written change order was prepared. On May 18, 1971, six months after completion of the project, Linneman wrote to Montana-Dakota stating that the total balance due under the contract was $43,510.65. This amount represented the retainage due and a minor amount for extras that had already been approved. Linneman filed a claim for the $460,000 some 10 months after completion of the job. This figure ($460,000) represented the total cost over and above the unit price already paid by Montana-Dakota for the 160,145 ft of mains behind the curb.

Should Linneman be paid the $460,000? Was constructive notice ever given?

Exercise 6-3: Acchione & Canuso and PennDOT

On October 5, 1973, Acchione and Canuso, Inc., entered into a contract with the Pennsylvania Department of Transportation (PennDOT) for improvements to Roosevelt Boulevard in Philadelphia. Trenching work was necessary to replace conduit for traffic signal wiring at existing intersections.

After reviewing the plans, Acchione's subcontractor, Tony DePaul and Sons, Inc., noted a discrepancy between its calculation of 16,658 linear ft of trenching and the bid documents, which called for 13,131 linear ft. As a result, DePaul telephoned PennDOT's engineers and was informed that PennDOT had directed its engineers to assume that some of the existing conduit would be reusable, thus requiring no trench excavation. Relying on its own calculations and this oral representation, DePaul calculated a unit price of $24 per linear foot. This figure was the average cost for earth excavation, pavement, and roadway trenching.

The contract contained the following provision:

> The contractor . . . shall not be entitled to present any claim . . . to the Board of Arbitration of Claims . . . for additional compensation . . . unless . . . it shall have given . . . the Engineer a 10-day due notice in writing of . . . its intention to present a claim . . .
>
> The "due notice in writing", as required above, must have been given to . . . the Engineer within 10 days of the time the contractor performed such work or any portion thereof.

As the project progressed, PennDOT discovered that most of the conduit could not be reused and that the engineers had failed to include a contingency item in the contract to cover this situation. PennDOT orally directed DePaul to perform the necessary trenching and then executed a written work order calling for an additional 17,433 ft to be paid for at the contract unit price of $24 per foot. According to Acchione, the additional trenching materially changed the character of the work and the unit cost. When the job was finished, Acchione was paid at the rate of $24 per foot for all work performed, 34,738.70 linear ft.

On August 4, 1975, after substantially completing all the trenching, Acchione wrote to PennDOT requesting an adjustment in the contract unit price. Acchione had worked on the trenching under the written work order for more than eight months. Acchione now argued that it did not realize that its costs had increased until July 1975.

Should Acchione be paid? Was constructive notice given?

Exercise 6-4: McKeny Construction Co. and Rowlesburg, WV

A flood that occurred in late 1985 severely damaged water and sewer facilities owned by the Town of Rowlesburg, West Virginia. To correct the damage, the town contracted with Lennon, Smith and Souleret (LSS) Engineering Co., Inc., to design new facilities. LSS developed three project proposals: one for sewer repair, one for a river intake, and one for the construction of a chlorine tank. McKeny Construction Co., Inc., proposed on all three projects and was awarded all three separate contracts.

The relevant contract provisions were the same in all three contracts and stated,

> No claim for an addition to the contract sum shall be valid unless
> so ordered in writing.

and

> Neither the contractor nor the surety shall be entitled to present
> any claim or claims to the Owner either during the prosecution
> of the work or upon completion of the Contract, for additional
> compensation for any work performed which was not covered
> by the approved Drawings, Specifications, and/or Contract, or
> for any other clause, unless due notice of his intention to present
> such claims as hereinafter designated.
>
> The written notice, as above required, must have been given to
> the Owner, with a copy to the Engineer, prior to the time the Con-
> tractor shall have performed such work or that portion thereof
> giving rise to the claim or claims for additional compensation or
> shall have been given within ten (10) days from the date the Con-
> tractor was prevented, either directly or indirectly, by the Owner
> or his authorized representative, from performing any work pro-
> vided in the Contract, or within ten (10) days from the happening
> of the event, thing, or occurrence giving rise to the alleged claim.

As the work progressed, McKeny was directed to make a number of
changes and to do a substantial amount of additional work. As a result
of the changes, eight change orders were issued as per the contract, and
the contract sum was modified accordingly. McKeny failed to give notice of certain additional work, and no written change orders were ever
issued for that work. The town refused to pay for the additional work
for which it had received no notice or for which no change order had
been issued.

Is McKeny entitled to compensation for its additional work?

Exercise 6-5: Northern Improvement Co. and South Dakota State Highway Commission

On August 15, 1968, Northern Improvement Co., Inc., entered into a
contract with the South Dakota State Highway Commission to construct a highway improvement project on U.S. Highway 212 in Cod-

dington County, South Dakota. The scope of Northern's work involved earthmoving, grading, gravel, and asphalt work. The project was to be completed in 150 working days. The contract contained the following language:

> 4.4 EXTRA WORK—Work not originally contemplated, or work made necessary by minor alterations of the plans, which is deemed necessary or desirable in order to complete fully the work as contemplated, and for which no provision or compensation is provided in the proposal or contract, shall be deemed extra work, and shall be performed by the Contractor in accordance with the specifications as directed by the Engineer; provided, however, that before any extra work is started a supplemental agreement (change order) shall be entered into, or an extra work order issued . . . . Claims for extra work which have not been agreed to beforehand in writing will be rejected.

> 5.14 CLAIMS FOR ADJUSTMENT AND DISPUTES—In any case where the Contractor deems that, as defined herein, the Contractor shall notify the Engineer in writing of his intention to make claim for such extra compensation before he begins work on which he bases the claim. If such notice is not given, and the Engineer is not afforded proper facilities by the Contractor for keeping strict account of actual cost as defined for force account then the Contractor hereby agrees to waive the claim for such extra compensation.

Part of the work required of Northern was the construction of a detour. Because of a deficiency in the roadbed material, the specified quantity of gravel placed at the detour location was inadequate to produce a stable roadway in the shoulder areas. To further complicate matters, the gravel contained insufficient binder clay to enable Northern to lay and compact it properly. Northern immediately orally requested that the engineer allow it to add clay as a binder at no additional cost to the state. However, the engineer refused to order any deviation from the specifications. Northern then orally advised the engineer that the instability of the detour would result in substantial additional maintenance cost for which Northern expected to be compensated. The engineer responded by ordering Northern to dig up the various soft spots and dry the material. Northern performed the work but under oral protest. Subsequent maintenance problems did in fact arise continuously throughout the project because the detour design was defective, and these problems cost Northern valuable time, in addition to the cost of the maintenance work itself. Northern was required to assign an entire crew whose sole job was to maintain the detour and repair it as it broke.

Northern repeatedly requested written supplemental agreements and written work orders, but these requests were consistently denied. On several occasions, Northern discussed the problem with the resident engineer, the district engineer, and department engineering personnel in Pierre, South Dakota. The engineers acknowledged that the detour was underdesigned. When the project was finally completed, Northern submitted a claim for its additional costs. The claim was denied because Northern had not submitted written notice as required by the contract.

Should Northern be paid? Was constructive notice given?

Exercise 6-6: D. Federico Co. and Commonwealth of Massachusetts

The D. Federico Co., Inc., entered into a contract with the Commonwealth of Massachusetts to perform certain development work at a campsite at Wompatuck State Park in Hingham, Massachusetts. The contract contained a typical notice provision requiring the contractor to notify the engineer of his intention to file a claim before he began work on items in question.

A dispute arose during the work because the excavation and borrow requirements for the roads were substantially underestimated by the consultant retained by the Commonwealth. No test borings or soil analyses were done in the area of the roads. Instead, the designer based his estimate on an assumption that, because roads had been constructed earlier in the vicinity (the campsite had been used earlier as a naval ammunition depot), it would not be necessary to excavate beyond an average depth of 12 in. In fact, excavation revealed substantial amounts of unsuitable organic material and nested boulders, all of which had to be removed. Instead of excavating 85,650 yard^3 of material, the amount the designer had estimated, Federico was required to excavate 215,492 yard^3 of material. The amount of ordinary borrow required to be brought in from other locations amounted to 278,493 yard^3, rather than the 30,150 yard^3 that had been estimated.

Federico was paid at the unit price rate in his contract, which amounted to $595,430.90. However, as the job neared completion, he realized that the contract price did not adequately compensate him for the work and that fair compensation should be $913,522.40. Federico now seeks compensation for the difference of $318,091.50. His argument is that the earthmoving operation was an entirely different type of job in scope and magnitude from the one he offered to perform in his bid.

Should Federico be paid extra compensation? Was constructive notice given?

References

American Institute of Architects (AIA). (1987). *General Conditions of the Contract for Construction*, AIA Document A201. American Institute of Architects, Washington, DC.

Black, H. C. (1979). *Black's Law Dictionary*, 5th Ed., West Publishing Co., St. Paul, MN, 1752.

Blankenship Construction Co. v. North Carolina State Highway Commission, 222 S.E.2d 452, 462, 28 N.C.App. 593 (1976).

Bramble, B. B., and Callahan, M. T. (1987). *Construction Delay Claims*, John Wiley & Sons, Inc., New York, 30.

Buchman Plumbing Co., Inc., v. Regents of the University of Minnesota, 215 N.W.2d 479, Sup.Ct., Minn. (1974).

Chaney and James Construction Co., Inc., v. United States, 421 F.2d 728 (Ct.Cl., 1970).

C. H. Leavell & Co. (Appeal), ASBCA No. 16099,73–1 B.C.A. (CCH) para. 9781 (1973).

Crane Construction Co., Inc., v. Commonwealth, 290 Mass. 249, 195 N.E., 110, (1935).

DOT v. Fru-Con Construction Corp. 426 SE.2d 905 (1993).

E. C. Ernst, Inc., v. General Motors Corp., 482 F.2d 1047 (5th Cir., 1973).

E. C. Ernst, Inc. v. Koppers Co., Inc., 476 F.Supp. 729, 626 F.2d 324 (3rd Cir., 1980).

Ginsburg, G. J., and Bannon, B. A. (1987). *Calculating Loss of Efficiency Claims*, Federal Publications, Inc., Washington, DC, 31.

Guest, A. G. (1984). *Anson's Law of Contract*, 26th Ed., Clarendon Press, Oxford, U.K., 436–437.

Hoel-Steffen Construction Co. v. United States, 456 F.2d 760 (Ct.Cl., 1972).

Lane-Verdugo (Appeal), ASBCA No. 16327, 73–2 B.C.A. (CCH) para. 10,271 (1973).

Loulakis, M. (1985). "Contract Notice Requirements." *Civ. Engrg.* January, 29.

McDevitt & Street Co. v. Marriott Corp., 713 F. Supp. 906 (1989).

Nat Harrison Associates, Inc., v. Gulf States Utilities Co., 491 F.2d 578 (5th Cir., 1974).

New Ulm Building Center, Inc., v. Studtmann, 225 N.W.2d 4, Sup.Ct., Minn. (1974).

Plumley v. United States, 226 U.S. 545, 33 S.Ct. 139, 57 L.Ed. 342 (1913).

Powers Regulator Co., GSBCA Nos. 4668, 4778, 4838, 71,302.

Sante Fe, Inc., VABCA Nos. 1898 and 2167.

Schnip Building Co. v. United States, 645 F.2d 950 (Ct.Cl., 1981).

State of Indiana v. Omega Painting, Inc., 463 N.E.2d 287 (Ind.App. 1 Dist., 1984).

Sweet, J. (1989). *Legal Aspects of Architecture, Engineering, and the Construction Process*, 4th Ed., West Publishing Co., St. Paul, MN.

Transportation Research Board. (1986). "Enforceability of the Requirement of Notice in Highway Construction Contracts." *Research Results Digest*, 152, 7.

Vanderlinde Electric v. City of Rochester, 54 A.D.2d 155, 388 N.Y.S.2d 388 (1976).

Watson Lumber Co. v. Guennewig, 226 N.E.2d 270, 276, 79 Ill.App.2d. 377 (1967).

Weeshoff Construction Co. v. Los Angeles County Flood Control District, 88 C.A.3d 579, 152 Cal.Rptr. 19 (1979).

Additional Cases

The following are additional cases related to issues associated with notice requirements. The reader is invited to review the facts of the case, apply the decision criteria in the flowchart, reach a decision, compare it with the judicial decision, and determine the rationale behind the judicial decision.

Dan Nelson Construction v. Nodland & Dickson 608 N.W.2d 267.

Dugan & Meyers Construction Co., Inc. v. State of Ohio. Ohio Court of Appeals.

Fru-Con v. U.S., 43 Fed. Cl. 306 (1999).

Chapter 7

Extra Work and Oral Change Orders

Oral communications are common to all construction projects. Oral directives do not normally create problems unless the directive is misunderstood or the design professional or owner later refuses to pay for an orally directed change.

Contract Language

All contracts contain language similar to Article 12.1.1 of AIA A201 (1977) requiring that all changes be in writing,

> A Change Order is a written order to the Contractor signed by the Owner and Architect, issued after execution of the contract, authorizing a change in the Work or an adjustment in the Contract Sum or the Contract Time. The Contract Sum and Contract Time may be changed only by Change Order.

and in AIA A201 (1987), Article 7.2.1,

> A Change Order is a written instrument signed by the Owner, Contractor, and Architect...

Some elements of contract clauses related to written change order requirements are

- Only persons with proper authority can direct changes.
- The directive must be in writing.

- The directive must be signed by a person with proper authority.
- Procedures for communicating the change are stated.
- Procedures for contractor response are defined.

Similar clauses in other contracts differ primarily in who signs the change order.

Background

Published literature is available describing certain legal aspects of oral change orders (Transportation Research Board 1986). However, much of this material is vague and confusing and is not readily understood by many contract administrators. The literature and case law are difficult to reconcile and can easily lead one to believe that the outcome of a dispute cannot be predicted with reasonable certainty.

Purpose of Written Change Orders

The purpose of the requirement that all contract modifications be in writing is similar to the requirement for written notice (Thomas et al. 1990; Transportation Research Board 1986, p. 3), which is

- the owner has the right to know the nature and extent of owner promises and liabilities, and
- the written change order protects the owner from unknowingly incurring a liability in the course of routine interpretations of the contract documents or normal interactions with the contractor.

Judicial Attitude toward Oral Changes

To understand how courts view oral modifications despite an agreement to do so only in writing, consider the following points of law.

Appraising the importance of limiting the form of change orders to written directives, the Supreme Court of Iowa, quoting from the *Corpus Juris Secundum* (1963), stated,

Such a provision (requiring change orders to be in writing), however, is not of the *essence of the contract, but is a detail in the performance* ... (Berg v. Kucharo Construction Co.; emphasis added)

The 4th Circuit Court of Appeals went beyond the Supreme Court of Iowa by stating that oral modifications may still be valid even though there may be specific contract language prohibiting oral changes. The reasoning is that because parties can validly make an oral agreement, they can also orally agree to change the requirement for a written directive.

Other courts have also affirmed the validity of oral changes. For example, in the case of Illinois Central Railroad Co. v. Manion, the court stated,

Though the written contract has a clause forbidding such oral alteration, and declaring that no change in it shall be valid unless in writing, such provision does not become a part of the *law of the land*; it is like another agreement which is superseded by a new one. So that in spite of it an oral alteration may be validly made. (emphasis added)

In Bartlett v. Stanchfield, the court noted,

Attempts of parties to tie up by contract their freedom of dealing with each other are futile. The contract is a fact to be taken into account in interpreting the subsequent conduct of the plaintiff and defendant, no doubt. But it cannot be assumed, as *matter of law*, that the contract governed all that was done until it was renounced in so many words, because the parties had a right to renounce it in any way, and by any mode of expression they saw fit. They could substitute a new oral contract by conduct and intimation, as well as by express words. (emphasis added)

Requirements for written contract modifications are valid conditions until the parties renounce the requirement by mutual consent or conduct. Thus, a prerequisite for the owner to insist on the written requirement is that all dealings with the contractor must be consistent with the requirement of the contract. Where evidence of a modification or conduct to the contrary exists, the requirement is set aside and common-law rules are applied. From a review of cases, it is evident that courts quickly default to common law where there is any hint of inconsistent conduct.

Rules of Application

This section presents the common-law rules used by the courts in deciding oral change order disputes. The criteria are based on a detailed examination of more than 70 appellate court decisions. The results show that the courts have dealt with the issue in a consistent manner and that the criteria can be arranged in a hierarchical manner, as shown in Fig. 7-1.

Primary Issues Governing Oral Changes and Extra Work

In deciding disputes involving oral changes, four general inquiries must be answered:

- Is there a statutory requirement?
- Does the changes clause apply?

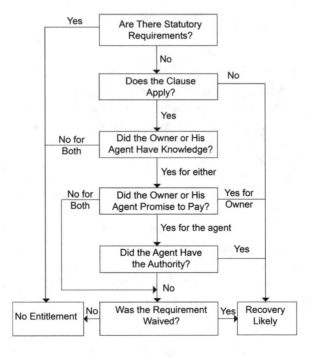

Figure 7-1. Common-law Rules Used by the Courts in Deciding Oral Changes.

- Was there a valid contract alteration?
- Was the requirement waived?

Is There a Statutory Requirement?

Where statutes and regulations require written directives, the requirement is not set aside. This result was affirmed by the District Court of Alaska, quoting the *Corpus Juris Secundum* (1963, 17 C.J.S. Contracts, §377, at 865, n. 36) and stated,

> A written contract may, in the absence of statutory provisions requiring a writing, be modified by a subsequent oral agreement. (U.S. v. Slater)

This criterion must be carefully weighed when performing work for public agencies, municipalities, and authorities. Where such a statute or regulation exists, there will probably be no recovery unless there is compliance with the statute. Examples exist where statutes were violated and the contractor was precluded from recovering. (Montgomery v. City of Philadelphia; Earl T. Browder v. County Court)

Does the Changes Clause Apply?

In examining change order disputes, the initial consideration is whether the disputed work is covered by the changes clause. Courts tend to limit the coverage or applicability of various terms in ways perhaps not intended by the drafter of the contract.

A case illustrating this point was heard before the U.S. Supreme Court in 1886 (R. D. Wood & Co. v. City of Fort Wayne). R. D. Wood and Co. was a contractor hired to construct a water distribution system for the City of Fort Wayne, Indiana. Pipe diameters ranged from 4 to 24 in. A river crossing was part of the scope of work. The contract gave the owner the right to alter the quantities and direct extra work or to make alterations in the extent, dimensions, form, or plan of the work. The changes clause stated that no claim for *extra work* would be considered unless performed in response to a written order from the owner. Finally, the contract specified that the engineer was responsible for determining the amount of work and materials authorized for payment and deciding all questions relative to the contract documents, and that the engineer's estimates and decisions were final. The original plans showed an under-river crossing at

Calhoun Street. After Wood was awarded the contract but before he began work on the under-river crossing, the city relocated the crossing to Clinton Street. The relocation resulted in an increase in water depth from 2 to 7 ft and from a solid bottom to a bottom described as quicksand. The city directed the engineer to accomplish the change but refused to issue any written directive or change order. The city promised the extra work would be taken care of at a later date but subsequently denied the contractor's claim because there was no written change order.

The court evaluated the contract provisions for extra work and alterations and determined that the work was an alteration rather than extra work; therefore, the disputed work was not covered by the changes clause; the contractor was allowed to recover.

The city's intent was probably to have all changes authorized only by written direction, including both extra work and alterations. However, a loophole existed in the contract that allowed the contractor to recover. It is for this reason that standard contract forms use all-encompassing language covering anything that may affect the time and cost aspects of the project.

Many recent decisions have consistently and narrowly defined extra work as something not called for in the contract. (In a unit-price contract, there would be no line item for this work.) Thus, from the judicial viewpoint, a contractor doing more or additional work would not necessarily qualify as performing extra work.

Suppose the language in the Wood contract was used for a commercial building contract. If the contractor is required to install more galvanized rigid conduit than is required by the contract, this is not extra work and the written requirement does not apply. But, if the contractor is required to install PVC conduit and the contract calls for none, this is extra work and the written requirement does apply.

Was There a Valid Contract Alteration?

In most instances, the changes clause applies and there is no statutory consideration. Thus, resolving a dispute is reduced to determining if there is a valid contract alteration or waiver.

Of primary importance is whether the contractor was orally directed to do the work or was acting as a volunteer. Quoting from the *Corpus Juris Secundum* (1963), the Appellate Court of Illinois stated,

> ...as a general rule, a builder or contractor is not entitled to additional compensation for extra work or materials voluntarily

furnished by him without the owner's request or knowledge that he (contractor) expects to be paid . . . (Watson Lumber Co. v. Guennewig)

In Watson Lumber Co. v. Guennewig, the Guennewigs contracted with Watson Lumber Co. (Watson) for the construction of a four-bedroom, two-bath house. The Guennewigs provided the plans, and Watson provided its bid and specifications based on these plans. After substantial completion, the contractor claimed extra compensation for no less than 48 items of labor and materials. In discussing the issue of payment for extras, the court established the following five conditions as prerequisites:

> The law assigns to the contractor, seeking to recover for "extras," the burden of proving the essential elements. That is, he must establish by the evidence that (a) the work was outside the scope of his contract promises, (b) the extra items were ordered by the owner, (c) the owner agreed to pay extra, either by his words or conduct, (d) the extras were not furnished by the contractor as his voluntary act, and (e) the extra items were not rendered necessary by any fault of the contractor. (Watson Lumber Co. v. Guennewig)

Apparently, the Guennewigs were aware that some of the items were not called for in the contract. Regarding whether the Guennewigs knew that Watson would later request compensation, the court stated,

> The evidence is clear that many of the items claimed as extras were not claimed as extras in advance of their being supplied. Indeed, there is little to refute the evidence that many of the extras were not the subject of any claim until after the contractor requested the balance of the contract price, and claimed the house was complete. This makes the evidence even less susceptible to the view that the owner knew ahead of time that he had ordered these as extra items and less likely that any general conversation resulted in the contractor rightly believing extras had been ordered. (Watson Lumber Co. v. Guennewig)

The Court of Appeals of Iowa decided a similar dispute, stating, "Waiver of written change order requirement does not entitle subcontractor to perform extra work without any approval whatsoever" (Central Iowa Grading v. UDE Corp.). In this case, Nelson-Roth was the owner, UDE was the prime contractor, and Central Iowa Grading (CIG) was an earthwork subcontractor on a low-income housing project. CIG was orally ordered by UDE to perform numerous changes. The requested work was done, but some of the work CIG performed was additional work that had not been requested. The claim for this extra work was rejected by both Nelson-Roth and UDE. Even though the owner and prime contractor had disregarded the contract provisions requiring written changes, the court determined that the owner was not liable for changes of which he was unaware because CIG was acting as a volunteer.

The issue of an oral directive was also involved in a case heard before the Supreme Court of Wisconsin. The case of Supreme Construction Co., Inc., v. Olympic Recreation, Inc., involved construction of a bowling alley for Olympic Recreation, Inc., in which Supreme Construction, as prime contractor, abandoned the project. A subcontractor, Christfulli Co., installed the electrical wiring, lighting, and equipment in the building. The subcontractor sought compensation from Olympic (the owner) and payment for extras performed at the oral direction of the prime contractor, Supreme. The changes in electrical work were not readily apparent, and it was unlikely that Olympic would have been aware that they were being done. The testimony presented was inconsistent, and the court was unable to establish a sound basis for compensation for the extras. The court placed the burden of proof on the subcontractor claiming the extra, and Christfulli was unable to prove that Olympic knew of the extra work. Therefore, recovery was denied.

The next inquiry for a valid oral directive is whether the owner knew the contractor was expecting additional compensation, or if the owner expressed or implied agreement to pay for the additional cost of the extra work. An owner, through the course of administering a construction contract, may be aware of various conditions involving extras, yet the owner may not be aware that the contractor expects additional compensation. Courts have determined that mere owner knowledge of additional work is insufficient to ensure contractor recovery of extra costs, and a contractor may be precluded from recovering additional costs if the owner does not know that the contractor expects additional compensation. However, a contractor may be able to recover if there has been an express or implied promise to pay.

The owner's right to know that the contractor expects an equitable adjustment is directly related to the owner's right to control his or her liabilities. This requirement protects an owner against a contractor voluntarily going ahead with extra work and then charging the owner.

The contractor must make his position clear at the time the owner has to decide whether or not he shall incur extra liability. Fairness requires that the owner should have the chance to make such a decision. (Bartlett v. Stanchfield; emphasis added)

In Watson v. Guennewig, the contractor argued that by accepting the work, the Guennewigs implied that they would pay for the extras. To this, the court applied another important principle:

Mere acceptance of the work by the owner . . . does not create liability for an extra. . . . More than mere acceptance is required even in cases where there is no doubt that the item is an "extra" . . . (Watson Lumber Co. v. Guennewig)

Some of the extra items were orally agreed to beforehand, some were ratified after the fact, and others were not claimed by the contractor as extras until the project was substantially complete. Without considering any other points, the court disallowed items that were provided by the contractor without the owner's knowledge or consent.

Case Study

Approval of Shop Drawings and Observation by Inspectors Does Not Necessarily Create a Liability

Community Science Technology Corporation, Inc.
ASBCA No. 20244 (1977)

Community Science Technology Corporation, Inc., received a $4,219,320 contract to construct 200 multifamily housing units at Fort Gordon, Georgia. The specifications read in part,

Section 150—Heating, Air Conditioning and Ventilation

2. Detail Requirements

E. Thermostat shall be low-voltage cooling-heating type with switches for Fan-On-Auto and Cool-Off-Heat selection; adjustable heat anticipator, separate sensing elements and adjustors for cooling and heating. Heating side shall be 55 to 75 °F with factory set 75 °F maximum; cooling side shall be 75 to 90 °F with factory set minimum of 75 °F.

The contract also contained the following language related to shop drawings:

> c) Shop Drawings: . . . If the Contractor considers any correction indicated on the drawings to constitute a change to the contract drawings or specifications, notice as required under the clause entitled "Changes" will be given to the Contracting Officer . . . . The approval of the drawings by the Contracting Officer shall not be construed as a complete check, but will indicate only that the general method of construction and detailing is satisfactory. Approval of such drawings will not relieve the Contractor of the responsibility for any error which may exist as the Contractor shall be responsible for the dimensions and design of adequate connections, details, and satisfactory construction of all work.

As required by the contract, Engineering Form 4025 dated October 1971 was used to transmit the shop drawings. Instruction No. 5 on the reverse side of the form stated, "Submittals not in accordance with the plans and specifications will be accompanied by a written statement to that effect in the space provided for 'Remarks.'"

During construction, Community Science submitted shop drawings with Form 4025 for a Honeywell Thermostat T834. Separate sensing elements and adjustors for cooling and heating were not indicated as being included in the thermostat. Community Science did not advise the U.S. government via Form 4025 that the thermostats were not in compliance with the contract documents. The shop drawings were approved by the government. Government inspectors were present for an undetermined period of time during installation of the thermostats. Quite some time after the installation was complete, the government discovered that the thermostats were not in compliance with the contract. Community Science was ordered to replace the thermostats at their own expense.

Does the presence of the government inspector during installation relieve Community Science of its contractual obligation? Does the government's approval of noncomplying shop drawings relieve Community Science of compliance?

The Board of Contract Appeals did not relieve Community Science of its contractual obligations. The board pointed out that Community Science did not comply with the contract requirements, in particular Form 4025, and that there is no evidence that the government inspectors should have known or recognized that Community Science was installing nonconforming thermostats. The presence of inspectors or approval of shop drawings was not a promise to pay.

A case heard before the Federal District Court in Alabama illustrates the essential promise to pay criterion (Blair v. U.S.). The U.S. government contracted with Algernon Blair to dismantle certain prefabricated buildings located at Granada, Mississippi, and transport them to Key West, Florida. There, the buildings were to be reassembled by the contractor. During the reassembly phase, a hurricane struck Key West and caused considerable damage to some of the buildings. The contracting officer directed the contractor to protect the work and repair the damage. In ordering the work, no promise or suggestion to pay was made to pay for the additional cost, but the question of liability for that cost was deferred until later, an action that could have potentially constituted a waiver. Ten days after the oral direction, the government sent a letter to Blair stating in part,

> In view of the foregoing, you are advised that the Contractor will
> be expected to complete the project in accordance with the terms
> of the Contract, without any additional cost to the Government
> as a result of damage caused by the hurricane. (Blair v. U.S.)

On completing the work, Blair requested additional compensation based on the oral direction. However, the government had never agreed to pay for the additional work. In deciding for the government, the court stated,

> Where contract provides that there shall be no charge for extra
> work unless a written agreement is made therefor, the builder
> cannot recover compensation as for extra work on account of al-
> terations made at the oral request or consent of the owner but for
> which no written agreement to pay additional compensation is
> made. (Blair v. U.S.)

An agreement to defer discussions about payment may be sufficient to put the owner on notice that the contractor is expecting additional monies. However, in the Blair case, the government never promised to pay.

Another case involving a promise to pay is Berg v. Kucharo Construction Co. The project was for the construction of more than 250 apartment buildings and houses for the U.S. Federal Housing Administration. Berg, the subcontractor, had a contract with Kucharo Construction Co., the prime contractor, which stated that no oral agreement would be honored, and only extras directed in writing and agreed to before construction would be recognized for additional compensation. Numerous defects in material and other items affecting Berg's work were brought to Kucharo's attention. Kucharo repeatedly instructed Berg to complete the work and promised payment. However, Kucharo later refused to pay, citing the requirement for written directives. The court stated,

> The courts have adopted various theories of avoidance which may be classed as those of independent contract, modification or rescission, waiver, and estoppel.... Among the acts or conduct amounting to waiver are the owner's knowledge of, agreement to, or acquiescence in such extra work, a course of dealing which repeatedly disregards such stipulation, and a promise to pay for extra work orally requested by the owner and performed in reliance thereon. (Berg v. Kucharo Construction Co.)

The court concluded that the written contract was properly modified, and Kucharo's promise to pay could not be rescinded.

Another case involving a promise to pay was heard by the Supreme Court of Oklahoma (Kenison v. Baldwin). The case involved the construction of a residence where the homeowner orally requested changes to the contract. The changes were performed by the contractor, and at the time, the owner acknowledged that there was additional cost involved. The contractor finished the house, and the work was accepted by the owner. However, the owner refused to pay for the extra work, contending that the directives were not in writing. The court determined that the contract had been orally modified and directed the owner to pay the contractor, even though the owner never expressly promised to pay. Knowing that the contractor expects additional compensation and not taking appropriate measures to stop the work is likely to be interpreted as an implied promise to pay.

Most contract forms expressly state who is authorized to order changes. However, at least for nongovernmental owners, authority may also be apparent or implied from the responsibilities or conduct of an individual or party. If circumstances lead a contractor to reasonably believe that an unauthorized person has the authority to direct changes, authority may be imputed to that individual. Contractors need to know at all times who is authorized to direct changes.

In a complicated case, Flour Mills of America, Inc., v. American Steel Building Co., the Supreme Court of Oklahoma considered the claims and counterclaims of the parties involved. The contract was for the construction of a building addition to grain storage facilities in Alva, Oklahoma, owned by Flour Mills. Problems arose during construction, including moisture damage to grain that was already stored. Additional work was also directed. The contractor claimed compensation for extras orally directed by the owner. The court specifically determined that those who had ordered the extra work had been authorized to do so by Flour Mills. The court also stated,

> The same principle . . . is recognized by this court in Jackson Materials Co. v. Grand River Dam Authority, supra, at page 560 of the Pacific report of the opinion, but was not applied therein because the person who made the subsequent oral agreements involved therein had not been authorized to do so and his action in doing so had not been ratified by the only entity authorized by statute to make such agreements. (Flour Mills of America v. American Steel Building Co.)

The Flour Mills case illustrates an often-applied condition that apparent authority cannot be extended to someone who does not possess authority unless there is some positive action (or inaction) by the person who actually possesses the authority (Sweet 1989, p. 36), i.e., apparent authority must be ratified by words or actions. Typical action is where an owner knew of unauthorized directives but failed to take appropriate action (Sullivan v. Midwest Sheetmetal Works; Weeshoff Construction Co. v. Los Angeles County Flood Control District).

A case heard before the 4th Circuit U.S. Court of Appeals illustrates apparent authority (*Sappho*). The contract called for overhauling a steamer ferry, the *Sappho*. The overhaul included significant timber replacement, remetaling, recaulking and plumbing, and straightening. The contract contained a provision stating that no extra work of any kind would be considered unless it was submitted beforehand and was approved and signed by the chairman of the board of the ferry company. When the work began, the condition of the ferry was found to be much worse than anticipated. At a conference that included the contractor, the master of the steamer, the president of the ferry company, and the inspector, additional repairs were agreed on orally. These repairs were subsequently performed by the contractor. The court found that there was confusion concerning what was said and what was intended by the various parties at the conference. However, at the conclusion of the conference, the plaintiff was told to "go ahead." The court stated,

> Work was immediately begun on the hull, under the direction of Capt. Cherry as superintendent, who stayed at the work, and directed personally what rotten wood and timbers should be taken out and what work should be done, and how it should be done, until the steamer was completed. (*Sappho*)

The court also noted that the president and various officers of the ferry company were frequently at the steamer and allowed the repair work to continue. Accepting another's performance that is in violation of the contract may be interpreted as a waiver. Although not provided for by the contract, Capt. Cherry possessed apparent authority to direct the additional work, and the officers of the company made no attempts to limit his directing the work. This inaction by the ferry company was sufficient to lead the contractor to believe that the ferry captain had the authority to direct the changes.

The problem of authority can be particularly troublesome when the owner is a public agency. Apparent authority may not be sufficient, and statutes may restrict actual authority to a single person or a limited group. With respect to statutory limitations that are inconsistent with what is stated in the contract, Simon states (1982, p. 11.6-1),

> Determination of the authority is not as easy as it may appear on the surface. In Blum v. City of Hillsboro, supra, the Mayor, City Council and Architect all approved the change. They are proper parties and have apparent and actual authority; however, external limitations (the bidding statutes) placed a different form of prohibition on that authority. This might be called an artificial limitation on authority, but to those involved in the construction process, when they are unable to be paid for what otherwise appears to be a properly authorized, issued, and executed change order, that is not an artificial barrier. It is very real.

Thus, statutes always prevail over the contract language, except in unusual circumstances, and the prudent contractor should be fully aware of the local statutes and regulations that specify who has authority to bind the public agency.

Was the Requirement Waived?

Waiver is created by words, actions, or inactions of the owner that result in the abandonment of a contract requirement. The following case provides an excellent example of the waiver of the written changes requirement.

The case of Reif v. Smith, heard before the Supreme Court of South Dakota, involved construction work on a log home. Partly because of the inadequacy of the plans, numerous changes were made during the construction. Section 15 of the contract specified that all changes had to be ordered in writing and any change in contract price had to be settled before beginning the work. Relative to Section 15, the court stated,

> Generally, provisions like Section 15 prevent contractors from re-covering for alterations or extras not subject to a written order . . . Such provisions, however, are impliedly waived by the owner where he has knowledge of the change, fails to object to the change, and where other circumstances exist which negate the provision; i.e., the builder expects additional payment, the alteration was an unforeseen necessity or obvious, subsequent oral agreement, or it was ordered or authorized by the owner . . . Additionally, repeated or entire disregard for contract provisions will operate as a waiver of Section 15. (Reif v. Smith)

The trial court record indicated that the Smiths (owners) were on the job site repeatedly, had knowledge of certain changes and authorized others, and made several progress payments after the changes were made. The court found the Smiths' actions inconsistent with the contract terms.

> It is incongruous that the owner, author of the contract with the written change order requirement, can come into court and acknowledge that he authorized the changes without a single written change order, admit liability for two or three specific items, but escape liability on the balance on the assertion that he understood that they were "tit for tat." (Reif v. Smith)

The Smiths' conduct was apparently inconsistent and resulted in several oral changes. Because the conditions of change order creation were identical, the Smiths could not acknowledge some and disavow others. The court determined that the Smiths, by their conduct, had waived the requirement for written change orders, and they therefore allowed the contractor to recover.

Other Issues

Several issues, although not directly included in the basic rules, can be important considerations in certain instances.

Consideration

Valid contracts require consideration or an exchange of something of value. Generally, courts are reluctant to enforce change orders where there has been no exchange of something of value. Sweet argues that parties should not be restricted from making agreements lacking consideration if the agreements are made voluntarily (Sweet 1989, p. 431).

Clauses Specifically Precluding Oral Direction

Some contracts contain language expressly prohibiting oral directives. However, by attempting to limit the owner's exposure to additional costs, the owner's flexibility to make changes in immediate situations is also reduced.

Form of Directive

Changes are usually directed by issuing a signed, written order. Some courts insist that the order be in writing, whereas others may simply insist that there be something written and an order. Simon explains (1979, p. 114) thus:

> Since most contracts do not specify the format for the writing, various documents might, in the judge's discretion, constitute the writing so as to fulfill the "written" portion of the clause requirement. That writing might be found in letters, transmittal notices, revised plans and specifications, notations on shop drawings, job minutes, field records, daily reports, signed time and material slips, internal memoranda, or other documents . . . . The next consideration is to determine whether the words "writ-

ten order" require a "written order" or a "written" "order." If they are read together as a single phrase (which they are not), the owner's furnishing a sketch, revised drawing, or a new plan, along with the oral directive to perform the work "or else," would not fulfill the technical requirements. However, if the words are interpreted to mean that both writing and an order must exist, the sketch and oral directive would suffice.

The literature suggests that courts tend to apply the less restrictive interpretation and require a "writing" and an "order." Thus, an oral direction must be supported by evidence that the owner directed, had knowledge of, and approved the change, and that some written record of the authorized change exists.

Theory of Equity

If a contractor cannot justify performance based on a valid oral direction or show that the requirement was waived, the contractor has little recourse other than to seek recovery on a theory of equity. However, the topic of equity receives little discussion in the literature, and the case-law review indicates that courts seldom render decisions in favor of the contractor based solely on equity considerations. Thus, it would appear that a contractor has little chance of recovering on the basis of equity alone without one of the other exceptions to the written order requirement being present.

Illustrative Example

The scenario described below is based on Metro Insulation Corp. v. Robert Leventhal.

Statement of Facts

Metro Insulation Corp. was a sub-subcontractor to the Frank Sullivan Co. for the construction of a federally assisted housing project for the elderly owned by the Boston Housing Authority. Beacon Construction Co. was the general contractor. The dispute involved insulation work performed by Metro of certain hot-water piping in the seventh floor ceiling and of certain cold-water piping in the pump and boiler rooms of the building.

Before June 13, 1967, the engineer employed by the housing authority's architect inquired of Metro about the type of insulation Metro proposed to use in the upper floors of the building. In a letter to the engineer dated June 13, 1967, Metro replied that the specifications did not call for the insulation of such piping and that Metro did not include anything for such insulation in the bid that had been submitted to the Frank Sullivan Co., the plumbing subcontractor. At the engineer's request, by letter on June 30, Sullivan requested Metro to submit a price "for the hot water piping not covered by the specifications." In a letter dated July 17, Metro submitted separate prices to Sullivan for the insulation of hot-water piping in the seventh-floor ceiling and cold-water piping in the pump and boiler rooms. Acting on a request from the housing authority, Sullivan instructed Metro to send a breakdown of its quotations "priced in accordance with the contract specifications." Metro did so, and Sullivan in turn sent the revised quotations to Beacon on August 21. On August 22, Beacon wrote Sullivan that the architect had "verbally authorized that this insulation was to be installed and that it was subject to a change order" and concluded by saying that "in the meantime, it appears that we are authorized to proceed with this change." Sullivan, by a letter of August 24, sent a copy of Beacon's letter to Metro.

Division 28.21 of the specifications (Pipe Covering and Insulation), as amended by Addendum No. 3, called for

(1) All hot water supply and return piping, branches, risers, and also including cold water piping throughout the first floor ceiling shall be insulated with ... (3) Insulate all water lines in first floor ceiling with ...

The specifications also provided, in the article titled "Changes in the Work,"

a. The Authority may make changes in the work of the Contractor by making alterations therein, or by making additions thereto, or by omitting work therefrom, without invalidating the Contract...

c. Except in an emergency endangering life or property, no change shall be made by the Contractor unless he has received a prior written order from the Authority, countersigned by the Architect, and approved on its face by the...Department of Housing and Urban Development (HUD) authorizing the change, and no claim for an adjustment of the contract price or time shall be valid unless so ordered.

The article titled "Disputes" provides in part as follows:

a. All disputes...arising under this Contract or its interpretations, whether involving law or fact, or both, or extra work,... shall within 10 days of commencement of dispute be presented to the Architect for decision.... Such notice need not detail the amount of the claim but shall state the facts surrounding the claim in sufficient detail to identify the claim, together with its character and scope. In the meantime the Contractor shall proceed with the work as directed. The parties agree that any claim not presented within this subsection is waived...

b. The Contractor shall submit in detail his claim and his proof thereof.

Following Metro's receipt of the last two letters, a conference was held with Metro, Sullivan, Beacon, the architect, the engineer, and the housing authority, at which Metro was "informed by all those present that a change order was forthcoming and that Metro was to proceed with the work immediately so as not to hold up" the project. Metro thereafter completed the work.

The housing authority, acting under the provisions of Article 10 of the General Conditions, submitted a change order request to the U.S. Department of Housing and Urban Development (HUD) for its written approval. HUD refused to approve the change order for the insulation work. Metro did not receive notice of such disapproval until November 14, 1967, by which time all the insulation work had been completed.

Analysis

Are There Statutory Requirements?

This issue is not applicable in this dispute because it is a prime-subcontractor dealing.

Does the Changes Clause Apply?

The changes clause is applicable in this dispute.

Did the Owner Have Knowledge?

The owner, in this situation is the housing authority, and it is obvious that the housing authority and all other relevant parties were aware that insulation of piping in the upper floors was not called for in the contract documents.

Did the Owner Promise to Pay?

This is the central issue in this dispute. The housing authority should not have directed Metro to do the work if it had not been approved by HUD. It was not Metro's responsibility to confirm that HUD approval had been given. When the housing authority issued the directive to do the work, Metro had the right to rely on that order and assume that HUD approval had been given. If not, the housing authority should not have issued the order.

Synopsis

The judicial outcome in this dispute was consistent with the flowchart analysis: Metro was paid for doing the work. In making their determination, the court said,

> If the Authority had requested us to consider the effect of the provisions of Article 10 of the General Conditions, we would have regarded them as waived by the Authority's actions, prior to seeking HUD approval of the change order, in directing the work to proceed and agreeing that it should be paid for.

Sullivan paid Metro. Sullivan was reimbursed by Beacon. Beacon was reimbursed by the housing authority. However, the housing authority did not receive any reimbursement from HUD because HUD did not approve the change; that was their prerogative.

Exercise 7-1: Long and Lazer Construction Co.

James and Kathy Long consulted Lazer Construction Co., Inc., about building a new home. Lazer, with the Longs' approval, drafted a contract on forms published by the American Institute of Architects. The contract stated a guaranteed maximum price (GMP) of $99,500, subject to written change orders prepared according to the contract. The contract also stated that the work was to be performed according to the owners' plans and specifications and provided for a contractor's fee of 5% over all costs of the project, unless the project, in the absence of approved changes, ran over the guaranteed maximum cost, in which case the owner was to pay only the direct costs.

The contract at the time of execution was based, in part, on an initial set of plans. Several weeks after construction began, Long changed the plans and a new set of plans was submitted. As construction continued, certain items were orally added, certain items were deleted, and certain items were paid for directly by Long. Lazer expended $140,771.50 on the project and billed Long for that amount. By this time, Long had paid $116,713.57. Long refused to pay any more monies to Lazer.

The extras included items that were added and deleted. Certain items were paid for directly by Long. All indications are that Long authorized all changes and Lazer provided a list of authorized changes amounting to $41,271.50. Long claimed that these cost should be audited, but he never requested an audit.

Long argued that Lazer was not entitled to additional payments because Long had already paid more than the GMP of $99,500. He also relied on contract provisions indicating that this maximum price could be changed only by preapproved written change orders.

Were the orally directed changes a valid contract alteration? Does the contract language entitle Lazer to the additional monies requested? Is one obligated to pay for orally directed changes, even though the contract says otherwise? How did Long waive the GMP? The written change order requirement?

Exercise 7-2: Henry's Electric Co. and Marilyn Apartments

In 1970, Henry's Electric Co. was an electrical subcontractor on the construction of an apartment complex known as Glenbrook Apartments. Marilyn Apartments, Inc., and R. C. Cunningham II were the contractors (hereinafter called Marilyn). Atrium Corp. was the owner.

The contract between Marilyn and Atrium called for a total project cost not to exceed $766,013. In arriving at this amount, the electrical work was shown as an "allowance" in the amount of $36,000. An allowance is merely an estimate and is not a figure based on a firm bid. The three principal parties involved knew that this figure was too low. Henry's electrical bid, which was thereafter submitted and accepted, was considerably in excess of the $36,000 figure. By written Change Order No. 1, the electrical work allowance figure in Marilyn's contract was increased by $17,850.

Marilyn, not Atrium, verbally directed Henry to obtain and install certain light fixtures and further directed Henry to bill Marilyn for the same. The light fixtures were not included in the contract and were extras. The subcontract between Henry and Marilyn contained the following provision regarding charges for extra work or material:

> 12. It is mutually agreed that no charges by Sub-Contractor for extra work or material under this sub-contract or for any other work on the above project shall be made or will be recognized or paid by Contractor unless agreed to in writing by Contractor before such work is done or the material furnished.
>
> In this connection, it is further mutually agreed that in any case where work is to be done or materials furnished "as directed" the said work shall not be done or materials furnished unless Sub-Contractor receives written direction therefor but Contractor shall not be required to order any such work or materials.

Marilyn paid to Henry a part of the total expense of obtaining and installing the light fixtures but refused to pay the remaining balance due, which amounted to $7,181.48. Marilyn's argument was that there were no written change orders and there was no written notice given.

Can Marilyn escape liability because of the absence of a written change directive?

Exercise 7-3: Laramee and Care Systems

Edward Laramee contracted with Care Systems, Inc., to construct an addition to a home owned by Laramee in Sarasota, NewYork. The total price was $28,797.38 for materials and labor. A change order of $3,221.61 was included in that price. The contract stated that any modification or alteration had to be in writing. The work was to be performed according to plans provided by Care Systems (the owner).

During the course of the project, Laramee brought to Care Systems' attention that a foundation, earlier constructed by Care Systems, had been improperly laid out. The plans had to be revised no fewer than five times, and no fewer than 15 change orders were requested by Laramee. Nevertheless, Care Systems insisted that Laramee proceed anyway. Care Systems orally directed Laramee to correct the problem that necessitated that Laramee perform extra work. Care Systems now refuses to pay for the extra work because there was no written directive as required by the contract.

Was the work an alteration or a modification? Is Care Systems responsible for payment?

Exercise 7-4: Owens Plumbing & Heating and Bartlett, Kans.

Owens Plumbing and Heating signed a contract with the city of Bartlett, Kansas, for a water distribution system. The contract price was $21,303. The contract called for the installation of underground water lines, fire hydrants, and meter hookups. The contract contained the following provision:

> 16. EXTRA AND/OR ADDITIONAL WORK AND CHANGES.—
> Without invalidating the contract, the owner may order extra work or make changes by altering, adding to or deducting from the work, the contract sum being adjusted accordingly. All the work of the kind bid upon shall be paid for at the price stipulated in the proposal (unless such prices were rejected at the time the proposal was accepted), or at the lump sum agreed upon between the Owner and Contractor, and *no claims for any extra work*

or materials shall be allowed unless it is ordered in writing by the Owner or its authorized representative. If the extra work shall be of the kind for which no price was stipulated in the Proposal, and the Owner and the Contractor cannot agree as to the fair value of such work prior to its performance, such work will be performed and the Contractor will be paid . . . ; *but no claim for such extra work will be allowed unless the same was done pursuant to a written order as aforesaid.* (emphasis added)

As the work progressed, work was orally requested by the mayor, including an extra fire hydrant, 600 ft of 2-in. pipe, and footings and lines to connect a water tower, all unit-price items in Owens's contract.

Both the city and the contractor encountered difficulties in the course of the project. The contractor encountered rock that had to be broken up and removed from the ditches. The rock was outside the scope of Owens's contract. Dirt and sand had to be hauled in for bedding under the water lines (these items were not called for in the original contract). The city ran short of money after issuing the maximum amount of revenue and general obligation bonds. The city fell behind in its progress payments, which caused Owens to suspend work for three weeks.

To get the work started again, the city orally agreed that it would rent and furnish special equipment to remove the rock and that the rock removal would be paid for as an extra to the contract. Throughout the course of his dealings with Owens, the mayor kept the city council advised, and the council in their meetings ordered and directed payment thereof.

After the water distribution system had been completed, Owens was paid for the extras associated with the hydrant and related items. However, the city refused to pay Owens for the additional difficulties encountered in rock removal. Their rationale was that no written directive was ever issued, and the work was part of the trench work for which the city had contracted.

Is Owens entitled to extra compensation? How did the city waive its obligations to not pay for rock removal?

Exercise 7-5: Utah DOT and Thorn Construction Co.

On March 27, 1973, the Utah Department of Transportation (DOT) contracted with Thorn Construction Co., Inc., for the construction of an access road at Rockport State Park, near Wanship, Utah. Before submitting its bid, several representatives of Thorn and a low-level Utah DOT representative, Virgil Mitchell, toured the site. During the tour, Thorn was taken to the Utelite borrow pit site where he was told the Utelite pit was available and could be used. Although other pits in the area were available, they were not visited on the tour.

The standard specifications in Sec. 105.17 contained the following language:

> If, in any case, where the contractor deems that additional compensation is due him for work or material not clearly covered in the contract or not ordered by the Engineer as extra work as defined herein, the contractor shall notify the Engineer in writing of his intention to make a claim for such additional compensation before he begins the work on which he bases the claim.

During the course of the project, the project engineer orally requested that Thorn widen a turning area. This work was clearly not part of the original contract work. The project engineer also agreed to pay Thorn for the extra expenses, although no costs were discussed. On completion of the project, the Utah DOT refused to pay Thorn because there was no written directive in accordance with the contract provisions.

Should Thorn be paid?

Exercise 7-6: Security Painting Co. and PennDOT

In March 1971, Security Painting Co. signed two contracts with the Pennsylvania Department of Transportation (PennDOT) for painting 12 bridges in Potter and Clearfield Counties. The total amount for the two projects was $23,580. As per the contract, Security was required to sandblast the metal surfaces on the bridges before painting. Security began sandblasting and was soon informed by PennDOT inspectors that to fully comply with the specifications, it would have to remove substantially all of the sound and adherent old paint from all of the surfaces. Before this development, Security had intended only to remove loose, excessively thick, or inflexible paint. Security acceded to the inspectors' demands and as a result, incurred considerable expense beyond their bid price. After completion of the 12 bridges, Security sought an additional sum of $49,703.58.

Should Security be paid?

References

American Institute of Architects (AIA). (1977). *General Conditions of the Contract for Construction*, AIA Document A201. American Institute of Architects, Washington, DC.

AIA. (1987). *General Conditions of the Contract for Construction*, AIA Document A201. American Institute of Architects, Washington, DC.

Bartlett v. Stanchfield, 148 Mass. 394, 19 N.E. 549, 2 L.R.A. 625 (1889).

Berg v. Kucharo Construction Co., 21 N.W.2d 561, 567 (1946).

Blair v. United States, 66 F.Supp. 405, 405 (1946).

Central Iowa Grading, Inc., v. UDE Corp., 392 N.W.2d 857, 858 (1986).

Corpus Juris Secundum. (1963). Contracts, Vol. 17, American Law Book Co., Brooklyn, NY.

Earl T. Browder, Inc., v. County Court, 143 W.Va. 406, 102 S.E., 2d., 425 (1958).

Flour Mills of America, Inc., v. American Steel Building Co., 449 P.2d 861, 878 (1969).

Illinois Central Railroad Co. v. Manion, 113 Ky. 7, 67 S.W. 40, 101 Am. St. Rep. 345 (1902).

Kenison v. Baldwin, 351 P.2d 307 (1960).

Metro Insulation Corp. v. Robert Leventhal, 294 N.E., 2d, 508 (1973).

Montgomery v. City of Philadelphia, 391 Pa. 607, 139 A.2d. 347 (1958).

R. D. Wood & Co. v. City of Fort Wayne, 119 U.S. 312, 321 (1886).

Reif v. Smith, 319 N.W.2d 815, 817 (1982).

The *Sappho*, 94 Fed. 545, 549 (1899).

Simon, M. S. (1979). *Construction Contracts and Claims*, McGraw-Hill, Inc., New York.

Simon, M. S. (1982). *Construction Law Claims and Liability*, Arlyse Enterprises, Inc., Butler, NJ.

Frank Sullivan v. Midwest Sheetmetal Works, 335 F.2d 33 (1964).

Supreme Construction Co., Inc., v. Olympic Recreation, Inc., 7 Wis.2d 74, 96 N.W.2d 809 (1959).

Sweet, J. (1989). *Legal Aspects of Architecture, Engineering, and the Construction Process*, 4th Ed., West Publishing Co., St. Paul, MN.

Thomas, H. R., Smith, G. R., and Wright, D. W. (1990). "Resolving Disputes over Contract Notice Requirements." *J. Constr. Engrg. and Mgmt.*, 116(2), 738–755.

Transportation Research Board. (1986). "Enforceability of the Requirement of Notice in Highway Construction Contracts." *Research Results Digest*, 152, 7.

United States v. Slater, 111 F.Supp. 418, 420 (1953).

Watson Lumber Co. v. Guennewig, 226 N.E.2d 270 (1967).

Weeshoff Construction Co. v. Los Angeles County Flood Control District, 88 Cal. App., 3d, 579 (1979).

Additional Cases

The following are additional cases related to issues associated with changes, oral changes, and extra work. The reader is invited to review the facts of the case, apply the decision criteria in the flowchart, reach a decision, compare it with the judicial decision, and determine the rationale behind the judicial decision.

Fox v. Mountain West Electric, Inc.
Gill Construction, Inc., v. 18th Vine Authority of Kansas City, Missouri.
Town of New Ross v. Ferretti, Court of Appeals of Indiana, 9/22/04.

Chapter 8

Rules for Contract Interpretation

When the parties disagree as to whether certain work is required by the contract, contract interpretation rules must be applied.

Contract Language

Several general clauses pertain to interpretation issues. For instance, Art. 1.2.2 of AIA A201 (1986) states,

> The intent of the Contract Documents is to include all items necessary for the proper execution and completion of the Work. The Contract Documents are complementary, and what is required by one shall be as binding as if required by all. Work not included in the Contract Documents will not be required unless it is consistent therewith and is reasonably inferable therefrom as being necessary to produce the intended results.

and in Art. 3.2.1,

> The Contractor shall carefully study and compare the Contract Documents with each other and with information provided by the owner . . . and shall at once report to the Architect any error, inconsistency or omission discovered.

Some contracts include an order-of-precedence clause. The clause lists various contract documents. In the event that there is a discrepancy or ambiguity between the provisions of two documents, the clause specifies that the provision of the document cited first will govern. The order-of-precedence clause is of no value if the ambiguity occurs in the same document.

Rules of Application

Primary Issues Governing Interpretation

The interpretation process involves four primary inquiries shown in Fig. 8-1. These are the following:

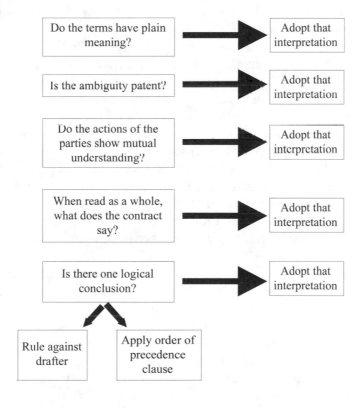

Figure 8-1. Decision Diagram for Disputes Involving Interpretation.
SOURCE: Thomas et al. 1994, ASCE

- Do the terms have plain meaning?
- Is the ambiguity patent?
- Do the actions of the parties show mutual understanding?
- When read as a whole, what does the contract say?

Do the Terms Have Plain Meaning?

Sweet points out that when the words have plain meaning (Table 4-1 in Chapter 4, Principle 5), courts may not look beyond the document itself to determine what the parties meant. He further states that rarely is the language sufficiently plain in meaning to preclude other inquiries from being made (Sweet 1989). Although the plain meaning rule holds a significant position in the rules of hierarchy, it would appear that it is seldom conclusive.

An ambiguous word or phrase may be the focal point of the dispute. Where this situation arises, courts determine all possible meanings of the word(s) in question.

When establishing meanings, courts determine the trade meaning (Framlau v. U.S.) and examine the context in which the word is used (Monroe M. Tapper & Associates v. U.S.). Conclusive, technical or trade meanings will be assigned to ambiguous terms, even though they may have a relevant nontechnical meaning (Lewis v. Jones).

In Columbus Construction Co. v. Crane Co., a gas line contractor (Columbus) entered into a contract with a local piping mill (Crane) to supply 8-in. threaded pipe with collars that were "tight in line." After installation, leaks were discovered in the pipe. Crane contended that "tight in line" should be taken in its literal sense as nonleaking. Columbus' position was that "tight in line," taken in the strict sense, was impossible, and that minor leaks were acceptable and according to industry standards were "tight in line." The evidence was clear that "tight in line," as defined by the contractor was impossible under the design stipulated in the contract (Columbus Construction Co. v. Crane Co.). The court found in favor of Columbus. Custom and usage were applied as the justification for favoring Columbus because there was no other way to resolve the dispute and custom and usage were conclusive.

Generic terms or phrases of the construction trade are normally defined by the context in which they are found. For instance, a contract may allow time extensions for labor strikes, unusual weather, and other events beyond the contractor's control. If a delay occurs because of a problem with a vendor, a time extension may not be included within this clause because one could argue that vendor problems were not similar to the items specifically mentioned. The "catchall" phrases only capture events closely related to those enumerated. Without a "catchall" phrase, the event must be specifically listed.

Is the Ambiguity Patent?

When a contract clause is found to be ambiguous, courts next seek to determine if the ambiguity is patent (obvious) or latent (hidden). The determination of patent ambiguity is based on whether at the time of the bid, the ambiguity was so obvious that any reasonable bidder needed the ambiguity resolved before its bid could be completed.

Where an ambiguity is patent, a duty is imposed on the contractor to inquire (WPC Enterprises v. U.S.). Courts look unfavorably on a party that recognizes (or should have recognized) an ambiguity in a contract and does not seek clarification (WPC Enterprises v. U.S.). The knowledge, or reasonable knowledge, that the other party has a different understanding of the contract, without bringing it to their attention, could imply bad faith and unfair dealings. Courts have shown support for the contractor who seeks clarification, particularly if the owner fails to respond to the inquiry (U.S. v. Rich).

The rule of patent ambiguity is not often applied and is usually applied only when it involves an essential element of the work and resolution is necessary for the contractor to prepare the bid. When the ambiguity involves obscure details in a voluminous set of contract documents, it has rarely been applied.

The following illustrates a patent ambiguity. Hensel Phelps Construction Co. contracted with the U.S. government to repair a missile facility (Hensel Phelps Construction Co. v. U.S.). The specifications required 18 in. of fill beneath concrete slabs. However, the drawings showed 36 in. of fill. The contractor noticed this discrepancy in preparing his bid but resolved the conflict by applying the order-of-precedence clause, which favored the specifications. The bid was submitted based on 18 in. Before placing concrete, he was told that 36 in. of fill was required. The contractor sought an equitable adjustment of $100,983. The government prevailed because the knowledge of the fill depth was an essential component to the bid.

The concept of being essential to the bid is also illustrated by the case of Newsom v. United States. Newsom was contracted to renovate medical preparation and janitor rooms at a Veterans Administration (VA)

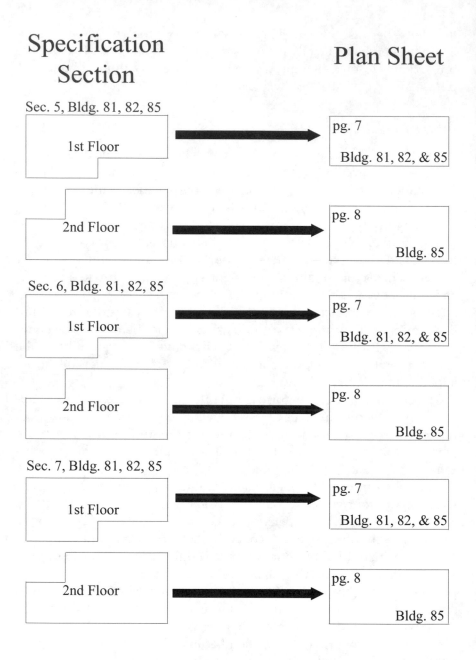

Figure 8-2. Relationship Between Specifications and Plan Sheets,
Newsom v. United States.

hospital in Knoxville, Iowa. As shown in Fig. 8-2, sections 5, 6, and 7 of the specifications each described buildings 81, 82, and 85. This figure shows the relationship between the plan sheets and specifications and clearly shows that through the relationship with the specifications, work was contemplated by the owner on the second floors of buildings 81 and 82. The first paragraph in each section described the work on the first floor and the second paragraph referred to the second floor. Page 7 of the plans showed the work on the first floor, and page 8 showed the second-floor work. The title block of page 7 referred to buildings 81, 82, and 85, but the title block of page 8 only referred to building 85. Newsom made no inquiry as to the proper interpretation and, instead, based his bid on doing both floors of building 85 and the first floor only of buildings 81 and 82. It was not until 4½ months later that the parties realized that there was a discrepancy between what the VA intended and what Newsom had understood. The court reasoned that the court "must recognize the value and importance of a duty of inquiry in achieving fair and expeditious administration of Government contracts" (Newsom v. U.S.). The judgment favored the government.

In a third example, Beacon Construction Co. of Massachusetts v. United States, a dispute arose between the federal public housing administration and Beacon Construction Co. Beacon was the low bidder on a 1963 project to construct 275 dwellings and 7 laundry units. The dispute involved a requirement to provide weather stripping for the windows. The contractor's position was that weather stripping was not required because the specifications only mentioned weather stripping for doors. The government insisted that weather stripping was part of the contract and should have been supplied. The relevant part of the specification read,

> Weather strips for entrance doors shall be brass, bronze, zinc, or stainless steel strips not less than .017 inches thick, 1 or 2 member, manufacturers standard type providing a weather tight seal on all four edges of doors and casement and *double hung sash.* They shall adjust themselves to the swelling and shrinking of the *sash* and frames without impairing their efficiency or the easy operation of the *sash* and doors. . . . Weather strips shall be provided for all doors, opening out, in the service building. (Beacon Construction Co. v. U.S.; emphasis added)

The court used the plain meaning rule to conclude that the word "sash" was a generic term in the building industry associated with windows. The drawings also contained clear indications (i.e., notations with large red arrows pointing to windows) that weather stripping was required on the windows. The court noted that a prospective bidder could not help but notice that a difference of intent existed. Furthermore, Beacon's president stated that the discrepancy had been noticed before bidding. In summary, the court stated,

> In this case it is plain that, as we have found, the discrepancy was in actual fact, and in reason must have been, fully known to the plaintiff before it computed its bid. It had ample cause and opportunity to seek an interpretation from the government before consummating the agreement, but it did not do so.... Having failed to take that route, Plaintiff is now barred from recovering on his demand. (Beacon Construction Co. v. U.S.)

In the above cases, it is worth noting that all three contractors noticed the ambiguity before the bid, all three made their own interpretation without asking, and all three were not allowed to prevail. Clearly, a contractor who chooses to make an interpretation without asking does so at his or her own peril.

Do the Actions of the Parties Show Mutual Understanding?

The rule of practical construction examines the conduct of the parties during performance of the contract. Where mutual understanding or intent is expressed, that interpretation is adopted (Wiebner v. Peoples). The rule of practical construction is an important rule and has been applied in many disputes.

In Bulley and Andrews, Inc., v. Symons Corp., the owner, Symons, contracted with Bulley to construct a building using a concrete formwork system provided by the owner. A dispute subsequently arose over the type of tie rod to be used. The operative clause in the specification read,

> All ... form ties, form hardware, ... will be furnished by Symons Manufacturing Company to the contractor from his standard catalog items, a copy of which is attached to this specification. (Bulley & Andrews v. Symons Corp.)

The catalog pictured both a standard and a threaded-end tie rod. The threaded-end tie rod was the tie rod furnished by Symons. Bulley alleged that the standard tie rod was intended for use. The threaded tie rods were provided by the owner's field representative to the Bulley site superintendent, who accepted and used the tie rods without protest. Frequent inspections and visits by both parties' representatives brought no evidence to light that Bulley was having any difficulties or was incurring additional costs due to the use of the threaded tie rods. Nine months after substantial completion of the concrete phase of the project, Bulley submitted a claim for additional work, alleging that the threaded tie rods caused the cost overrun. Symons refused to pay the claim. In its findings, the court stated,

> Although there may have been an ambiguity in the contract as to the specified type of form tie rods, any doubt as to the type of equipment to be used on the job was dispelled when Symons's supervisor supplied Bulley & Andrews' field superintendent with the threaded tie rod ends. At that time, plaintiff (Bulley) knew what was intended by defendant (Symons) under the contract, but plaintiff failed to protest, negotiate, comment, or otherwise call to defendants' attention any doubts it may have had as to whether the contract called for the use of threaded form tie rods. We feel the ambiguity was resolved by plaintiff actions.... (Bulley & Andrews v. Symons Corp.)

Where one party acquiesces to another party's actions, a modification or waiver of the contract may be established. The waiver doctrine has been upheld by many court decisions (Bulley & Andrews v. Symons Corp.; Lewis v. Jones).

When Read as a Whole, What Does the Contract Say?

Courts examine the entire agreement to determine the contract meaning. Two basic principles guide the interpretation of a contract as a whole. First, conflicting pieces of language are reconciled by reading them as complementary and not contradictory, and second, every clause is given meaning and purpose. It has been a long-standing principle that no single word, phrase, or sentence is interpreted out of context with the rest of the contract. Each relevant provision is looked on as serving a distinct purpose and not an implied purpose. As such, it is interpreted in light of that express purpose (Lewis v. Jones; Corbetta Construction Co. v. U.S.).

Case Study

Clear and Unambiguous Language

James A. Cummings, Inc., v. Bob Young
589 So.2d 950 (1991)

On April 1, 1985, Young, a subcontractor, entered into a contract with Cummings, a general contractor, for demolition, site work, paving, and installation of storm water systems at Florida Power and Light's Miami district office building. The contract amount was $297,778.

The project resulted in several claims. One in particular dealt with the relocation of boulders in accordance with the landscaping plans. Paragraph 44 of the contract read as follows:

> Demolition includes removal of concrete terrace, footing, concrete canopy, aluminum railing, remove and relocate benches, existing sidewalks, removal only of flagpole, parking lot lighting fixture bases, removal and relocation of existing boulders, sewage lift station (disconnect by Cummings), precast bench, concrete slab fence, concrete curb, ramp, planters, catch-basin, soakage pit and pavement removal.

Young asserts that "remove and relocate" means simply remove and stockpile the boulders at the commencement of the job to prepare the site for construction. Cummings, however, interprets the language as requiring Young to remove and stockpile the boulders at the commencement of the job and to return at the end of the job to place them in accordance with the landscaping plans.

How should the language of the contract be interpreted?

The Third District Court of Appeals of Florida spent little time in discussing Young's interpretation. The Court agreed with Cummings's interpretation that the language was clear and unambiguous. Young tried to introduce parol evidence, but this too was dismissed because the language was subject to only one interpretation.

The contract contained two categories of work. Specific items of work in each category are as noted below.

REMOVE	REMOVE AND RELOCATE
concrete terrace, footing, concrete canopy, aluminum railing, flagpole, parking lot lighting fixture bases	benches, existing sidewalks, existing boulders, sewage lift station (disconnect by Cummings), precast bench, concrete slab fence, concrete curb, ramp, planters, catch-basin, soakage pit, and pavement

If Young's interpretation is correct, there would be no need to have two categories because everything would be included in the remove category. Young's interpretation was unreasonable.

Conflicts in a single specification clause usually arise because of poor structure. The steps to be followed in resolving conflicts are

1. Determine the intent of the clause by (a) identifying the heading of the clause and subclauses and (b) reading the general language of the ambiguous clause.
2. Divide the clause into its relevant parts.
3. Identify the purpose of each part.
4. Review each part to be certain that it is interpreted as to what it says, not what it was meant or intended to say.
5. Review the clause to ensure that it is consistent with the intent of the clause.
6. Repeat steps 2 through 5 for the interpretation of the other party.
7. Evaluate the reasonableness of the two interpretations using the standards of interpretation cited in Chapter 4 to determine if there is more than one logical interpretation.

To illustrate, in Lewis v. Jones, Lewis, the general contractor, brought suit against Jones, the owner, to recover the unpaid balance of the contract price for the construction of a building. Jones withheld the final payment, asserting that construction was not completed in the 125 "working days" allotted in the contract, but instead the work was substantially completed in 365 calendar days. The issue before the court was to determine what was meant by "working days." The pertinent section of the preprinted contract form read,

Article 2. Time of completion—The work to be performed under this contract shall be commenced not later than November 28,

1947. It shall be substantially completed in one hundred and twenty five (125) working days.... and from the compensation otherwise to be paid, the owner may retain the sum of $25 for each day thereafter (Sunday and whole holidays not included) that the work of the general contractor remains incomplete or unacceptable to the architects. (Lewis v. Jones)

The owner asserted that "working days" was defined as every day as expressed in Article 2. Sundays and whole holidays were working days, but liquidated damages would not be assessed for those days. The contractor's interpretation of "working days" was that it meant days he actually could work. This interpretation excluded Sundays and whole holidays, plus days he could not work because of bad weather and when there were delays by the owner. The court concluded that there was nothing ambiguous about what the parties meant. The clause was divided in two parts: the first described the time of completion and how it would be determined (working days), and the second defined the exceptions so that liquidated damages could be assessed. The court reached its decision by looking at the way the clause was structured. Jones's interpretation required that to find the definition of working days in the first part, one had to look to the second part. This meant that the second part had two purposes, which the court did not accept. The Court said that interpretation was unreasonable and Lewis prevailed.

A similar problem arose, but in a somewhat more difficult situation, in Metropolitan Paving Co. v. City of Aurora. The City of Aurora, Colorado, contracted with Metropolitan Paving Co. to lay a 55-mi water pipeline. The dispute arose over the backfill requirements. The contractor used concrete pipe, and Fig. 8-3 graphically shows the requirements. The work was divided into zones, and the disagreement was over the maximum particle size for Zone 3. The contract stated,

12.2.44 *Zone 3 Backfill Material.* Zone 3 backfill material shall consist of selected material from the trench excavation, free from frozen material and lumps or balls of clay, organic or other objectionable material. When compaction of Zone 3 backfill is called for the material shall be well graded and easily compacted throughout a wide range of moisture content. Alternatively, if flooding, jetting and vibration are to be used for placing or compaction, the material shall meet the additional requirements specified in paragraph *Zone 1 and Zone 2 Bedding Material*

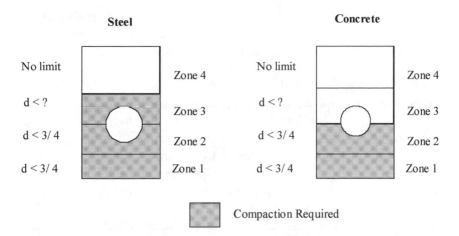

Figure 8-3. Graphical Representation of Compaction Requirements in Metropolitan Paving Co. v. City of Aurora.

Article 12.2.44 Zone 3
Backfill Material

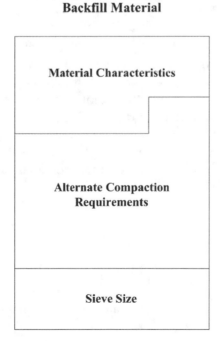

Figure 8-4. Rationalization of Compaction Requirement Decision.

for material to be placed and compacted by flooding, jetting and vibration. The maximum size shall pass a 2-inch U.S. Standard Series sieve. (Metropolitan Paving Co. v. City of Aurora)

The contractor argued that the 2-in. limitation applied only when compaction was done by flooding, jetting, or vibration. The city countered by stating that all Zone 3 material was subject to the 2-in. limit, regardless of the method of compaction. As a result of their disagreement, Metropolitan filed a claim for more than $3.5 million. The court studied the contract and listened to extensive expert testimony. The court found no ambiguity to exist and adopted the city's interpretation. Fig. 8-4 shows how this decision can be rationalized.

Most problems arising from ambiguities in a single clause occur because of poor structure. The preferred structure of a clause is for it to discuss subject A, then subject B. When clauses are interpreted by one party to cover topic A, B, then A or topic A followed by AB, that interpretation is usually rejected. Each clause or subpart thereof should have a single purpose.

Diagrammatic Details

Details and sketches can be intended to show details of contractor performance (detailed) or to show generalities (diagrammatic), like arrangements, relationships, or general layouts. It follows that it is important to an interpretation to determine the intent of details or sketches (e.g., is it intended to be diagrammatic or detailed?). A detail that is diagrammatic is not to be used to illustrate existing conditions or what the finished construction is to look like, except in general terms. The following case study illustrates the proper use of a diagrammatic detail in a dispute that never went past the negotiation phase.

Case Study

Interpretation of a Single Clause

Proceres Construction Co. signed a contract to rehabilitate the brick facade of three 19-story apartment buildings owned by Berkshire Properties, Inc. The contract documents called for brick to be removed at each floor level, steel angles to be anchored to the slab, and new brick to be installed in accordance with Detail 1 (Fig. 8-5). Along with Detail 1, the contract suggested the following sequence of operations:

1. Install protective fence enclosures. ...
2. Install protective plywood barricades on side of ground floor patio adjacent to each repair area. ...
3. Prior to fabricating bent plate relieving angle, contractor shall drill holes 24 inches on center to determine dimension from face of brick to edge of concrete slab. Horizontal leg of bent plate shall be ½ inch less than this dimension.
4. Submit shop drawings indicating all dimensions for relieving angles based on edge of concrete slab survey.

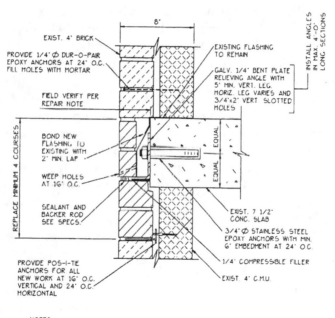

DETAIL   1

NOT TO SCALE

Figure 8-5. Detail 1 in the Proceres Construction Co. v. Berkshire Properties, Inc., Case.

5. Installation of relieving angles and lintel repairs shall begin at roof level and proceed toward the first floor.

The contract did not contain a concealed conditions clause but did contain the following clause:

21.2 Contractor acknowledges that it has fully inspected the job site, is fully familiar with all relevant conditions as the same may reasonably be expected to change during the performance of the Work and is fully familiar with all applicable state and local building codes. Based on such inspection and familiarity, Contractor agrees that the Contract Sum is just and reasonable compensation for all of the Work, including (a) all unforeseen and foreseeable risks, (b) inconsistencies between the Contract Documents and all state and local building codes, wherever arising, (c) hazards, (d) difficulties in connection therewith, and (e) any Work which is not required by the Contract Documents but without which the Work could not be completed. Contractor acknowledges, notwithstanding any provision of the Contract Documents to the contrary, that the Owner does not warrant or guaranty the conditions which will be encountered during the performance of the Work. Contractor and Owner agree, not withstanding any provision to the contrary, that this Agreement will result in completion of the Work, as defined by the Contract Documents, with no Change Orders unless Change Orders are requested by Owner in accordance with Article 18 or as a result of hidden conditions.

When Proceres began work, it became known that because of poor construction tolerances maintained by the original contractor, the brick was located at varying distances from the edge of the slab, and in many instances, the brick was not supported by the slab. Proceres argued that Detail 1 showed the brick being supported and that he was entitled to additional compensation based on the last sentence of para. 21.2.

What are the relevant parts of Article 21.2, and what is the purpose of each part? What is the purpose of Detail 1?

Conflicts may occur between several clauses or several documents. The consistent goal is to seek harmony or concord among the clauses, not discord (Monroe M. Tapper & Associates v. U.S.). When interpreting conflicts between clauses or documents, the steps to follow are the following:

1. Resolve conflicts in individual clauses if necessary.
2. Determine the *main* purpose(s) of each clause.
3. Compare the paragraph headings with the purpose to ensure consistency.
4. Make sure that other relevant clauses or phrases have not been rendered meaningless or useless.
5. Check to make sure that the interpretation is consistent with the purpose of each individual clause.

Many problems arise when a phrase in a clause is used for an interpretation that is not consistent with the main purpose of the clause. Headings often provide the telling clue as to the purpose of the clause, although these can be incorrect, so caution is necessary. For example, wording in a clause found under the heading of "Concrete, Workmanship" is unlikely to be the place to find details about concrete material requirements.

Conflicts between plans and specifications are illustrated in the case of Unicon Management Corp. v. United States, where there was a conflict between a provision of the specification and the plans. Unicon was constructing a floor in equipment room No. 1 of the missile master facilities building near Pittsburgh, Pennsylvania. At issue was whether floor tile was to be laid on a ¼-in. steel plate and steel beams (Fig. 8-6). The work involved embedding two steel beams in a concrete floor so that a pallet could be freely moved. The floor also contained cable trenches with a ¼-in. steel cover plate. The technical specifications under a section titled "Miscellaneous Metals" read as follows:

> *17-23 Equipment Room No. 1: AAOC Main Building.* Floors shall have steel beams embedded in the concrete floor so that they run transverse to the equipment lengths. The flange surfaces shall be flush with the finished concrete. Two beams shall be used, one at each end of the pallet, as shown on the drawings. These beams will be used to anchor the equipment pallets, to level the equipment palettes and can be used for references for concrete finishing tools during pouring of the floor. These beams shall be anchored to the building grounding system so that they serve as a grounding means for the equipment pallets. The beams shall be

Unicon v. United States

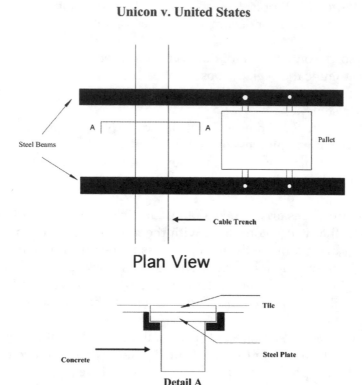

Figure 8-6. Floor Detail, Unicon Management Corp. v. United States.

one continuous length, with portions removed to provide clearance for cable troughs. (Unicon Management Corp. v. U.S.)

The contractor based his position on clause 17-23 and the order-of-precedence clause, which read that specifications governed over plans and argued that tile was not required over the steel plate, despite the fact that the drawings in one detail showed the floor tile on top of the plate.

The court reasoned that clause 17-23, read by itself, could be interpreted to implicitly envision an all-concrete floor; however, when viewed as a whole, the intent of paragraph 17-23 was to call attention to the beams set in the floor to support the pallets. Another telling aspect to support this conclusion is the heading of the section containing paragraph 17-23, which was "Miscellaneous Metal." The order-of-precedence clause did not have effect.

Headings play an indicative role in defining the purpose of the clause or detail. Headings can be particularly helpful when interpreting ambiguities between the contract plans and the specifications (Newsom v. U.S.).

Frequently, the general canons are used in conjunction with the rule of interpretation as a whole. The most common canons used are

- Specific terms control over general terms.
- A contract is read as a whole.
- Negotiations may be examined to explain, but not enlarge, the contract.
- Ordinary meaning of words is preferred over technical meanings.
- Plain contract language overrides trade usage.
- Specifications and drawings are read together.
- Specifications control over drawings, if in conflict.
- Performance specifications control over design specifications.
- Design specifications are generally warranted as correct.
- A lawful result is preferred over an unlawful one.
- Typewritten text controls over preprinted language.
- Writings control over figures (and specifications over plans).

The following cases illustrate their use. Construction Services Co. brought suit against the U.S. government for withholding payments on a dam project (Construction Services Co. v. U.S.). The contractor was required to place reinforced concrete in a spillway bucket section, where (he alleged) the government intended gravity concrete to be placed. Included in the invitation for bids was a unit price schedule containing estimates of material quantities for two types of concrete: gravity and reinforced concrete. Payment was to be based on the actual volume of concrete placed within the pay lines as indicated on the drawings. The original drawings contained no pay lines. However, the U.S. Army Corps of Engineers sent to all prospective bidders an addendum to the specifications stating that only three portions of the dam would consist of reinforced concrete. A drawing with the addendum had the following note attached:

> All concrete will be paid for as gravity concrete except spillway and bucket training walls and Gate House above Elev. 836.0 which are classed as Reinforced Concrete for payment.

The drawing showed pay lines that defined the parts that were to be paid as gravity concrete. After substantial completion, the plaintiff contended that all of the spillway bucket concrete was to be paid for as reinforced concrete. The contractor pointed to the original bid documents as the basis for his position. The government argued that the spillway bucket concrete was part of the spillway and bucket training walls, as defined in the addendum.

The court noted that before the addendum was issued, the contract was ambiguous as to what was to be classified as reinforced concrete, but went on to say,

> The crucial fact, however, is that this omission was cured by the addendum. The relevant section of the addendum specified the parts which were to be labeled "reinforced concrete" and provided an explanatory note. (Construction Services Co. v. U.S.)

Finding for the government, the court stated,

> By limiting attention to the note set forth in the addendum, plaintiff attempts to find an ambiguity. The short answer to this is that when all pertinent provisions of the addendum are considered it is clear that the spillway bucket was excluded from the category of reinforced concrete. (Construction Services Co. v. U.S.)

The contractor was denied recovery because the addendum overruled the original documents on this issue. In this instance, the more specific document prevailed. If the specifications prevailed over the plans, then pay lines would be of no value.

Sometimes, the canon that words overrule figures is used. Caution must be used when applying this principle, because drawings may give more specific or detailed information than the applicable specifications (Unicon Management Corp. v. U.S.). A determination must be made as to which document is more specific.

Rule against the Drafter

Common law provides that courts rule against the drafter when there is more than one reasonable meaning after all the primary rules and canons have been exhausted (Peter Kiewet Sons' Co. v. U.S.). In reality, the rule against the drafter criterion is used as a tiebreaker only when all other rules have proven inconclusive or unpersuasive.

Blount Brothers Construction Co. v. United States is an excellent example. The case involved a Navy contract to construct a ship-maneuvering basin. During performance, a dispute arose as to whether a concrete slab was required to rest solely on concrete fill and natural rock or if the slab could rest on soil. The contractor asserted that concrete fill was only required when overexcavation had occurred, and otherwise, portions of the slab could rest on soil. The government's position was that the bottom slab would only rest on concrete fill or natural rock. The court concluded that the ambiguity was not patent, and both plaintiff's and defendant's interpretations were reasonable. The governing case law was established in WPC Enterprises, Inc., v. United States, which it quoted:

> It is precisely to this type of contract that this court has applied the rule that if some substantive position of a government drawn agreement is fairly susceptible of a certain construction and the contractor actually and reasonably so construes it, in the course of bidding or performance, that is the interpretation which will be adopted.

In practice, the rule against the drafter is often argued and is often applied by the judiciary as a tiebreaker rule. This is true even when standard form contracts are "proposed" by one of the contracting parties. The party proposing a standard form may be considered the "drafter."

The Order-of-Precedence Clause

Many government and some state DOT contracts have order-of-precedence clauses that establish an order of priority among the various documents when an ambiguity occurs between documents (Reed v. U.S.). Many of the previous cases in this chapter had such clauses, but the court refused to adopt them (Beacon Construction Co. v. U.S.; Unicon Management Corp. v. U.S.; Construction Services Co. v. U.S.). They are used as a substitute for the rule against the drafter.

The positive application of the order-of-precedence clause is shown in the case of Franchi Construction Co. v. United States. The Franchi Construction Co. was awarded a government contract to install partitions and floor tile in an office building. The controversy arose about a year into performance over the sequence in which vinyl–asbestos floor tile was to be installed in relation to gypsum wallboard partitions. Should the tile abut the partitions after the partitions were erected on the concrete floor, or should the tile be placed first with the partitions on top of the tile? The specifications made reference to the sequence of the work and said that the tile should be placed after other work that might cause harm to the tile was finished. The drawings showed the tile being placed first. Both requirements were explicit and inconsistent with one another. After much discussion, the court was able to reconcile the matter by resorting to the order-of-precedence clause, stating that "the plaintiff is entitled to take the Government-sponsored order-of-precedence clause at face value" (Franchi Construction Co. v. U.S.).

Illustrative Example

This example is based on Monroe M. Tapper & Associates v. United States. The contractor brought action arising from constructing a post office facility in Worchester, Massachusetts.

Statement of Facts

The dispute involved the material makeup of the backfill to be placed in utility trenches. Specifically, did the specifications permit the use of excavated earth in backfilling utility trenches? Tapper said it did; the government said it did not and required the contractor to use gravel instead.

The applicable contract language was as follows:

SECTION 0220—EARTH WORK

3. MATERIALS

. . .

C. Material for Filling: All backfills and fills for the building, retaining walls, walks, roads and other surfaced areas from grades existing after removal of overlying materials shall be made with gravel (except lawns and planting areas).

4. WORKMANSHIP

. . .

D. EXCAVATION

...

(6) Do all excavation within the building area and for utilities such as water lines, underground electrical lines, sanitary lines, storm water lines, catchbasins, and manholes. *Refer to site utilities section for more detailed information.* (emphasis added)

E. BACKFILLING

...

(1) Do all backfilling with selected granular material specified properly compacted. Excavated material on the site, if gravel and free from loam, sticks, rubbish, and other foreign matter, may be used, if it meets the specified requirements but if the quantity is insufficient or the quality unsuitable for the particular purpose for which it is intended then the specified gravel shall be provided.

SECTION 0255—SITE UTILITIES

4. WORKMANSHIP

...

I. Backfilling. Backfill trenches only after piping has been inspected, tested (if water) and locations of pipe and appurtenances have been recorded, backfill by hand around the pipe and for a depth of one foot above the pipe; use earth without rock fragments or large stones and tamp firmly in layers not exceeding six inches in thickness, taking care not to disturb the pipe or injure the pipe coating. Compact the remainder of the backfill thoroughly with a rammer of suitable weight or with an approved mechanical tamper, or if the soil is granular by flooding, provided that under pavements, walks and other surfacing, the backfill shall be tamped solidly in layers not thicker than 6 inches. Exclude all cinders and rubbish from trenches in which metal pipes are laid.

5. SITE UTILITIES

A. Water lines: Excavation, installation and backfilling for water lines within the street and up to the property line shall be done by the City of Worcester Water Department and the cost of same shall be charged to the Contractor. Excavation and backfill for water lines within the property lines shall be done under Section 0220, Earthwork.

B. Sanitary lines: Excavation, installation and backfilling for sanitary lines within the street to the property line shall be the responsibility of the plumbing contractor. The installation of this line within the street shall be done by a licensed drain layer. The installation of sanitary lines within the property lines shall be done by the plumbing contractor and excavation and backfilling for sanitary lines within the property line shall be done under Section 0220, Earthwork.

C. Storm water lines: Excavation, installation and backfilling for storm water lines from a point 5'-0" outside the building foundation wall to the street connection and including the street connection shall be the responsibility of the Contractor and done under Section 0220. The installation of this line within the street shall be done by a licensed drain layer.

D. Underground electrical, telephone, and fire alarm lines: Excavation and backfill for the underground electrical, intercom, telephone, and fire alarm lines shall be done under Section 0220, Earthwork. Also excavation and backfilling for underground cables or conduit to walk lights, etc., shall be done under this same section, 0220.

Analysis

Plain Meaning Rule

The plain meaning rule is first applied to Section 0220, 3.C, Material for Filling. When the special associated words canon for language is used, it is clear that site utility trenches are excluded because it is not part of the list: "building, retaining walls, walks, roads and other surfaced areas." (The contractor did not argue that gravel was required in utility trenches beneath these listed areas.) However, the plain meaning rule is not sufficient in this instance because there are other operative clauses.

Patent Ambiguity

There is no patent ambiguity in this instance, even though patent ambiguity was argued by the government.

Actions of the Parties

The parties displayed actions that were consistent with their opposing positions stated above. This issue is not relevant in this dispute.

Interpret as a Whole

Fig. 8-7 was prepared to assist in this complex analysis. The purpose of Section 0255—Site Utilities is to describe materials and methods for backfilling site utility trenches (earth), and the purpose of Section 0220—Earthwork is to describe materials and methods for backfilling all other areas. Paragraph D.6 of Section 0220 actually refers to Section 0255. Even though the reference appears in a paragraph under the heading Excavation, it is clear that Section 0255 describes backfilling. Overall, there is general consistency with the paragraph headings, although the description of the use of gravel is under the heading of earthwork. Paragraph 5 of Section 0255 discusses who is responsible for the work and then refers back to Section 0220, only to establish who is responsible. This interpretation is consistent with the meanings of the paragraphs. Furthermore, if the government's interpretation is adopted, there is no reason to have the materials portion of Section 0255.4.I. This part is useless because there are no other site utility trenches. If interpreted that the site utility trenches require earth, then the entire specification is in harmony. Section 0220 describes the use of gravel in all trenches except those listed in paragraph 3.C. Section 0255 describes the use of earth in all site utility trenches.

Synopsis

Based on the above analysis, when read as a whole, the contract contemplated earthen backfill in many areas. Tapper should recover monies for the added cost of backfilling with gravel.

The court summarized its findings for the contractor by saying,

> The court is here presented with a contract containing a detailed section entitled site utilities which specifically details how backfilling needs to be accomplished, i.e., with earth. We have other provisions of the contract which do not mention site utilities and only refer to the use of gravel in backfilling in specific

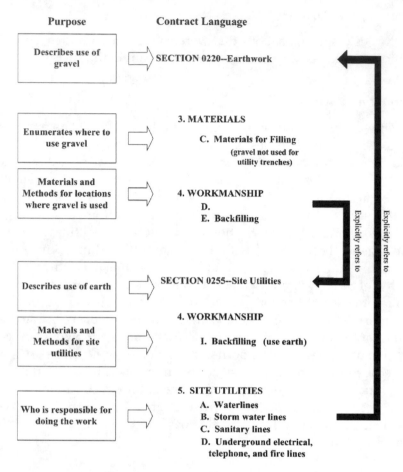

Figure 8-7. Framework of Contract Language in Tapper v. United States.

circumstances. It is clear that the contract viewed as a whole is not ambiguous in regard to backfilling of utility trenches. (Monroe M. Tapper & Associates v. U.S.)

Exercise 8-1: Barash and New York

M. Barash, a contractor, contracted with the state of New York for the rehabilitation of room 202 in the state Capitol building. The disputed work involved the refinishing of the wood ceiling. The dispute arose when the state architect informed the contractor that cabinetwork on the ceiling was required. Barash contended that cabinetwork on the ceiling was not required by the contract.

The contract provisions in question were Section 19A, Mill and Cabinet Work: Addendum No. 2 to the specifications, and Article 12, Paragraph 50 of the General Conditions. Section 19A detailed the rehabilitation for the wood ceiling, in particular, the repair of beams, moldings, and panels, and other related work. "Addendum No. 2 to Specification No. 17469, construction work for rehabilitation of existing room No. 202 ..." stated, "Disregard the entire Section 19A as contained in the specification and use the following in lieu thereof." The addendum completely omitted the work required in the original Section 19A that related to the ceiling. However, the contract drawings still showed the detail of the wood ceiling as to beams, moldings, and panels that was referred to in the original specification, Section 19A.

The state architect decided that cabinetwork on the ceiling was required pursuant to Article 12, Paragraph 50, which gave him the authority to determine

> ... the true meaning of the drawings or specifications on any point concerning the character acceptability and nature of any kind of work or materials or construction thereof ...

Does Article 12, Paragraph 50, give the architect the right to order anything he or she wants? What does the contract say? Is the ceiling work reasonably inferable? Should Barash recover his additional cost?

Exercise 8-2: Long and Lazer Construction Co.

In 1985, James B. Long approached the Lazer Construction Co., Inc., about building a new home. Lazer, with Long's approval, drafted a contract on forms published by the American Institute of Architects, stating a guaranteed maximum price (GMP) of $99,500, subject to preapproved written change orders prepared according to the contract. The contract also stated that the work was to be performed according to the owner's plans and specifications and provided for Lazer's fee to be 5% over all costs of the project, unless the project cost, in the absence of approved changes, ran over the GMP, in which case the owner was to pay only the direct costs. Although the contract stated a GMP, other provisions of the contract indicated that this limit was flexible. Most notably, the contract specifically provided that in the absence of approved changes, Long would pay only direct costs exceeding the GMP.

The contract at the time of the execution was based, in part, on an initial set of plans. Several weeks after construction began, Long changed the plans, and a new set of plans was submitted. As construction continued, certain items were added, certain items were deleted, and certain items were paid for directly by Long. Long verbally approved the changes. Lazer spent $140,771.50 on the project and billed Long for this amount. Long paid Lazer $116,913.57 and then refused to pay any more.

Long later argued that Lazer was not entitled to additional payments because he had already paid more than the GMP of $99,500. He also asserted that the contract provisions stated that the maximum price could be changed only by preapproved written change orders. Furthermore, Long asserted that he should have received credits totaling $33,886 for the costs of certain items deleted from the original plans and items he paid for directly instead of buying them through Lazer. Ignore issues related to changes, and instead focus on the contract language.

Does the rule of practical construction play a role in this dispute if it was learned that Long had already paid Lazer the sum of $116,913.57? Does the contract allow Lazer to recover the $33,886 if the changes were validly made?

Exercise 8-3: W.H. Armstrong & Co. and U.S. Government

W. H. Armstrong and Co. entered into a contract with the U.S. government to build 2 field officers' quarters and 11 company officers' quarters at Bolling Air Force Base in Washington, D.C. The walls were to be of brick. Outside exposed surfaces of the walls were to be of facing brick, and the basement and the inside portion of the walls behind the facing brick were to be of common bricks.

The contractor was to furnish all labor and material to complete the job except that as per Article 38 of the specifications, "Brick, Common will be furnished by the U.S. at the location indicated by the contracting officer." Before submitting its bid, Armstrong inspected the stockpile of common brick from which the contracting officer proposed to furnish. The stockpile contained between a million and a million and a half salvaged red shale common bricks in excellent condition. Armstrong took several bricks as samples. On the basis of this inspection, Armstrong made its estimates as to the amount of labor and mortar to be used in laying the common bricks.

Before the laying of bricks was completed, the contracting officer orally directed Armstrong to discontinue the use of bricks from said stockpile. The reasoning behind this directive was because the government wished to use these bricks for facing work on other contracts in progress at the same site. At this time, Armstrong had used 210,820 bricks from the stockpile.

Armstrong was directed to use bricks salvaged from a nearby dismantled steel plant. These bricks had been used in furnaces and were not common bricks, but firebricks. These bricks were of varying sizes and shapes such as circle bricks, wedge bricks, arch bricks, oversize bricks, split bricks, and brickbats. They were more porous than common brick. The irregularities required the use of more mortar and labor at an increased cost to Armstrong.

Armstrong protested to the contracting officer for having to use these bricks. The parties orally agreed that Armstrong would continue to use the brick from the steel mill and at a later date would submit the extra cost to the contracting officer for audit. Armstrong subsequently used 386,500 firebricks.

When the work was complete, Armstrong submitted the extra cost to the contracting officer for audit. By this time, there was a replacement contracting officer, who refused to pay for the extra cost because there was no written change order.

Ignoring the issue of an orally directed change, does the contract provide relief to Armstrong?

Exercise 8-4: R. B. Wright Construction Co. and U.S.

The R. B. Wright Construction Co. entered into three contracts with the U.S. government in 1985 for miscellaneous repairs, including painting, to approximately 200 World War II–era barracks and office buildings. The three contracts contained drawings identifying the areas to be painted. The painting schedule detailed the surface preparation, type of paint, and number of coats. Most of the surfaces to be painted had been previously painted.

Each contract contained a specification section, 9P2A, Painting, General. Paragraph 14 of section 9P2A read,

> 14. SURFACES TO BE PAINTED: Surfaces listed in the Painting Schedule, other than those listed in paragraphs SURFACES NOT REQUIRING PAINTING and SURFACES FOR WHICH PAINTING IS PROHIBITED, will receive the surface preparation, paints, and number of coats prescribed in the schedule.

Paragraph 18 provided,

> Painting schedule: The PAINTING SCHEDULE prescribes the surfaces to be painted, required preparation, and the number and types of coats of paint.

The painting schedule contained five columns, which listed, for each of the various surfaces, the type of surface preparation and specified the type of paint for three coats. A typical example was as follows:

SURFACE	PREPARATION	1ST COAT	2ND COAT	3RD COAT
Exterior wood surfaces not otherwise specified	As previously specified	MIL-P-28582 (which the record shows was a primer coating)	Exterior oil paint or TT-P-19 or TT-P-1510	Exterior oil paint or TT-P-19 or TT-P-1510

For a number of surfaces, however, the schedule stated "None" under the listing for the "3rd Coat." For example, one item read,

SURFACE	PREPARATION	1ST COAT	2ND COAT	3RD COAT
Exterior ferrous (metal) surfaces, exposed, unless otherwise specified	As previously specified	TT-E-489 Class A	TT-E-489 Class A or TT-E-1593	None

The painting schedule had 38 separate items for different surfaces, 26 of which specified "None" for the third coat. The surfaces listed in the painting schedule included wood, ferrous surfaces, concrete, and plaster.

After the work on the first building had been completed, the government inspector discovered a previously painted wall that, in its opinion, had not been properly repainted (its old color, purple, could still be seen). The government learned that Rembrandt, Wright's painting subcontractor, had applied only one coat of paint to that wall and all other walls that had been previously painted. The government then ordered Wright to (1) apply three coats of paint to every surface specified in the painting schedule without regard to whether it had been previously painted, or (2) give the government credit for any work not performed. Wright's subcontractor argued that the painting schedule was ambiguous and did not apply to previously painted surfaces, only to unpainted ones. The basis of his argument was that it was customary in the painting business that three coats of paint are not applied to previously painted surfaces.

Assuming that Rembrandt was paid for all extra coats where the contract called for two coats or less, is the contract ambiguous as Rembrandt argues? Is Rembrandt entitled to additional monies? What rule from Chapter 4 is applied to resolve this dispute?

Exercise 8-5: Lyon Metal Products and Kaiser Construction Co.

Lyon Metal Products, Inc., is a corporation engaged in the business of "furnishing manufactured material for the installation of lockers, wardrobe cabinets, bookcase units, etc. in public buildings, industrial plants, army barracks and warehouses."

Lyon had a subcontract with the prime contractor, Hensel Phelps Company, to supply and install the lockers, wardrobe cabinets, and bookcase units in barracks at Fort Carson, Colorado. Lyon first subcontracted with the Holm Heating and Sheet Metal Co. to assemble and install these units. In the latter part of May 1966, at the specific request of Lyon's Kansas City district sales manager, Richard Brown, Kaiser Construction Co. sent two of its supervisory employees, Fred Eshelman and Carl Darby, to Fort Carson to aid the Holm Co., which was having difficulty with the quality and progress on the project. For two weeks, Eshelman and Darby worked with Holm and then surveyed the job with respect to the Lyon–Holm subcontract. Their findings were submitted to Mr. Kaiser, the principal of Kaiser Construction, who apparently had been approached by Lyon regarding a takeover of the Holm subcontract. Holm voluntarily surrendered its contract, and on June 3, 1966, Lyon entered into a written contract with Kaiser, under which Kaiser, for the sum of $62,000, agreed in part to

> ... unload, distribute, erect, install locks, adjust doors and equipment in place; touch up all mars and scratches and dispose of all crates, boxes and packing ...

associated with the lockers, wardrobe cabinets, and bookcase units. These items were specifically identified by reference in the contract to numerous numbered erection orders, and totaled 310 wardrobes and 240 bookcase units for each of the 10 barracks.

The contract also required that all changes be authorized in writing.

When Kaiser began work, they found that they had to remove concrete that had been spilled on the floor and that the floors and ceilings were not plumb, which required "shoring up." The materials did not arrive at the site on schedule, and they were often required to store it in one building and then move it to another when the construction had progressed to such point that it could be used. On occasion, it was necessary to store part of the units outside on the ground, and this required laying a floor of 4-in. x 4-in. timbers and a covering of tar paper. Much of the material was damaged from either improper handling or from the elements and was sometimes rusty, requiring repainting or refinishing. To fit the cabinets and lockers squarely and evenly against the uneven ceilings and floors, it was necessary to install angle iron clips. The difficulties in construction were defects that are usually eliminated by the prime or building contractor and are not customarily left to be performed by the contractor that is charged only with the duty of installing lockers, bookcases, and wardrobes. No written change order was ever executed.

Throughout the work, the government site representatives were telling them "to do this and to do that" and get the barracks done. Mr. Eshelman for Kaiser was told by Clint Peterson, Lyon's Denver representative, to "do anything you have to do to get these (government) people satisfied." Mr. Kaiser, the company president, also discussed the matter with Mr. Peterson, and was told, "We'll have to do it. We've got to maintain schedule." With respect to replacing material that had been damaged through storage on the ground or otherwise, Mr. Kaiser said that Mr. Brown, Lyon's supervisor on the job, told him to "use everything possible—refinish and repaint" and said that "We would be taken care of, that it was a thing that had to be done in order to sell the job." Mr. Swartz, Kaiser's supervisor during the latter part of the work, discussed the difficulties with the progress of the work with Mr. Brown on occasion, and also with Al Corredor of the Lyon office, and they told him to go ahead and do whatever was necessary and assured him that they would be taken care of.

Aside from the issue of an oral change order, does the contract offer the Lyons any avenue for recovering their added expenses? What was the intent of the Lyon–Kaiser contract? Is it necessary for the Lyons to show that a valid change was made (see Chapter 7) before they can argue contractual relief? Are the Lyons entitled to an equitable adjustment?

Exercise 8-6: Lancaster Area Sewer Authority and Environmental Utilities Corp.

In the late 1960s, the Lancaster Area Sewer Authority developed a plan for an extensive and interconnected sewer system throughout various townships and boroughs in the Lancaster (Pennsylvania) area. Huth Engineers, Inc., was retained to prepare the plans and specifications for the project and to supervise construction. Huth divided the system into several parts, and for each, a separate contract was awarded and executed. Environmental Utilities Corp. was awarded Contract No. 10 in the area of the borough of Mountville. Four months after Environmental Utilities commenced work on this project, Samuel Berlanti, a vice president of Environmental Utilities, took charge of the project. He also prepared bids on Contracts No. 20 and 21, in the area of West Hempfield Township, while Environmental Utilities was performing the work on Contract No. 10.

The terms of the three contracts were identical in most material respects, especially in the areas designated as "Information for Bidders," "Technical Provisions (Standard)," and "Technical Provisions (Detailed)." The contractor was to be paid by the linear foot for the installation of pipe, by the cubic yard for the installation of backfill, and by the square yard for the replacement of paving. Furthermore, in the last two categories, backfill and paving, there were constraints known as "pay widths," which limited the amount of backfill and paving for which the sewer authority would pay. Thus, when 8-in. pipe was laid, the backfill pay width was limited to 2 ft, and the paving pay width for nonstate highways was limited to 38 in. and 48 in. for state highways.

The specifications also provided that if extra work became necessary, a prescribed procedure was to be followed. Section 17 of the Supplemental General Conditions of the Specifications of Contract No. 10 reads in part,

> Without invalidating the contract, the Owner may order extra work or make changes by altering, adding to or deducting from the work, the contract sum being adjusted accordingly. All the work of the kind bid upon shall be paid for at the price stipulated in the proposal, and no claims for any extra work or materials shall be allowed unless the work is ordered in writing by the Owner or its Engineer, acting officially for the Owner, and the price is stated in such order.

Section 22 of the General Conditions of the Specifications of Contract No. 10 read as follows:

> No claim for extra work or cost shall be allowed unless the same was done in pursuance of a written order of the Architect/Engineer approved by the Owner, as aforesaid, and the claim presented with the first estimate after the change or extra work is done. When work is performed under the terms of sub-paragraph 17(c) of the General Conditions, the Contractor shall furnish satisfactory fills, payrolls, and vouchers covering all items of cost and when requested by the Owner, give the Owner access to accounts relating thereto.

In addition to this language regarding alteration of work and payment, there is language regarding the condition of the work sites. The language of all three contracts is identical to that in Contract No. 10 contained in the "Information for Bidders" section; it reads as follows:

> 11. CONDITIONS OF WORK
> Each bidder must inform himself fully of the conditions relating to the construction of the project and the employment of labor thereon. Failure to do so will not relieve a successful bidder of his obligation to furnish all material and labor necessary to carry out the provisions of his contract. Insofar as possible the Contractor, in carrying out his work, must employ such methods

or means as will not cause any interruption of or interference with the work of any other contractor.

19. SITE CONDITIONS

Where information as to soil conditions, test borings, test piles and existing underground and overhead structure locations is shown on the Engineer's plans, specifications or drawings, or in preliminary reports prepared by the Engineer, such information is for the Owner. The correctness of such information is not guaranteed by the Owner or the Engineer, and in no event shall be considered as part of the contract, an inducement to bidding or as a factor for computation of bids. If such information is used by a bidder in preparing his proposal, he must assume all risks that conditions encountered in performing work may be different from the approximation shown. If any bidder so desires, the Owner will afford him an opportunity, at his own expense, to make borings or soundings, to drive test piles or to dig test pits properly refilled to the satisfaction of the Owner.

The Contractor shall satisfy himself, by careful examination, as to the nature and location of the work, the character of equipment needed preliminary to and during prosecution of the work, the general and local conditions, and all other matters which can in any way affect work under this contract.

20. APPROXIMATE ESTIMATE OF PROPOSAL QUANTITIES

The bidder's attention is directed to the fact that in contracts based on unit prices the estimate of quantities of work to be done and materials to be furnished under these specifications, as shown on the proposal form and in the contract, is approximate and is given only as a basis of calculation upon which to determine the lowest bidder. The Owner does not assume any responsibility that estimated quantities shall be maintained in the construction of the project, nor shall the Contractor plead misunderstanding or deception because of such estimate of quantities, or the character of the work or location, or other conditions pertaining thereto. The Owner reserves the right to increase or diminish any or all of the above mentioned quantities of work or to omit any of them, as it may deem necessary, and such increase or decrease of the quantities given for any of the items shall not be considered as sufficient grounds for granting an increase in the unit prices bid.

During the work, Environmental Utilities had to excavate at a width greater than the pay width because local governmental regulations, incorporated into the contract, require sheeting and shoring in certain types of soil. A trench box was an acceptable alternative to sheeting and shoring. To use such a trench box, Environmental Utilities had to excavate a trench 6 ft wide; therefore Environmental Utilities' work method required greater quantities for backfill, shoulder restoration, and paving than the authority specified as the pay width. Environmental Utilities disputed the work but never received a change order for any of the disputed work. Furthermore, Environmental Utilities did not submit written claims for extra work with any of its monthly estimates, or within 15 days of the alleged occurrence giving rise to the claim. In fact, Environmental Utilities waited more than two years before filing claims for work under Contract No. 10 and almost 18 months before filing claims for work under Contracts No. 20 and 21.

What is the intent of the contract relative to "pay width"? Ignoring Environmental Utilities' shortcomings on submitting its claim and lack of timeliness, are they entitled to extra compensation?

Exercise 8-7: Metro Insulation Corp. and Boston Housing Authority

Metro Insulation Corp. was a sub-subcontractor to the Frank Sullivan Co. for the construction of a federally assisted housing project for the elderly owned by the Boston Housing Authority. Beacon Construction Co. was the general contractor. The dispute involved insulation work performed by Metro on certain sprinkler piping in the first-floor ceiling of the building.

The General Conditions in Division 1.2.3 stated,

> The intent of the Contract Documents are to include all items necessary for the execution and completion of the Work. The Contract Documents are complementary, and what is required in any one shall be as if required by all. Work not covered in the Contract Documents will not be required unless it is consistent therewith and is reasonably inferable therefrom as being necessary to produce the intended results...

Division 28.21, Pipe Covering and Insulation, discussed piping insulation in general terms, stating,

(1)All hot water supply and return piping, branches, risers, and also including cold water piping throughout the first floor ceiling shall be insulated with . . .

(2)Insulate all water lines in first floor ceiling with . . .

The specification was silent with respect to the insulation of the sprinkler piping. The article titled "Disputes" read as follows:

All disputes . . . arising under this contract or its interpretations, whether involving law or fact, or both, or extra work, . . . shall within 10 days of commencement of dispute be presented to the Architect for decision. A copy of the notice of the dispute shall be sent to the Authority and the Regional Office of . . . [HUD]. Such notice need not detail the amount of the claim but shall state the facts surrounding the claim in sufficient detail to identify the claim, together with its character and scope. In the meantime, the contractor shall proceed with the work as directed. The parties agree that any claim not presented within this subsection is waived.

The housing authority realized that the pipe was not being insulated and directed Metro to do so. Metro objected by asserting that it is not the custom in the insulation trade to insulate sprinkler piping unless that insulation is clearly specified. A contract drawing clearly distinguished between the water piping and the sprinkler piping to be installed in the first-floor ceiling. A jobsite meeting was held with Metro, Sullivan, Beacon, the architect, the engineer, and the housing authority present. At that meeting, Metro was informed by all those present that a change order for the extra insulation work was forthcoming and that Metro was to proceed with the work immediately so as not to hold up the project. Metro proceeded and completed the work.

Meanwhile, the housing authority submitted the request for a change to HUD, whose approval was required as per the contract. HUD refused to approve the change, and the housing authority subsequently decided not to pay Metro. The housing authority's objection to payment was that the work in question was already required under Division 28.21.

Is Metro entitled to additional compensation? What is the basis?

Exercise 8-8: Foothill Junior College District and Jasper Construction Co.

Foothill Junior College District of Santa Clara County, California, awarded Jasper Construction Co., Inc., a contract to construct the Calvin C. Flint Center for the Performing Arts, located on the De Anza Community College campus in Santa Clara. Jasper was the general contractor and Foothill, the owner. The design was performed by Kump, Masten and Hurd, a joint venture consisting of two separate architectural firms.

Under the terms of the contract signed on May 17, 1968, Jasper was to be paid $3,307,403 and was required to complete the project within 600 calendar days or by January 7, 1970. Jasper agreed to pay as liquidated damages the sum of $400 for each calendar day that the project was delayed.

Jasper received extensions of 545 days from Foothill, which would have made the completion date June 11, 1970. Requests totaling 223 days of extension were denied. Notice of completion was not in fact filed until June 10, 1971, or 363 days beyond the completion date. Foothill thus withheld $145,200 in liquidated damages because of the delay.

With regard to the construction joints in the reinforced concrete walls, the contract specified the following:

11. CONSTRUCTION JOINTS

(A) Location and details of construction joints shall be as indicated on the structural drawings, or *as approved by the Architect*. Relate required vertical joints in walls to joints in finish. In general, approved joints shall be located to least impair the strength of the structure.

When Jasper began work, he noted that the location of the construction joints was not shown on the drawings, but certain structural drawings indicated to Jasper that the steel reinforcement was from floor to floor and that, therefore, the concrete would be poured horizontally from floor to floor. This understanding was consistent with the way Jasper bid the job. Jasper began pouring the basement from floor to floor, but the architect instructed Jasper to keep the pour joints vertical from wall to wall. Jasper objected to this method of pouring because it resulted in many wooden concrete forms that could not be reused. Jasper alleged that having to use different methods of pouring cost $500,000 in "extra material and extra labor."

What rule from Chapter 4 is instrumental in deciding this claim? Ignoring issues of time, is Jasper entitled to extra monies?

Exercise 8-9: Western Contracting Corp. and Georgia State Highway Department

In the late 1960s, Western Contracting Corp. entered into a contract with the Georgia State Highway Department to construct 5.049 mi of grading, dredging, and hydraulic embankment on Interstate 95 in Glynn and McIntosh Counties. The highway department provided plans designating certain borrow pit areas adjacent to the proposed right of way from which Western was to pump the hydraulic material for the embankments. After the work began, Western learned that certain of the designated, privately owned borrow areas were unavailable for use because owner consent had not been obtained. Western incurred additional expenses for having to use other borrow pits.

Of particular importance was Sheet 22 of the plans, titled "Pit Location Sketches Hydraulic Borrow." Sheet 22, along with other portions of the plans, shows that the proposed highway embankment traverses several streams and rivers (i.e., the Altamaha, Champney, and Butler Rivers and the Darien and Cathead Creeks). Sheet 22 showed five clearly marked and numbered areas where the U.S. Army Corps of Engineers had agreed to permit dredging. Furthermore, Sheet 22 was drawn to scale and showed a certain amount of precision with respect to the location and boundaries of the borrow pits in question. It further showed for each pit location the type of borrow material available and whether it was usable with or without "muck removal." The sheet showed for each pit the volume of material available, although with regard to Pit No. 5, the volume available was denoted as "approximate." It further showed where in the project in delineated lengths and widths and in what volume the material taken from *each respective pit* was to be placed to create the embankment. Finally, Sheet 22 showed the owners of the property through which the embankment and the various waterways run. Sheet 22 also contains a conspicuous note stating the following: "These pits are shown as possible sources of material."

Other provisions in the contract include Special Provision Sec. 108, which specifies that

> The quantity of material shown on the plans as available in the borrow pits is not guaranteed by the State Highway Department. Payment for material from sources where such material is privately owned will be made by the Contractor directly to the property owner.

The standard specifications, Vol. 1, Sec. 2.06, required the bidder to examine the site of the work, the proposal, plans, specifications, etc., and to satisfy himself or herself as to the conditions to be met, as to the character, kind, and quantities of work to be done, specification requirements, etc. Standard Specifications Vol. 1, Sec. 6.01b provides,

> The sources of local material will be shown on the Plans, and the amounts of royalties and other costs and conditions of acquisition of the material from the owner will also be so shown.... The Department will obtain all necessary options from the owners.... The Department does not guarantee that the quantity or quality of acceptable material required can be obtained from any designated source, and the failure of such designated sources to contain material in sufficient quantities of acceptable quality shall not be the basis for any claim...

Finally, the Supplemental Specifications in Sec. 5.04 contain an order-of-precedence clause stating that in the event of discrepancies in the contract documents, the specifications overrule the plans.

Identify and reconcile, where possible, conflicting provisions of the contract. What is the intention of the contract? Is Western entitled to more monies?

Exercise 8-10: Jacksonville State University, Dawson Construction Co., and Bob Roberts Co.

On March 10, 1970, Jacksonville State University contracted with Dawson Construction Co. to construct a women's dormitory. Dawson made subcontracts with Bob Roberts Co., Inc., to install exterior exposed aggregate panels and with Copeland Glass Co. to finish and install windows and doors. Among the provisions of the contract were that Dawson was to construct the building according to plans and specifications as prepared by the architect, Hofferbert-Ellis and Associates. In the specifications, the architect specified the materials that were to be used in constructing the exterior aggregate panels and the performance requirements the panels should meet. Windows were specified to the same level of detail. According to the specifications, the aggregate panels were to be composed of insulation board on exterior wall studs, lathing, epoxy or cementitious matrix, exposed aggregate embedded in matrix, sealer coat, caulking, zinc accessories, and related items. The aggregate panels were specified to be waterproof.

Rather than specifying a product name for the cementitious or epoxy matrix, the architect listed the various properties the matrix should have. From the listing of properties, Roberts interpreted that specifications required a specific product, Boncoat, made by the M. D. Corp.

The project started in 1970. After the outside aggregate panels were installed, water began to come through the panels, causing many of them to buckle away from the wall. As per manufacturer's recommendation, Roberts put a scratch coat of cement between the matrix material and the metal lathe. This did not solve the leakage problem. It was also discovered that water was leaking around the windows installed by Copeland Glass. Roberts recaulked the window jambs, but this did not stop the leakage. There was also evidence of corrosion on the metal portion of the windows and in the retainer channels. Although there were leakage problems with the windows, it was subsequently determined that the main source of leakage was from the panels. In particular, the leakage was the result of a defect in the Boncoat.

The building was completed in May 1971, and the dormitory was occupied in June 1971. It was not until December 1972 that all repairs were considered acceptable.

Some time after December 1972, the architect issued a directive to Dawson to correct stains on the aggregate panels. Dawson advised the architect that his guarantee period had expired and that the architect would have to contact either Roberts or Copeland Glass. In October 1973, Jacksonville sued Dawson for breach of contract because of defective windows and aggregate panels.

Part of the applicable language can be found in Paragraphs 34 and 41, which say,

34. Subcontracting

c. The contractor shall be fully responsible to the Owner for the acts and omissions of his subcontractors, . . .

41. Conflicting Conditions

   Any provisions in any of the Contract Documents which may be in conflict or inconsistent with any of the paragraphs in these General Conditions shall be void to the extent of such conflict and inconsistency.

And in Part 2, Division 16b, Paragraph 11, the Exposed Aggregate Panels:

   a. All portions of the exposed aggregate panel installation, including necessary caulking, shall be done by one subcontractor and he shall be completely responsible for the entire installation.

   b. To insure against leaks, this subcontractor shall furnish a 3-year Indemnity Bond on the Exposed Aggregate Panel Installation to be completely watertight and free from leakage, cracking or other defect in the work.

Paragraph 34 can be viewed as being in conflict with Part 2, Division 16b, paragraph 11. Dawson contends that based on paragraph 11, Jacksonville needs to contact the Bob Roberts Co. because the company was solely responsible for the work. Jacksonville counters that paragraph 34 puts Dawson in charge.

   Whom is Jacksonville to contact, Roberts or Dawson? Ignore the possibility that the Boncoat may be defective.

Exercise 8-11: Security Painting Co. and PennDOT

In March 1971, Security Painting Company signed two contracts with the Pennsylvania Department of Transportation (PennDOT) for the painting of 12 bridges in Potter and Clearfield Counties. The total amount for the two projects was $23,580. As per the contract, Security was required to sandblast the metal surfaces on the bridges before painting. Security began sandblasting and was soon informed by PennDOT inspectors that it would have to remove substantially all of the sound and adherent old paint from all of the surfaces. Before this development, Security had intended only to remove loose, excessively thick, or inflexible paint. Security acceded to the inspectors' demands and as a result, incurred considerable expense beyond their bid price. After completing the 12 bridges, Security sought an additional sum of $49,703.58, arguing that it was required to perform work beyond that which was required in the contract.

The operable contract language is contained in the following clauses:

> **Extent of Work.**—The work on all bridges shall consist of the removal of all soil, cement spatter, drawing compounds, salts or other foreign matter prior to cleaning and *complete commercial blast cleaning (Method A) of all surfaces of the structures* including open steel mesh decking. Each structure shall be painted with one (1) full coat of Sandstone Paint (No. 8) and one finish coat of Antique Bronze Paint as specified in Section 1070.
>
> **Cleaning of Surfaces.**—Amending Section 1073.01(a), the contractor shall remove all debris or buildup of foreign materials on the tops of abutments and piers. Such materials shall be disposed of prior to performing any painting operations.

The contract also provided,

> **5.**—The contractor further covenants and warrants that he has read and is completely familiar with and understands thoroughly the General Conditions, Specifications, . . . contained in and governing the performance of this contract . . .

The standard specifications are contained in the following description of Method A:

> **(B) Method A — Commercial Blast Cleaning.**[1]—Surfaces of metal shall be cleaned by use of abrasives propelled through nozzles or by centrifugal wheels. The method of propelling the abrasives

1. For a partial description of some of the various blast cleaning methods which may be used to accomplish the requirements of this specification, reference to Steel Structures Painting Council Surface Preparation Specification: No. 6 Commercial Blast Cleaning (SSPC-SP6) is recommended.

and the type and size of abrasives used shall be such that all oil, grease, dirt, rust scale and foreign matter have been completely removed from all surfaces and *all rust, mill scale and old paint have been completely removed except for slight discolorations caused by rust stain, mill scale oxides or slight, tight residues of paint or coatings that may remain. At least two-thirds of each square inch of surface area shall be free of all visible residues and the remainder shall be limited to the light discoloration, slight staining or light residues mentioned above.*

Security wanted to use a method other than Method A. The PennDOT inspectors required Method A, contending it was required by the specifications. The basis of Security's argument is that the footnote to Method A allows alternate methods of blast cleaning to be used.

Do the specifications allow an alternate method to be used? What rule(s) from Chapter 4 are instrumental in resolving this dispute?

Exercise 8-12: D'Annunzio Brothers and NJT

In October 1984, D'Annunzio Brothers, Inc. (DBI), obtained bid documents from the New Jersey Transit Corporation (NJT) for excavation and foundation-related work at NJT's rail maintenance facility in Kearny, New Jersey. A site inspection and prebid conference were held that month. In at least two places, the bid documents advised the bidders to seek prebid correction or clarification by NJT of any perceived contract, drawing, or specification obscurities, errors, or discrepancies.

Before submitting DBI's bid for the November 28 opening, DBI's chief estimator, who prepared the bid, was aware of an apparent discrepancy regarding the amount of excavation and backfill involved in the contract and the method of payment. There were two separate unit-price bid items, for structural excavation and structural backfill, which NJT estimated at a total of 90 yard^3. DBI read the items as including the excavation and backfill for all of the hundreds of structures in the contract, even though DBI recognized that the estimate of 90 yard^3 would have been wildly inaccurate for all of the work.

Another possible reading of the bid documents was that the cost of excavation and backfill associated with each of the many structures to be installed was to be included in the cost calculated for each structure, and that the separate 90-yard^3 item was for a particular structure in an area where work by another contractor made it impossible for NJT to estimate the amount of necessary work accurately.

DBI rejected this interpretation without contacting NJT, even though the chief estimator and company president recognized the discrepancy. DBI's project manager independently recognized the discrepancy and discussed it with the chief estimator before the bid was submitted. Instead, DBI included the cost of excavation, which eventually totaled more than 11,000 yard³, into the line item for structural excavation and backfill for which NJT estimated 90 yard³. The estimated price for that item was $100 per cubic yard.

NJT's estimated cost of the project was $6.58 million. DBI's bid was $6.72 million. The second and third low bids were $6.94 and $6.98 million.

The discrepancy was brought to NJT's attention after the contract was signed. DBI was informed that its interpretation was incorrect. DBI was paid for 90 yard³ of structural excavation and backfill at $100 per cubic yard but was not paid for the structural excavation and backfill for the hundreds of other structures at the $100 per cubic yard rate. DBI now seeks to recover $1,100,000 in unreimbursed costs caused by ambiguous specifications.

Did DBI have a duty to inquire? Is DBI entitled to more monies? Does it matter that the second and third low bidders were close to DBI's bid?

Exercise 8-13: Blake Construction Co. and U.S. Navy

In 1990, Blake Construction Co., Inc., contracted with the U.S. Department of the Navy to construct replacement medical facilities at the San Diego Naval Regional Medical Center. The contract called for the construction of four new structures, including a mechanical equipment building, a warehouse, an auxiliary building, and a nursing tower. A corridor of approximately 1,000 ft ran most of the length of the ground floors of the buildings and was designed to enable people, supplies, and utilities to move among the buildings. The contract called for electrical conduit to be installed along this corridor as part of an electrical feeder system.

Specifications governing the installation of the electrical feeder system were prepared by a joint venture architect and engineering firm. Electrical power coming into the hospital from outside high-voltage lines was to be reduced to lower voltages by transformer units and then distributed throughout the buildings via a branching series of smaller cable and conduit. The series of drawings pertaining to the installation of the electrical conduits within and between the buildings depicted the conduits as installed overhead along the west side of the corridor, hanging either exposed from utility racks or hidden from view by a dropped ceiling. These drawings also included notes describing the drawings as "diagrammatic." On some drawings, the notes stated, "All feeder details & sections are diagrammatic. And the contractor shall relocate any/all conduits as per existing conditions to coordinate with all other trades." On other drawings, the notes similarly stated, "All feeder locations are diagrammatic. Contractor shall relocate feeders as per existing conditions and shall coordinate with other trades."

After Blake was awarded the contract and before any construction of the buildings had commenced, its electrical subcontractor, Steiny and Co., Inc., began installation of the electrical feeder system in an underground concrete duct bank along the path of the corridor. When the Navy challenged this installation method, Steiny asserted that the contract's diagrammatic notes permitted the contractor to relocate the electrical conduits so as to avoid conflict with other trades, such as mechanical and plumbing, which were also to be installed in the corridor. The Navy issued a stop work order informing Blake that the underground duct bank did not comply with the specifications and directed that the conduits be installed overhead as shown in the drawings. Blake notified the Navy that it considered this directive to be a constructive change to the contract. Steiny removed the underground duct bank under protest.

After the corridor was constructed and ready for utility installation, a conflict arose between Steiny and the mechanical subcontractor over the location of their respective trades. To alleviate this interference, Steiny agreed to move to the east side of the corridor. In one 30-ft section, the electrical conduit had to be placed outside the corridor, and in another they were located outside the planned route of the feeders. Both changes were approved by the Navy. In general, Steiny had to weave the conduits between and around the other utilities in the corridor.

As a result of the alleged change in installation method, Steiny sustained an estimated $1,679,000 in damages, and a claim was submitted to recover these damages.

What is the meaning of "diagrammatic" relative to this dispute? What was the intent of the contract relative to conduit locations? What rule from Chapter 4 can be used in the resolution of this claim? Does Blake have a valid claim?

Exercise 8-14: Farrell-Cheek Steel Co. and Forest Construction Co.

Farrell-Cheek Steel Co. contracted with Forest Construction Co., Inc., for the preparation, construction, and landscaping of a parking lot. The unit-price contract called for the contractor to furnish 1 ton of asphalt at $180/ton. The asphalt was to be manually applied. The contract required the owner to order changes in the work in writing, with the amount and method of compensation to be determined at the time of ordering.

After the work began, the parties determined that additional asphalt should be used to level severe depressions, remove high spots, and ensure proper drainage. After 130 tons of asphalt was delivered and placed, a written change order was executed. It called for Farrell to pay $180/ton for 130 tons of asphalt. The asphalt was placed via paving equipment in lieu of manual methods. Farrell then ordered an additional 381 tons, which Forest delivered and placed. There were no discussions about the price. Subsequently, a dispute arose when the owner refused to pay the contractor $180 per ton for the additional 381 tons. The owner asserted that a reduced unit rate would reflect a more reasonable price based on economies of scale.

What was the intention of the parties? What can be said about Farrell's actions? Is Forest entitled to $180/ton?

Exercise 8-15: Granger Contracting Co. and Chiappisi Brothers

Granger Contracting Co., Inc., entered into a contract to build a high school. Granger made a subcontract with the Chiappisi Brothers under which Chiappisi agreed to furnish all labor and materials for completing the lathing and plastering work covered in paragraph 19 of the specifications. By the subcontract, Chiappisi agreed to be bound to Granger by the plans and specifications, including all general conditions. The general conditions of Granger's contract included Article 16, which reads,

> If the contractor claims that any instructions by drawing or otherwise involve extra cost under this contract, he shall give the Architect written notice thereof within a reasonable time of receipt

of instructions or otherwise, and in any event before proceeding to execute the work, except in emergency endangering life or property .... No such claim shall be valid unless so made ...

and Article 37:

> The contractor agrees to bind every sub-contractor and every sub-contractor agrees to be bound by the terms of the ... General Conditions, the Drawings, and Specifications as far as applicable to his work including the following provisions of this article .... The sub-contractor agrees ... to make all claims for extras ... to the Contractor in the manner provided in the General Conditions of the Contract ... for like claims by the Contractor upon the Owner, except that the time for making claim for extra cost is one week.

The work to be done by Chiappisi included the installation of Zonolite spray insulation on the metal flute deck over the science area of the building. The flutes were to be filled with insulation and then a 1¼-in. coat of Zonolite was to be applied over the entire surface. Chiappisi, from a drawing that dealt with the science area of the building, measured the size of the flute openings as shown on that plan. From these measurements and the area of the deck, an estimate was made of the materials and labor necessary to fill the flutes and cover the surface. Chiappisi used these computations in bidding the job. The drawing that was used by Chiappisi in making the calculations, if measured to scale, shows the flute openings in the metal deck to be 1¼ in. wide at the mouth. This drawing was, however, intended by the architect who prepared it to be schematic or symbolic (diagrammatic), rather than an actual diagram of the roof deck installation. The drawing does not disclose this intention clearly, but the drawing contained no dimensions. The drawing used by Chiappisi may have been inconsistent in some respects with other drawings and paragraph 19 of the specifications for the metal deck.

The metal deck was installed in accordance with the manufacturer's instructions referred to in paragraph 19 and the shop drawings (approved by the architect). The shop drawings were never shown to Chiappisi. As installed, the roof deck had flute openings of a width of 4¼ in. If installed with one surface toward the roof, the metal flute spaces to be filled would be 1¼ in. wide and ½ in. deep. If exposed with the reverse surface toward the roof, the flute spaces to be filled would be 4¼ in. wide and of the same depth. The architect's drawing relied on by Chiappisi showed the narrow flute spaces on the underside of the flute deck (The diagram showed the metal deck inverted.) They were installed with the narrow flute spaces pointing upward and the broad flute spaces pointing downward, as the manufacturer intended.

While the work was in progress, more material was needed than Chiappisi had estimated. Alphonse Chiappisi went to the school and found that the flute openings that were being filled were wider than had been estimated. The orientation of the flutes was opposite to the plan sheet on which Chiappisi had relied. Before Chiappisi left the area, he went to the office and protested the installation to Gillette, the job superintendent for Granger. He told Gillette that the work on the flutes was extra because he was having to use more material than estimated and he wanted to be paid for it. Chiappisi ordered more Zonolite, and his men continued with the science area roof deck and completed filling the flutes. In a bill sent to Granger on April 29, 1963, after the work in the science area was completed, Chiappisi made no reference to any extra work in the science area roof deck. On May 23, 1963, Chiappisi wrote Granger asking for $2,927 for added work involved in connection with the roof deck.

Did Chiappisi not seeing the shop drawings have any bearing on the outcome of this dispute? Is Chiappisi entitled to more monies?

Exercise 8-16: J. A. Jones Construction Co. and U.S. Army Corps of Engineers

J. A. Jones Construction Co. was awarded a contract in 1954 by the U.S. Army Corps of Engineers for the construction of five multipurpose hangars with shells (roofs, siding, and sliding doors) of galvanized steel sheeting of prescribed gauges for the U.S. Air Force at the Ernest Harmon Air Force Base in Newfoundland, Canada. J. A. Jones subcontracted this work to Capitol Steel and Iron Co. All but the furnishing of the sheeting was sub-subcontracted to the Western Steel Erection Co.

Paragraph 14-04 of the technical specifications for roofing and siding provided,

> GALVANIZED SHEETS shall be the manufacturer's standard commercial type, having corrugations approximately 2½ inches wide, and shall be tight-galvanized with not less than ¾ ounce zinc coating per square foot on each side of the sheet. Sheets shall

be of plain open-hearth 20 gage for roofing and 24 gage for siding. All interior corrugated metal work shall be galvanized sheets.

A dispute arose regarding the requirement for 3/4 oz zinc coating "on each side." Paragraph 14-02 of the technical specifications listed a series of standard specifications generally applicable to roofing and siding metal, and included Federal Specification QQ-I-716, Iron and Steel, Sheet, Zinc Coated, Galvanized. This standard specification set forth requirements for the *nominal weight of zinc coating* for 20 and 24 gauge galvanized sheets. It provided,

	CLASS A (OZ/FT²)	CLASS B (OZ/FT²)	CLASS C (OZ/FT²)
20 gage	2.75	2.50	1.75
24 gage		2.50	1.50

QQ-I-716 also set forth the minimum weight of zinc coating:

NOMINAL WEIGHT (OZ/FT²)	MINIMUM WEIGHT AS DISCLOSED BY	
	SHEET WEIGHT TEST (OZ/FT²)	DIAGONAL TRIPLE SPOT TEST (OZ/FT²)
2.75	2.45	2.30
2.50	2.20	2.00
2.00	1.80	1.60
1.75	1.55	1.35
1.50	1.30	1.15

The weights set forth in this specification are total weights; that is, the total weight of the zinc coating *applied to both sides of a square foot of a galvanized sheet*.

Reading the tables together, one can see, for example, that where 20-gage, Class C sheeting is required, the nominal weight of zinc coating specified in the first table is 1.75 oz per square foot. The corresponding nominal-weight figure in the second table would show the contractor that for the coated sheets to be acceptable, the minimum ounces per square foot weight of the coating would have to be 1.55 under the sheet weight test and 1.35 under the diagonal triple spot test. The manufacturing industry routinely uses the diagonal triple spot test in testing for compliance with this specification.

According to the contractor, there are several problems inherent in paragraph 14-04. These include the absence of a prescribed test method, an alleged difficulty in measuring the weight of coating if it is not a total for both sides, and the requirement for a "tight" coat. In the industry, a tight coat is one that will not flake if the metal is bent. The ability to produce a "tight" coat is associated with the coating thickness.

The contractors read the specifications and determined in their own minds that the government really wanted a zinc coating totaling 1½ oz of zinc on both sides. They did this because of their experience in steel work, "the information of the industry," and because in their experience a thickness for single sides had never before been specified. They related this 1½ oz requirement to the Class C 24-gage, although 20-gage was also to be supplied.

In July 1954, the contractor ordered corrugated galvanized sheets for the hangar roofing and siding and for the hangar doors. While specifying in the order the lengths, widths, and gages, the contractors did not specify in writing the class or weight of zinc coating required. They claim to have stated orally that they desired Class C, and that is what was delivered.

In the normal government procurement involving the use of galvanized sheeting, the invitation and contract specify the class of sheeting to be used, and from this the contractor can readily ascertain exactly what is needed by consulting the pertinent tables in Federal Specification QQ-I-716 (see above). Indeed, paragraph 1-5 of that specification required that government invitations and contracts state the class of coating desired. Both users and suppliers in the galvanized sheeting industry subscribe uniformly to such classifications.

During the course of two conferences with the government representatives before the award, the language or application of paragraph 14-04 of the technical provisions of the specifications was not questioned, nor did the contractor inquire whether it affected Federal Specification QQ-I-716. The question of what class of sheeting was contemplated was not raised.

The galvanized sheeting that Jones ordered was produced in late July or early August, and after a stay in a Houston warehouse arrived by sea transport at the job site late in November 1954. Installation began in early December. The government made no inspection of the sheeting as it arrived at the job site, as the job specifications provided, but it would have been difficult to inspect without removing the protective paper wrappers in which the material was delivered. The wrappers were left intact to protect the material while it was in open storage.

In January 1955, the government inspector reported to the contractor's representative that he had noticed some discoloration of the sheeting. On March 14, 1955, the government's representative wrote to Jones that many of the siding sheets and a lesser number of the roofing sheets being installed on hangar No. 4 indicated improper workmanship in the galvanizing process. It appeared that the sheets had some bare or improperly coated areas or defects. This representative recommended that installation of sheets with apparent defects be discontinued, and Jones was advised that samples would be submitted by the government to a testing laboratory for determination of the cause of the defects and the weight of the zinc coating.

On April 11, 1955, Public Service Testing Laboratories, Inc., reported that tests disclosed less than ¾ oz per square foot of zinc coating on each side, contrary to paragraph 14-04. The four test samples represented sheets in poor, fair, and average conditions. Corrosion was detected in these sample sheets and was deemed to be the result of temperature and humidity conditions while in storage. On April 13, 1955, the government issued a stop order for further installation of substandard galvanized metal siding sheets pending corrective action by the contractors.

The contract also contained the following clause:

> ARTICLE III. *Specifications and Drawings.*—. . . In any case of discrepancy either in the figures, in the drawings, or in the specifications, the matter shall be promptly submitted to the Contracting Officer, who shall promptly make a determination in writing. Any adjustment by the Contractor without this determination shall be at his own risk and expense.

Jones contends that the sheet weight test was used, and the actual weight per sheet if the triple spot test was used exceeded the specifications. Jones filed a claim for damages due to unjustified rejection of the galvanized sheets.

What role does the canon "specific over general" play in this dispute? How do you reconcile conflicting test results? Can custom and usage be helpful? Does Jones have a valid claim?

References

American Institute of Architects (AIA). (1986). *General Conditions of the Contract for Construction*, AIA Document A201. Washington, DC.
Beacon Construction Co. of Massachusetts v. United States, 314 F.2d 501 (1963).
Blount Brothers Construction Co. v. United States, 346 F.2d 962 (1965).
Bulley & Andrews, Inc., v. Symons Corp., 323 N.E,2d, 806 (1975).
Columbus Construction Co. v. Crane Co. 98 F. 946 (1900).
Construction Services Co. v. United States, 357 F.2d 973 (1966).
Corbetta Construction Co., Inc., v. United States, 461 F.2d 1330 (1972).
Framlau v. United States, 568 F.2d 687 (1977).
Franchi Construction Co. v. United States, 609 F.2d. 984 (1979).
Hensel Phelps Construction Co. v. United States, 314 F.2d 501 (1963).
Lewis v. Jones, 251 S.W.2d 942 (1952).
Metropolitan Paving Co. v. City of Aurora, Colorado 449 F.2d. 177 (1971).
Monroe M. Tapper & Associates v. United States, 602 F.2d 311 (1979).
Newsom v. United States, 676 F.2d. 647 (1982).
Kenneth Reed v. United States, 475 F.2d 583 (1973).
Sweet, J. (1989). *Legal Aspects of the Architecture, Engineering, and Construction Process*, 4th Ed., West Publishing Co., St. Paul, MN.
Thomas, H. R., Smith, G. R., and Mellott, R. E. (1994). "Interpretation of Construction Contracts." *J. Constr. Engrg. and Mgmt.*, 120(2), 321–336.
Unicon Management Corp. v. United States, 375 F.2d. 804 (1967).
United States v. F. D. Rich, 434 F.2d 855 (1970).
Wiebner v. Peoples, 142 P' 1036 (1914).
WPC Enterprises v. United States, 323 F.2d 874 (1964).

Additional Cases

The following are additional cases related to issues associated with interpretation. The reader is invited to review the facts of the case, apply the decision criteria in the flowchart, reach a decision, compare with the judicial decision, and determine the rationale behind the judicial decision.

Elter, S.A., ASBCA 52327.
Jowett, Inc., v. United States 29 Fed.Appx. 584 (2002; trade practices and customs).
Maryland State Highway Administration v. David A. Bramble 717 A.2d 943 (1998).

Chapter 9

Differing Site Conditions

Annually, considerable monies are spent in pursuing differing site condition (DSC) claims because the various parties do not understand what must be proven or they misunderstand the roles of the soil report, disclaimers, and site visit requirement.

Contract Language

The federal DSC clause is as follows:

(a) The Contractor shall promptly, and before the conditions are disturbed, give a written notice to the Contracting Officer of (1) subsurface or latent physical conditions at the site which differ materially from those indicated in this contract, or (2) unknown physical conditions at the site, of an unusual nature, which differ materially from those ordinarily encountered and generally recognized as inherent in work of the character provided for in the contract.

(b) The Contracting Officer shall investigate the site conditions promptly after receiving the notice. If the conditions do materially so differ and cause an increase or decrease in the Contractor's cost of, or the time required for, performing any part of the work under this contract, whether or not changed as a result of the conditions, an equitable adjustment shall be made under this clause and the contract modified in writing accordingly.

(c) No request by the Contractor for an equitable adjustment to the contract under this clause shall be allowed, unless the Contractor has given the written notice required; provided, that the time prescribed in (a) above for giving written notice may be extended by the Contracting Officer.

(d) No request by the Contractor for an equitable adjustment to the contract for differing site conditions shall be allowed if made after final payment under this contract.

Additionally, the site visit clause may also be important. Typically it states,

Bidders should visit the site to ascertain pertinent local conditions readily determined by inspection and inquiry, such as the location, accessibility and general character of the site, labor conditions, the character and the extent of existing work within or adjacent thereto, and any other work being performed thereon.

Background

In theory, using a differing site condition (DSC) clause in construction contracts reduces the cost of construction because contractors do not have to include contingencies to cover the costs of hidden or latent subsurface conditions (Stokes and Finuf 1986). As stated by the court in Al Johnson Construction Co. v. Missouri Pacific Rail Co.,

The purpose of the changed conditions clause is thus to take at least some of the gamble on subsurface conditions out of the bidding. Bidders need not weigh the cost and ease of making their own borings against the risk of encountering an adverse subsurface, and they need not consider how large a contingency should be added to the bid to cover the risk. They will have no windfalls and no disasters. The government benefits from more accurate bidding, without inflation for risks which may not eventuate. It (the owner) pays for difficult subsurface work only when it is encountered and not indicated in the logs.

Whether the owner actually saves any money as a result of including a DSC clause in a contract is a subject of intense debate. There always seems to be a contractor willing to include no contingency monies, even without a DSC clause.

When a typical DSC clause exists, the contractor needs to show that the actual conditions were materially different from those reasonably indicated or suggested by the contract documents. Courts have held that there is a difference between having a clause and not having one. Without a DSC clause, entitlement must be based on the theory of misrepresentation, which is a breach of contract. The situation can be sensitive to the type of work being constructed.

> In misrepresentation, the wrong consists of misleading the contractor by a knowingly or negligently untrue representation of a fact or a failure to disclose where a duty requires disclosure.... Some degree of culpability—either untruth or such error as is the legal equivalent—must, however, be shown.... The claim based upon the modern changed conditions clause is very much different, though it may arise from the same facts and be joined with a claim for misrepresentation.... Misrepresentation is not the issue ... the changed conditions clause eliminates the factual elements of misrepresentation and any need to impose a burden on plaintiff to prove those elements. (Foster Construction v. U.S.)

Most owners use the federal DSC clause in which the DSC clause is divided into two parts, commonly called type I and type II. A type I condition allows the contractor to recover his additional expenses if the conditions found differ materially from those *indicated* in the contract documents. A type II condition allows the contractor to recover additional expenses when the actual conditions differ from what could be reasonably expected for work of the character contemplated in the contract.

Because the federal form of clause is used in many contracts, federal precedent is heavily relied on in reaching a judicial decision. Furthermore, courts have ruled that when the language is similar to the federal clause, federal precedent may be used to decide the dispute (Metropolitan Sewerage Comm. v. R.W. Construction; Town of Longboat Key v. Carl E. Widell & Son; Sornsin Construction Co. v. State of Montana). If a different clause is used, the exact wording must be carefully evaluated.

More detailed discussions of the DSC clause can be found elsewhere (Currie et al. 1971, Parvin and Araps 1982).

Rules of Application

Fig. 9-1 summarizes the rules that courts have used in deciding DSC disputes when the contract contains a DSC clause. Each is discussed below.

In general, seven selected questions are highly relevant:

- What does the contract say? (Is the contract silent?)
- Who bears the risk? (What did the contract indicate?) (type I only)

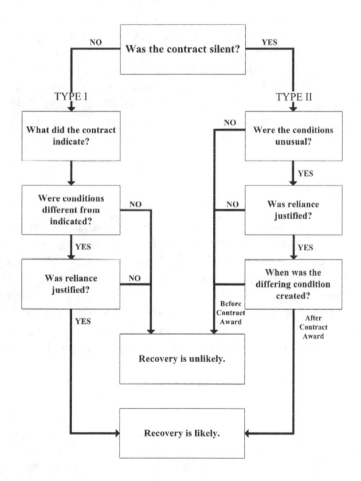

Figure 9-1. Decision Tree for Disputes Involving Differing Site Conditions.
SOURCE: Thomas et al. 1992, ASCE

- Were conditions different from those indicated? (type I only)
- Were the conditions unknown and unusual? (type II only)
- Was the contractor misled to his detriment?
- Was the contractor justified in relying on the information provided?
- When was the differing condition created? (type II only)

The first step in the decision process is to determine whether what the contractor is really arguing is different. Although this may seem a trivial inquiry, it is nonetheless important. Assertions made in a contractor claim cannot always be relied on, so an independent analysis may be needed.

Next, the entire contract must be evaluated to determine if there was an indication in the contract of what conditions would be encountered. This is the key differentiation between type I and type II disputes. If there is no indication of the conditions, then the contract is considered to be silent on those conditions, and the contractor must base a claim on a type II condition. If the contract is not silent, it is a type I claim. There may be instances when both a type I and type II condition may be claimed. Most claims are type I.

Type I Condition

A type I condition occurs when subsurface or latent physical conditions at the site differ materially from those indicated in the contract. Examples of a type I condition are encountering a fine silty substance where the borings indicated clay (Blount Bros. Construction Co.), finding fat clays and organic soils where the contract documents showed lean clay soils (Bergman Construction Corp.), and encountering a large crevice in a river bed where the contract documents indicated only small openings that an ordinary grout program could have handled (Farnsworth & Chambers Co. v. U.S.).

What Did the Contract Indicate?

With a DSC clause, the standard of proof is an indication or suggestion that may be established through implication and inference. "An *indication* may be proven, moreover, by inferences and implications..." (Foster Construction v. U.S.). "A contractor cannot be eligible for an equitable adjustment for changed conditions unless the contract *indicated* what those conditions would supposedly be..." (North Slope Technical Ltd. v. U.S.).

As stated by the court in Foster Construction v. United States,

> ...it is not necessary that the "indications" in the contract be explicit or specific; all that is required is there be enough of an indication on the face of the contract documents for a bidder reasonably not to expect "subsurface or latent physical conditions at the site differing materially from those indicated in this contract."

Currie et al. further reinforces this important concept by stating,

> To recover for either type of changed condition you are *not* required to prove *fault* or *bad faith* on the part of the Govt. Even if the Govt. *innocently* misled you, or *honestly* failed to discover (and reveal to you) the actual conditions, you are entitled to relief if you simply prove that the conditions you encountered differed materially... (Currie et al. 1971, p. 2)

and

> It is not required that the *indications* in the contract (upon which you are entitled to rely) be *affirmatively expressed* on plans or specific contract provisions. Instead, such indications may be inferred from reading the *contract as a whole*. (Currie et al. 1971, p. 3)

Indications or suggestions may be found in the contract documents, such as borings, profiles, design details, and other contract clauses.

Contract indications are normally found in the plans and specifications. If the actual conditions are different from the *indications* contained therein, the DSC clause allows the contractor to recover the additional costs of performing the work.

The changed conditions clause makes it clear that bidders are to compute their bids, not upon the basis of their own preaward (subsurface) investigations, but on the basis of what is *indicated* and shown in the specifications and on the drawings (Al Johnson Construction Co. v. Missouri Pacific Railroad Co.). (emphasis added)

Information about borings included in the contract documents is a particularly valuable source in establishing contract indications, depending on the conditions at issue. The court in United Contractors v. United States said,

> Borings are nevertheless considered the most reliable reflection of subsurface conditions.... The most reliable and specific indicator—the borings—had shown that water would not interfere with excavation.

and in Woodcrest Construction Co. v. United States, the court stated,

> ...the main purpose of such borings is to indicate subsurface conditions which would not otherwise be discovered.

Design details may also provide indications of the subsurface conditions. In Vann v. United States, the contract drawings showed the ocean floor to be rock even though a soil report was not prepared. The contractor found the actual bottom to be a spoil pile instead of rock. The spoil pile greatly hindered the pile-driving operation. The court allowed the contractor to recover his additional expenses because the contract *indicated* rock and rock did not exist (Vann v. U.S.).

In Foster Construction v. United States, the court stated that the contract documents, viewed as a whole, provided sufficient *indications* of the subsurface conditions.

> ...the court is of the view that the other *indications* in the contract of an impermeable subsurface permitting excavation in the dry—the notation as to the types of concrete; the direction that "all concrete shall be placed in the dry"; the omission from the concrete provisions of the documents of any provision for a concrete seal or a class of concrete of which seals are made; and the so-called "6 tons" note—are sufficient in themselves, without the logs, to sustain the determination that a changed condition was encountered. (Foster Construction v. U.S.; emphasis added)

Although courts have not been precise on the role of the soil report, when a DSC clause is present, courts have shown a strong willingness to go beyond the boundaries, to look at the four corners, as it were, of the contract to examine the soil report and determine if it contains an indication or suggestion. This situation often arises when the soil report is referred to in the contract itself, even though it may not be a part of the contract documents. Depending on the language used, referring to the soil report in the contract may actually make the soil report a part of the contract documents. A number of cases are similar to the Woodcrest case in that courts place great weight on the contents of the core-boring logs (Woodcrest Construction Co. v. U.S.). In Ruff v. United States, the soil report contained the test results from test pits showing yellow clay only in the excavation area. The contractor found rock and claimed for the additional expenses. The court, in ruling for the contractor, said that the soil report and test pit results gave the contractor the *indication* that the entire subsurface was yellow clay. Thus, a valid claim for type I DSC existed.

In other cases, courts have examined blow counts (Dravo Corp. v. Commonwealth of Kentucky) and test results for rock samples (Stock & Grove v. U.S.) as an aid to reaching a decision. These and other cases point to the dangers of owners feeling secure simply because the soil report is not part of the contract documents. However, if the contract contains clear indications, the soil report is not likely to be consulted (Foster Construction v. U.S.).

Groundwater issues are a cause of many DSC disputes. Sometimes these disputes are caused by the water table not being shown on the soil report or in the contract documents. In the 1954 case of Ragonese v. United States, the U.S. Court of Claims stated,

> The plans and specifications set out the character of the soil disclosed by these borings, but said nothing one way or another about subsurface water. It, therefore, cannot be said that the contractor encountered subsurface or latent conditions materially different from those specifically shown on the drawings or indicated in the specifications.

The court reasoned that because the contract was silent on the groundwater condition, a type I DSC could not be claimed. However, the same court later modified what constitutes silence on the subject of groundwater indications in soil borings. In United Contractors v. United States, the water table was not shown. The court, while discussing the Ragonese decision, stated,

> But United (contractor) claims that the plans furnished bidders not only failed to indicate the unusually high water table, but

showed the water table to be at or below grade.... Our conclusion is that the drawings (borings), properly viewed, did speak "one way about subsurface water." ... Carefully read, the (soil) profiles in this contract indicated that water would not be encountered in meaningful amounts in excavating for the project. (United Contractors v. U.S.)

Another case reinforced the latter view that a water table, not shown, is an indication that the water table is below the lowest elevation of the borings. In Woodcrest Construction Co. v. United States, the water table was again omitted from the borings. The court stated,

Although no actual representation was made by the government that there was no ground water, and thus, we cannot say there was a warranty, the effect upon the contractor of furnishing core boring logs without indicating the groundwater shown by such borings may be the same as if a representation had been made. (Woodcrest Construction Co. v. U.S.)

Courts now hold the view stated by United and Woodcrest that if the water table is not shown, it is an indication that the water table is below the level of the borings.

Were Conditions Different from Those Indicated?

The actual conditions must be materially or significantly different from those indicated. Minor and inconsequential differences are not sufficient to sustain a DSC claim. Although no clear judicial rules can be stated for what is a material difference, statements such as the following are common: "These changes were plainly substantial modifications of the work to meet changed conditions" (Foster Construction v. U.S.). Therefore, a difference that does not significantly affect the contractor's work would not seem to be a valid basis for a claim. The case of United Contractors v. United States brings the importance of this step to light.

Since the actual conditions must differ materially from those expected, the initial inquiry is whether United (contractor) ran into significant amounts of water in excavating. If no such factual finding has been or can be made, plaintiff's (contractor's) case fails at the outset. (United Contractors v. U.S.)

Was Reliance Justified?

The issue here is whether the contractor reasonably interpreted the contract indications. There are times when the contractor may not be justified in relying on the indications. Conditions that may reduce the contractor's reliance on the contract indications include other contract language, a site visit, or the contractor's experience. If any of these conditions signify that the indications are incorrect, then the contractor may not be justified in relying on those indications. Disclaimers may also negate a contractor's interpretation.

To determine if contractor reliance was reasonable, courts place themselves in the position of a reasonable contractor.

> A proper technique of contract interpretation on this problem is for the court to place itself into the shoes of a "reasonable and prudent" contractor and decide how such a contractor would act in appellant's (contractor's) situation. (North Slope Technical Ltd. v. U.S.)

There may be other clauses that will reduce reliance on indications of the subsurface. For other contract clauses to override subsurface information, they must be specific statements, not general ones. In United Contractors, the court stated that the general statement that high groundwater existed could not negate the precise information given by the borings (United Contractors v. U.S.).

Contracts typically require contractors to become familiar with the construction site before submitting a bid. If readily apparent or obvious conditions contradict the contract indications, then the contractor cannot reasonably rely on those indications.

Courts have consistently affirmed that contractors are responsible for conditions that are readily apparent. In Mojave Enterprises v. United States, the contractor estimated, using the plans, the amount of rock it had to remove from a hiking trail. The contractor did not visit the site, although there were opportunities to do so. A site visit would have made it obvious that the drawings were not meant to reveal the amount of rock to be removed but merely to indicate that rock removal was part of the project (diagrammatic). The court denied the contractor's claim, stating that the contractor acted unreasonably.

A site visit must be performed professionally and to the standard of other reasonable contractors.

A reasonable site investigation is properly evaluated against what a rational, experienced, prudent, and intelligent contractor in the same field of work would discover. (North Slope Technical Ltd. v. U.S.)

The depth of inquiry may also be an issue. As stated by one court,

This is not to say, of course, that such (contract) indications would excuse a site inspection or that such site inspection need not discover patent (hidden) indications plainly, to a layman, contradicting the contract documents. (Stock & Grove v. U.S.)

Sometimes, owners deny a claim, stating that the site visit clause required the contractor to perform an independent subsurface investigation. However, courts do not generally agree.

In the cases arising under the modern changed conditions clause, caution continues to be observed that the duty to make an inspection of the site does not negate the changed conditions clause by putting the contractor at peril to discover hidden subsurface conditions or those beyond the limits of an inspection appropriate to the time available. The contractor is unable to rely on contract indications of the subsurface only where relatively simple inquiries might have revealed contrary conditions. (Foster Construction v. U.S.)

Generally, the contractor is not expected to perform a subsurface exploration unless the site visit clause specifically requires that it be done. However, in some instances, a reasonable site visit may necessitate investigation. For example, on trenching work where blasting is required, the contractor may need subsurface information to determine rock hardness characteristics, but in most instances, the following is the norm:

. . . we are not inclined to view the requirement that the contractor examine the construction site, under the circumstances of this case (presence of a DSC clause), as contemplating that the contractor make its own separate test borings before submitting its bid. (Town of Longboat Key v. Carl E. Widell & Son)

Owners sometimes seek to reduce their exposure to liability by disclaiming responsibility for the accuracy of the soil report and related information included in the contract documents, but of the more than 75 DSC cases reviewed for this book, no case was found where a court decided in favor of the owner because of the disclaimer. Rather, the outcome was based on one or more of the relevant issues detailed above. Thus, it is well documented that courts do not view disclaimers favorably, and normally, they are too general and nonspecific to override the provisions of the DSC clause. In United Contractors, the court stated,

> It is true that Provision 1-07 also provided that "the Government does not guarantee that materials other than disclosed by the explorations will not be encountered, or that the proportions of the various materials will not vary from those *indicated* by the logs of the explorations." But we have held, in comparable circumstances, that broad exculpatory clauses, identical in effect to this one, cannot be given their full literal reach, and do not relieve the defendant (government) of liability for changed conditions as the broad language would seem to indicate.... General portions of the specifications should not lightly be read to override the Changed Conditions Clause. It takes clear and unambiguous language to do that. (United Contractors v. U.S.)

The attitude of the courts is that owners cannot induce lower construction bids by including certain information plus a differing site condition clause that promises to pay for unforeseen events and then disclaim responsibility for inaccurate information when the event occurs.

Case Study

Disclaimer and Site Investigation

Town of Longboat Key v. Carl E. Widell and Son
362 So.2d 719 (1978)

Carl E. Widell and Son entered into a contract to construct a pumping and lift station as part of a central wastewater system for the town of Longboat Key, Florida. In making its bid, Widell relied on a subsurface report prepared by Ardaman and Associates. The report indicated that the excavations for both the pumping and lift stations could be kept dry by using a well point method of dewatering. Based on this report, the contract documents called for dewatering using well points. The foundation was to be caissons; and there was no tremie concrete.

The contract contained the following differing site condition clause:

> During the progress of the work should the contractor encounter, or the engineer or owner discover, subsurface or latent conditions at the site differing materially from those shown on the drawings or indicated in the specifications, or unknown conditions of an unusual nature differing materially from those ordinarily encountered and generally recognized as inherent in work of the character provided for in the drawings and specifications, the engineer's attention shall be called immediately to such conditions before they are disturbed. The engineer will thereupon promptly investigate the conditions, and if he finds that they do materially differ, with the written approval of the owner, the contract will be modified to provide for the increase or decrease of costs and difference in time resulting from such conditions.

The contract did not contain a clause requiring that all extra work was to be authorized only via written change order. A standard notice provision was included in the contract. The contract also contained a typical site visitation clause, which required the contractor to visit the site and become familiar with the requirements of the contract, to satisfy himself or herself as to conditions, including the nature of the groundwater table.

When Widell began excavation, he encountered a water level that rose and fell with the tides. Widell concluded that it was not possible to dewater the site and that caissons were infeasible. He communicated this situation to the project engineer. The engineer redesigned the foundation and submitted new plans to Widell, who then proceeded with the construction. No formal notice was given, nor was any written change order ever executed.

Will the lack of written notice prevent Widell from recovering the extra costs? What about the lack of a written change order to complete the work according to the revised plans? Were the conditions materially different from those indicated in the contract?

Regarding the notice and extra work requirements that communications must be in writing, the Second District Court of Appeals of Florida took the position that "It is clear that the only burden to the contractor under the changed conditions clause was to call the attention of the town's engineer to the unexpected conditions and the contractor did so." With respect to the issue of the disclaimer language, the Court said,

> It has been held that, where conditions are in fact materially different from those contemplated by the contract, relief is not to be denied because of the presence in the contract of one or more of various admonitory or exculpatory clauses, such as one requiring the contractor to examine the construction site. Furthermore, we are not inclined to view the requirement that the contractor examine the construction site, under the circumstances of this case, as contemplating that the contractor make its own separate test borings before submitting its bid.

The Court allowed Widell and Son to recover its additional costs.

The contractor's experience may alert him or her to an existing condition. If the contractor knows or should know that the actual conditions are different from those indicated, he or she cannot receive a windfall because of that incorrect indication. This test is discussed in some cases but is seldom given full weight, unless there is other evidence that the contractor was unreasonable. In Morrison-Knudsen Co. v. United States, the court stated that the contractor should have expected to encounter permafrost even though the borings did not show any in the area of construction. The court cited the contractor's experience in the immediate area, the fact that borings outside the construction area but included in the contract documents showed permafrost, and specifically worded contract clauses warning of permafrost.

If the contractor ignores information that is known to exist, the contractor may not claim that he or she was being reasonable. In Leal v. United States, the contractor found the water table higher than expected. The borings showed the water table with the abbreviation "WT" but did not define the abbreviation. The court denied the contractor recovery, stating,

> There was sufficient information in the drawings and specifications to indicate to an experienced operator the existence of a water table in the valley. (Leal v. U.S.)

Type II Condition

A type II DSC claim occurs when unknown physical conditions at the site, of an unusual nature, differ materially from those ordinarily encountered and generally recognized as inherent in work of the character provided for in the contract. Examples include buried timbers (Caribbean Construction Corp.), utility lines (Neale Construction Co.), and unusual and erratic soil behavior (Pacon, Inc.; Guy F. Atkinson Co.). A type II condition requires a variance between the actual site conditions and those reasonably expected.

Were the Conditions Unknown and Unusual?

This question is the central issue in type II disputes. The condition needs to be unknown and different from what a reasonable contractor would expect in doing the type of work involved in the contract. Unusual conditions may arise in many nontypical ways. For example, in Kaiser Industries Corp. v. United States, the contractor could not get rock of the correct size from the only quarry that the government "approved." The court reasoned,

> Certainly, encountering a condition in a "quarry"—let alone an "approved" quarry—which makes it not a usable quarry at all for the purposes involved, should, it seems clear, normally be considered an "unusual" one not "ordinarily encountered and generally recognized as inherent in" quarrying operations. Thus, it seems almost self-evident that plaintiff (contractor) would be entitled to an equitable adjustment under this plain language of the above quoted second part of Article 4 (the DSC clause).

Clearly, the above condition was not anticipated by either party when the contract was made, an important element in type II disputes. It would appear that the test is whether the condition was anticipated or contemplated by either party. If so, a type II condition cannot exist.

The condition need not be a freak, but merely unknown and unusual for the type of work contracted. In Western Well Drilling Co. v. United States, the court stated,

> The term "unusual" does not refer to a condition which would be deemed a geological freak but rather a condition which would

not be anticipated by the parties to the contract in entering into their initial agreement.

For example, where diversion of the river flow by an upstream contractor caused increased water flow at the site, a type II changed condition was established (Hoffman v. U.S.).

Was Reliance Justified?

Owners are not insurers of losses incurred by contractors. In Blauner Construction Co. v. United States, the contractor made a mistake about the type of material that was to be removed, and the court stated,

> The defendant (owner) is not an insurer of contractors against loss. Where a contractor has miscalculated, and, through its own negligence in not examining the site, has failed to take into consideration conditions which actually existed and which had been called to his attention in the specifications by a warning to visit the site, and sustains a loss, no claim arises.

Generally, if the conditions are truly unusual, the contractor should be able to argue successfully that he or she was not negligent. Furthermore, most of the same issues in determining if the contractor was justified in relying on an indication for a type I condition should be examined.

When Was the Differing Condition Created?

The DSC clause does not alter existing policy on acts of God that occur after the contract was awarded. In Arundel Corp. v. United States, the contractor sued for a higher unit price when required to remove only 70% of what the contract estimated. The court determined that the reduced quantities occurred because of a hurricane after the bids had been opened. In ruling against the contractor, the court stated,

> It is a general principle of law that neither party to a contract is responsible to the other for damages through a loss occasioned as a result of an act of God, unless such an obligation is expressly assumed. (Arundel Corp. v. U.S.)

As stated by Currie et al., "after-bid abnormal rainfall in and of itself is *not* a changed condition, but an unforeseeable, *pre-bid* subsurface condition *reacting to* unprecedented rainfall has been held to be a changed condition" (Currie et al. 1971, p. 7).

Other Issues

Having a DSC Clause or Not

Including a DSC clause in the contract alters the contract risks in several subtle but important areas (Thomas et al. 1992a). These areas are discussed below.

Without a DSC clause, the contract should contain positive factual statements that certain conditions exist (Thomas et al. 1992a). Including a DSC clause significantly reduces the burden of proof on a contractor because all that is needed is an indication or suggestion. To illustrate, without a DSC clause, boring data show what is in the borehole only, nothing more (Morrison-Knudsen Co. v. U.S.). The contractor must show that the boring data were incorrect. With a DSC clause, the contractor should be able to reasonably interpolate between boreholes. If the boring data are sufficiently conclusive to suggest that the contractor's interpretation is reasonable, then recovery should be allowed.

Exculpatory clauses and disclaimers are not favored by courts unless they are specifically worded. It appears that disclaimers are less favored with a DSC clause. Courts are concerned with owners inducing lower bids by including a DSC clause, then arguing that the accuracy of the information provided was not their responsibility.

Soil reports are seldom included in the contract documents. Courts may consult the soil report to help interpret the contract documents, but if there is no DSC clause, the contract documents must contain a positive factual statement before a contractor is entitled to recover additional costs. With a DSC clause, courts have shown a willingness to examine the soil report. In a number of cases, the suggestion or indication was found in the soil report rather than the contract documents. Currie et al. state that "you may compare actual conditions not only with the express representations in the contract documents but, also, with *all reasonable inferences and implications* that can be drawn from these documents" (Currie et al. 1971, p. 3, emphasis added). Thus, the role of the soil report seems amplified when there is a DSC clause.

In general, other readily available information is not often consulted. Without a DSC clause, contractors are expected to consider these sources. With a DSC clause, courts seem less reluctant to impose this as a condition; however, little case law addresses this issue.

Illustrative Example

The following example is based on the 1988 Armed Services Board of Contract Appeals, decision of William F. Wilke, Inc. The narrative below illustrates the use of the decision rules in Fig. 9-1.

Statement of Facts

William F. Wilke, Inc., was the low bidder on an excavation project for the U.S. Army Corps of Engineers. A standard form of general conditions was used, including a differing site conditions clause. The specifications included a warning that the site had low areas where surface water would pond. Under the contract, the contractor was responsible for site drainage. To alleviate some of the drainage problem, the contract plans showed an 18-in. culvert as a part of the work to provide for drainage for a large part of the site.

Once the work began, large amounts of rainfall caused water to accumulate on much of the site. After investigation, the contractor found that the plans showed incorrect elevations for the culvert. Rather than drain the site, the culvert actually allowed water from an adjacent site to drain into the contractor's work area.

Analysis

The contract did contain indications relating to the site drainage problem. Therefore, the contractor's claim must be evaluated as a type I condition.

What Did the Contract Indicate?

With a DSC clause, the contractor has the right to rely on the drainage conditions shown on the plans and to assume that the culvert, installed according to the elevations shown on the plans, would assist in draining the rainfall from the work area.

Were the Conditions Different from Those Indicated?

There is little doubt that conditions differed because rainfall drained into the site, not away from the site.

Was Reliance Justified?

The site visit, disclaimer, and contractor experience are three relevant issues under justified reliance. First, the site visit was unlikely to have alerted the contractor of the drainage characteristics of the site. Furthermore, the contractor is not expected to verify the proper culvert elevations. The disclaimer was general and should not be allowed to negate a specific contract indication. Last, the contractor's experience is not pertinent to this dispute.

The contract contained a *force majeure* clause specifically stating that the government was not responsible for acts of God. Ordinarily, rainfall is considered an act of God. The board addressed this point as follows:

> Although weather conditions are not of themselves differing site conditions for which price adjustments are allowable, such conditions may affect physical factors at the site so as to create compensable differing site conditions if the site factors were improperly shown in the contract documents and unknown to the contractor, as in the present matter . . . (William F. Wilke, Inc.)

Therefore, the contractor was justified in relying on the contract indications and was compensated.

Synopsis

The Wilke case illustrates an interesting point. If the plans had not shown the elevations of the culvert or had not shown the culvert at all, the contractor would have been solely responsible for site drainage as part of construction means and methods. The important elements of indication and reliance would have been missing. The only recourse for the contractor would have been as a type II condition. Most likely, his claim would have been unsuccessful because the drainage problem was entirely the result of acts of God.

Exercise 9-1: Blauner Construction Co. and U.S. Treasury

On July 12, 1937, Blauner Construction Co. entered into a written contract with the U.S. Treasury Department to furnish all labor and materials for the construction of a new post office building in Orange, Massachusetts, for the price of $47,740. The work was to be finished within 300 calendar days from the receipt of the notice to proceed, which was given on July 30, 1937. The completion date of the contract was to be May 27, 1938.

Before advertising for bids, the U.S. government dug four test pits around the site of the proposed building to determine the type of foundations that might be necessary. These test pits were 6 ft². Pit No. 1 was located about 10.5 ft from the northeast corner of the site; Pit No. 2, at the east approach and extending about 1 ft within the site of the building proper; Pit No. 3, about 8 ft from the northwest corner of the site; and Pit No. 4, about 6.5 ft from the southwest corner of the site. These locations are diagrammatically shown in Fig. 9-2.

Pit No. 1 was dug to a depth of 6 ft. A rock ledge was encountered 3.5 ft below the surface, at elevation 510.3, in an inwardly sloping po-

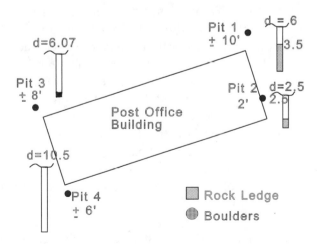

Figure 9-2. Diagram of Post Office Test Pits.

sition. The remainder was topsoil, sandy clay, coarse sand and gravel, and boulders.

Pit No. 2 was dug to a depth of 2.5 ft below the surface, at elevation 514.5, when a rock ledge was encountered and the pit could not be continued. The material excavated was topsoil, boulders, and sandy clay.

Pit No. 3 was dug to a depth of 6.07 ft below the surface when large boulders were encountered and the pit could not be continued. The material removed was topsoil, boulders, and dry sandy clay. No ledge was encountered.

Pit No. 4 was dug to a depth of 10.5 feet below the surface. No rock ledge was encountered. The material removed was topsoil, sandy clay, sand, gravel, and damp and closely packed clay.

This information was made available to the bidders in contract drawings Nos. 1 and 400. A part of the contract work was the backfilling of these test pits.

The specifications required bidders to visit the site and inform themselves as to all conditions and stated that failure to do so would not excuse performance. Before bidding, no representative of Blauner visited the site of the work.

Article 4 of the contract reads as follows:

> *Article 4. Changed Conditions.* Should the contractor encounter, or the Government discover, during the progress of the work subsurface and/or latent conditions at the site materially differing from those shown on the drawings or indicated in the specifications, or unknown conditions of an unusual nature differing materially from those ordinarily encountered and generally recognized as inhering in work of the character provided for in the plans and specifications, the attention of the contracting officer shall be called immediately to such conditions before they are disturbed. The contracting officer shall thereupon promptly investigate the conditions, and if he finds that they do so materially differ the contract shall, with the written approval of the head of the department or his duly authorized representative, be modified to provide for any increase or decrease of cost and/or difference in time resulting from such conditions.

Blauner began the excavation and struck hard blue granite ledge rock in places at higher elevations than ledge rock was found in any of the pits. On the first day of excavation, Blauner called the situation to the attention of the construction engineer and, at his suggestion, presented the matter to the contracting officer by long-distance telephone and by letter on September 2, 1937. The letter read as follows:

> We respectfully call attention to Contract, Article 4, Page 3, under heading Changed Conditions, stating that during progress of work subsurface or latent conditions encountered at the site, materially different from those shown on drawings or indicated in specifications; that the attention of the contracting officer shall be called to such conditions before they are disturbed. We wish to advise that we have today communicated with the Department by telephone and talked with Mr. Lund (contracting officer), calling the attention of the Department to the fact that we encountered rock, same being at a considerably higher elevation than shown on test pits. Wish to advise that we are proceeding with the work, in order to avoid delay, and we will at a later date submit itemized proposal of the extra cost involved in excavating rock. We trust that consideration will be given to said proposal.

When the excavation was completed on May 5, 1938, Blauner submitted to the contracting officer a claim for additional cost of excavating ledge rock over common (earth) excavation, above the elevations of ledge rock that it alleged were shown on the contract drawings.

What did the contract indicate relative to ledge rock? What was Blauner alleging as the basis of their claim? Was this a type I or type II condition? Is Blauner entitled to an equitable adjustment?

Exercise 9-2: U.S. Army Corps of Engineers and Stuyvesant Dredging Co.

On June 24, 1982, the U.S. Army Corps of Engineers awarded Stuyvesant Dredging Company a contract for maintenance dredging of the Corpus Christi Entrance Channel.

Maintenance dredging is done periodically to restore a channel to its so-called "acceptable prism," which is the original shape and size of the channel (Fig. 9-3). This shape always remains the same. Beneath the acceptable prism is an area known as the "prescribed prism." Because of

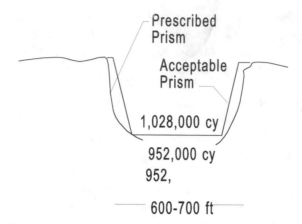

Figure 9-3. Diagram of Acceptable and Prescribed Prisms.

the inaccuracies inherent in the dredging process, material also may be removed from the prescribed prism.

All previous maintenance dredging of the channel had been performed by the Corps with its own dredges with the last regularly scheduled maintenance dredging done in 1978. However, after Hurricane Allen in 1981, emergency dredging was used, designed to open the channel to navigation until maintenance dredging could restore the channel to its acceptable prism. Stuyvesant was aware of these previous dredging operations.

Before bidding on the Corpus Christi Entrance Channel contract, Stuyvesant had bid on contracts to dredge two other channels: the Sabine-Neches Waterway Outer Bar Channel and Freeport Harbor Channel, both located on the Texas Gulf coast. In preparing bids for those two projects, Stuyvesant had reviewed the records in the Corps' offices regarding previous dredgings, examined physical samples of the material to be dredged, and for the Freeport project, which was the company's first dredging contract for the channels of the Texas Gulf coast, performed echo soundings.

Unlike its research on the other two sites, Stuyvesant did not review the records for the Corpus Christi Entance Channel, nor did anyone from Stuyvesant visit the site to take material samples or echo soundings. Instead, Stuyvesant concluded that the wording of the technical provisions in the Corpus Christi project were "very similar, almost identical" to the technical provisions of the Sabine-Neches and Freeport bid documents that it had previously reviewed and that it was not warranted to go to the expense of performing additional investigations. Stuyvesant also assumed that the 1981 emergency dredging had reestablished the acceptable prism and that it was required to remove only the material deposited since that time. It concluded that the information contained in the Corpus Christi documents therefore was sufficient to enable it to make a bid.

The contract provided for payment based on the amount of material removed from both prisms. Material dredged from outside the two prisms would not be paid for. The contract required an estimated 1,028,000 yard³ of material to be removed from the acceptable prism of the channel. The channel was 600–700 ft wide and 20,500 ft long. The prescribed prism included the acceptable prism and contained an estimated 1,970,000 yard³ of material (including the 1,028,000 yard³ in the acceptable prism). Technical provision 4-1.1 of the contract stated, in part,

> 4-1. Character of Materials.
>
> 4-1.1. The material to be removed to restore the depths within the limits specified in Construction Technical Provision 1-1. DESCRIPTION OF WORK, is that composing [sic] of shoaling that has occurred since the channel was last dredged, however, some virgin material [earth never before dredged in that particular channel] may be encountered in the prescribed prism, and/or side slope dredging. Bidders are expected to examine the site of work and the records of previous dredging, which are available in the Galveston District Office, 400 Barracuda Essayon Building, Galveston, Texas 77553 and Corpus Christi Area Office, No. 3 Science Park, Corpus Christi, Texas 78401 and after investigation decide for themselves the character of the materials.

Technical provision 4-1.1 also stated,

> In-Situ Densities. The following table details the results of a nuclear density survey conducted in the project area on 4 April 1979. The [sic] in-place density readings presented represent the average value of the density readings taken within the range indicated. The averaged values should not be interpreted as indicating the maximum or minimum density of material which may be encountered.

The table showed six places in the channel at which readings had been taken, and average densities ranged from 1.380 to 1.675 kg/L. The contract also contained a differing site condition clause.

Stuyvesant planned to remove approximately 60,000 yard^3 of material a day for 33 days, but after the contract was awarded, the Corps returned the contract to Stuyvesant to submit a new work plan based on the 120 days specified in the bid documents. Stuyvesant changed its work plan accordingly, but still intended to complete the project in 33 days.

For the first few weeks, Stuyvesant met or exceeded its plan. Thereafter, the rate of dredging fell drastically because Stuyvesant encountered large quantities of material that were difficult to dredge, which had densities of "up to, if not more than 1.9 kilograms" per liter. Stuyvesant believed that this material had not been previously dredged, but the Corps informed the contractor that this was the same type of material the Corps had encountered in the channel, and this information was shown in the records in the Corps' offices.

The work ultimately required 24 days more than Stuyvesant had anticipated and included approximately 302,500 yard^3 of material taken from the area outside the prescribed prism, for which the Corps refused to pay.

Were the densities encountered materially different from those indicated in the contract? Is Stuyvesant entitled to an equitable adjustment?

Exercise 9-3: Morrison & Lamping and State of Oregon

Morrison and Lamping contracted with the state of Oregon to construct 5.04 mi of highway in Wallowa County. A portion of the construction crossed an irrigated field owned by Lawrence Estes, one of the farmers in that area. The highway divided the Estes field in a generally east–west direction, and the field drained or sloped to the north. The land was irrigated by water taken from Wallowa Lake by the Silver Lake irrigation ditch, located south of the Estes property. Estes removed his share of the water from the Silver Lake irrigation ditch by another ditch which, in turn, had at least three lateral ditches traversing his field from south to north. The new construction divided these latter ditches and crossed over Prairie Creek, which was also on the Estes property (Fig. 9-4).

The changed conditions clause of the contract read as follows:

Should the contractor encounter, or the engineer discover during the progress of the work, [1] subsurface and/or latent condi-

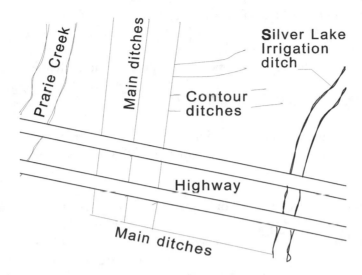

Figure 9-4. Diagram of a Portion of the Estes Fields in Wallowa County, Oregon.

tions at the site materially differing from those shown on the drawings or indicated in the specifications, [2] or unknown conditions of an unusual nature differing materially from those ordinarily encountered and generally recognized as inhering in work of the character provided for in the plans and specifications, the attention of the engineer shall be called immediately to such conditions before they are disturbed. The engineer shall thereupon promptly investigate the conditions, and if he finds that they do so materially differ, the contract may be modified by the engineer to provide for any increase or decrease of cost and/or difference in time resulting from such conditions.

The contract between the parties also included the following as part of the special provisions:

In performing the work, the contractor will be held responsible for all damage to crops. The ditch changes shall be built, in so far as is practicable, during the time the farmers are not using water from the existing ditches. When the irrigation ditch changes are constructed during the irrigation season, they shall be constructed, in so far as is practicable, before the existing ditches are closed or destroyed.

The contractor shall so arrange his work that the flow of water to the land will not be interrupted, and he shall cooperate with the users of the water to the greatest possible extent.

The contractor shall conduct his operations in such manner as will cause no interference with the flow of water in irrigation ditches and canals during the irrigation season. The contractor shall be liable for, and shall protect the State and the Federal Government against, any claims that may arise from any interruption of irrigation service caused by the contractor's operations or by his negligence.

Estes used a flood system of irrigation and irrigated the land between the main ditches by the use of contour ditches. The ditches were dammed off every 16 ft to spread the water over the ground. After one portion was irrigated, the dams were moved from place to place until the whole area was covered. Normally, Estes irrigated 5 to 10 acres per day by this method.

Mr. Morrison, one of the partners for the construction company, was familiar with irrigation problems, knew the irrigation season was in July and August, and made a careful investigation of the area before submitting his bid. In bidding the job, he took into consideration delays and shutdowns from the irrigation operation by allowing an additional 5 cents per yard[3].

On the day that Morrison and Lamping began construction, Estes advised Mr. Lamping, that irrigation would start in less than a week. As soon as the irrigation started, the right of way became wet and soggy and in some places covered with water. The contractor's equipment could not operate efficiently, and this process resulted in a one-month delay.

The plans called for the construction of siphon boxes with corrugated pipes to carry the water from the main ditches under the highway. Pending the construction of the siphon boxes, the water from the main ditches was to be transported across the highway by culverts. Actual construction began before the siphon boxes were installed. The plans also called for distribution ditches to redistribute the water to the north portion of the Estes property after it came through the siphon boxes. The contractor was unable to construct these ditches on the north side because the engineering was not complete, and the ownership of the property had changed. No provisions were made in the contract to keep the right of way free of the irrigation water coming from the Estes field south of the roadway. Because the fill material was wet and soggy, it was removed at the request of the state engineer and replaced by rock. The contractor was paid for the rock fill, but the state refused to pay for the removal of the wet fill material.

The contractor's position is that they expected irrigation water in the ditches, not in the fields, and they expected the irrigation water to cause some work delay and interruption and allowed for it in their bid, but they did not expect the volume of water or the extent of the delay. Thus, it was the contractor's position that it was the obligation of the state or the farmers to keep the water off the right of way. The basis of their claim was a DSC.

Is a DSC the correct basis for the claim? What is Morris and Lamping arguing that is the basis of their claim? Is this a type I or type II claim?

Exercise 9-4: Umpqua River Navigation Co. and Western Pacific Dredging Corp.

Umpqua River Navigation Co. (Umpqua), a general contractor, contracted with the Crescent City Harbor District to expand a boat basin in Crescent City, Louisiana. Umpqua subcontracted the dredging part of the work to Western Pacific Dredging Corp. (Western). The specifications were prepared by Swinc Engineering, Inc.

In the early 1970s, the Crescent City Harbor District decided to investigate the possibility of expanding its boat facilities. In June 1971, as part of its preliminary research, the Harbor District commissioned an independent firm, AAA Drilling, to drill eight test holes on the beach near the site of the proposed new boat basin to find bedrock levels. AAA subsequently prepared drawings showing the location of its borings, the depth of bedrock, and the materials found between the surface and bedrock. AAA made no underwater borings in the proposed dredging channel. Shortly after these drillings, the Harbor District informed Swinc that it had been chosen to design the new boat basin. The Harbor District turned the AAA results over to Swinc, and, after prolonged negotiations, the Harbor District and Swinc signed a formal agreement in May 1972. In August 1972, Swinc arranged for Harding, Miller, a soils engineering firm, to make two test borings. These holes were drilled in the area where the breakwater surrounding the boat basin was to be constructed, not in the dredging channel.

The results of the AAA and Harding, Miller borings were included in sheet 002 of the boat basin plans. Sheet 002 showed the presence of sandy clay, sand, clay, and gravel at the test hole sites. The term "gravel" was not defined in the plans, and nothing in the contract materials furnished to bidders interpreted or elaborated on this soil information.

In March 1973, the Harbor District invited sealed bids for the boat basin construction. The bidding materials included plans, specifications, and special and general conditions. Particularly relevant to this dispute were special conditions 3, "Examination of Site," and 4, "Soil Information and Pile Tests," and general condition 21, "Subsurface Conditions Found Different." Contractors were given 30 days to submit their bids.

Special conditions 3 and 4 and general condition 21 read as follows

3. EXAMINATION OF SITE

Each bidder shall thoroughly examine and be familiar with the site of the proposed project and submission of a Proposal shall constitute an acknowledgement upon which the Owner may rely that the bidder has thoroughly examined and is familiar with the site. The failure or neglect of the bidder to fully familiarize himself with the conditions at the project site shall in no way relieve him from or to the Contract. No claim for additional compensation will be allowed which is based upon lack of knowledge of the site.

4. SOIL INFORMATION AND PILE TESTS

The drawings show soil test logs and pile test logs reproduced from reports by the District's Soil Consultant. Copies of these reports are on file at the offices of the Engineers and at the District Office and may be examined by prospective bidders. Each bidder shall make his own evaluation of the information contained in the reports. Neither the Owner or the Engineers guarantee that the soil borings, pile test logs or other information shown are typical for the entire site of the work.

21. SUBSURFACE CONDITIONS FOUND DIFFERENT

Should the Contractor encounter subsurface and/or latent conditions at the site materially differing from those shown on the Plans or indicated in the Specifications, he shall immediately give notice to the Architect/Engineer of such conditions before they are disturbed. The Architect/Engineer will thereupon promptly investigate the conditions, and if he finds that they materially differ from those shown on the Plans or indicated in the Specifications, he will at once make such changes in the Plan

and/or Specifications as he may find necessary, and any increase
or decrease of costs resulting from such changes to be adjusted in
the manner provided in Paragraph 17 of the General Conditions.

Before bidding, Umpqua examined the results of the AAA borings reported on sheet 002 and concluded that they were unreliable except for
showing the depth of bedrock. Umpqua's project engineer found the test
results defective because the soil logs were inconsistent and because the
logs did not include a legend defining various soil terms. Umpqua did,
in compliance with special condition 4, request the reports underlying
the Harding, Miller soil logs. Those borings, however, were taken at the
location of the proposed breakwater and not in the dredging channel.

Western Pacific also conducted a prebid investigation. Robert Kaltsukis, a Western Pacific dredge captain, visited the project site. Kaltsukis, who had encountered rocks while dredging a nearby portion of
Crescent City Harbor in the mid-1960s, saw submerged rocks in an area
adjacent to the dredging site. As part of his on-site inspection, Kaltsukis
used a backhoe to dig test holes near two of the AAA boring locations.
The results of these tests were inconsistent with the information shown
on sheet 002, yielding sand and pea gravel, rather than only gravel as indicated in the AAA results. Despite these discrepancies, Western Pacific
did not comply with special condition 4 and attempted to obtain the
source data for sheet 002.

Umpqua was awarded the boat basin construction contract in May
1973 and shortly thereafter selected Western Pacific as its dredging subcontractor. Western's subcontract bid was based on an estimated dredging output of 10,000 yard³ per day, using hydraulic dredging methods.

Western began dredging on May 29, 1973. Almost immediately, the
dredge *Polhemus* encountered cobbles, boulders, and cemented sand. The
dredge also sustained damage, requiring extensive repairs. These difficulties arose during the dredging of the boat basin's access channel and turning area, areas in which no soil borings had been taken. On July 26,
Western notified Umpqua that it had encountered conditions materially
different from those indicated in the plans and that it was withdrawing
Polhemus from the project. Umpqua forwarded this notice to Swinc, which,
in turn, informed the Harbor District. Western brought in a second, larger
dredge, the *Herb Anderson*, which encountered difficulties similar to those
encountered by *Polhemus*. The parties offered conflicting explanations of
these problems. Umpqua contended that the damage to the dredges was
unprecedented, resulting from extreme and unforeseeable concentrations

of cobbles and boulders. The Harbor District and Swinc asserted that Western's dredging troubles were due to inefficient crews and improper maintenance of the dredges. Whatever the source of its dredging problems, Western Pacific, on August 25, again notified Swinc that it had encountered conditions different from those shown in the plans.

Swinc and the Harbor District did not formally investigate Western's claims until November 28. When representatives of Western, Umpqua, Swinc, and the Harbor District dug test holes at three locations designated by Western, these borings yielded a large quantity of sand and silt, some cobbles, and one boulder. The Umpqua and Swinc engineers decided the basin was dredgeable and directed Western Pacific to continue. Western Pacific completed dredging on December 21. In more than six months of intermittent dredging, Western rarely met its projected output of 10,000 yard^3 per day. The material dredged during the entire project was 80% sand, 15% gravel less than 3 in. in diameter, and 5% material exceeding 3 in. in diameter. Neither Umpqua nor Western Pacific obtained a change order, as required by general condition 21, or submitted a bill for extra costs at any time before June 6, 1974, when Umpqua reported that the boat basin project was completed.

On June 11, 1974, Western informed Umpqua that it had incurred additional dredging expenses of $523,402. This claim was forwarded to the Harbor District, which formally denied the claim on July 30, 1974. The Harbor District then paid Umpqua the contract price of 75 cents per cubic yard dredged, and Umpqua paid Western 65 cents per cubic yard.

What does the contract indicate? Ignoring issues over notice, how serious is Umpqua's failure to comply with special condition 4? Can Umpqua or Western submit a type I claim for increased cost in the boat basin's access channel and turning area, areas in which no soil borings were taken? Is Umpqua entitled to an equitable adjustment?

Exercise 9-5: Grand Forks, N.D., and Moorhead Construction Co.

The city of Grand Forks, North Dakota, entered into a contract with Moorhead Construction Co., Inc., to build a sewage treatment facility. The city had divided the construction project into two phases, each to be performed by separate contractors. Phase I was designed by the city's own engineering department and covered primarily the earthwork and site preparation for four aerated anaerobic treatment ponds, including installation of piping and appurtenances such as foundations for the compressor and meter buildings. The four ponds or earthen cells were to be formed by building earthen embankments in a square pattern divided into four large, square, watertight sections. The phase I contractor, Valley-Mayo, was scheduled to complete its work in September 1969, before phase II was to commence. The phase I contractor, however, did not substantially complete its contract until November 1970. The final acceptance by the city of the phase I work was not until October 1971, when the contractor was paid in full and discharged.

Phase II of the project was designed by Richmond Engineering, Inc., of Grand Forks, the city's agent and supervisor for the project. It consisted of completing the buildings, constructing manhole installations and access bridges into and over the ponds, and installing all electrical and mechanical equipment. When completed, the aeration equipment would treat the city's sewage primarily in the aerated cells; secondary treatment would occur in lagoons. The separate phase II contract was awarded to Moorhead in July 1969, with completion scheduled for October 30, 1970. The contract stated in part,

> F20. CHANGED CONDITIONS. Should the contractor encounter or the Owner discover during the progress of the work *subsurface or latent physical conditions at the site differing materially from those indicated in this contract, or unknown physical conditions at the site of an unusual nature, differing materially from those ordinarily encountered and generally recognized as inherent in work of the character provided for in the contract, the Engineer shall be promptly notified in writing* of such conditions before they are disturbed. *The Engineer will thereupon promptly investigate the conditions and if he finds they do so materially differ and cause an increase or decrease in the cost of, or the time required for performance* of the contract, an *equitable adjustment will be made and the contract modified in writing accordingly.*

At the time Moorhead bid on the phase II contract, the phase I earthwork had just commenced. Because an inspection of the site by Moorhead would not then have disclosed the difficult site conditions that it would later face due to excess moisture and lack of compaction, Moorhead in estimating its bid relied on the city to provide a construction site prepared in accord with the specifications of phase I. Those specifications called for 90% compaction of the soil embankments and cell bottoms.

As early as December 2, 1969, Moorhead expressed concern about increased costs related to the delay. On December 19, 1969, Moorhead wrote to the city,

> I am very deeply concerned and perturbed in regards to the contract we hold with the City. We bid the project under certain stipulations. We had to follow time limits. As of this date we have not even been given access to the project.

Before January 1970, when Moorhead was notified to proceed, its president inspected the site and refused to take responsibility for it.

> ...I went to inspect the project site and I definitely would not accept accessibility to the site and be responsible for it in its present condition.... We are going to incur additional costs, as to increased labor, material, sales tax, warehousing and scheduling material shipments and placement of our crews to complete this project.

The bottom surface of the lagoon cells was extremely soft. As a result of the unstable soil conditions actually encountered in the cell bottoms and on the embankments, Moorhead was forced to work by different, more expensive methods without heavy equipment. Most of the foundation footings for the mechanical installations and access bridges had to be redesigned and spread apart for greater support.

On February 17, 1970, Moorhead wrote to Richmond,

> It is very difficult to construct a job under existing information and complete the same when the conditions and time of availability are not the same as stated under the bidding plans and specifications.
>
> According to the specifications of Phase I, the bottom and slopes shall have 90% compaction. This definitely is not there.

On February 18, 1970, Richmond, relaying Moorhead's letter to the city, stated,

> As we all know, the soil conditions at the site are treacherous. It is entirely possible that 90% compaction by the Phase I Contractor was a physical impossibility considering the time of year the work was done.

On March 23, 1970, Richmond wrote to the city. In discussing the unstable soil matter with Moorhead, they again noted that the soil conditions were beyond Moorhead's control and that they bid the job under the premise that the phase I contractor would obtain 90% compaction in the cell bottoms and dikes. Richmond said that "since the dikes are not completely finished and are still frozen it is probably too early to comment on their density, but *the bottoms are definitely a changed condition.*" (emphasis added)

Due to adverse weather, soil conditions, and other delays, phase II was not completed until November 1971. Moorhead claims that the phase II job was entirely changed and greatly increased its construction cost. Moorhead filed a claim based on a type I DSC.

Can the city argue that Moorhead's contract did not promise 90% compaction density? Is Moorhead entitled to extra compensation?

Exercise 9-6: Western Contracting Corp. and California Department of Water Resources

Western Contracting Corp. (Western) was awarded a fixed-price contract with the state of California's Department of Water Resources for the construction of the Castaic Dam in Los Angeles County. The projected period of performance was four years. The contract provided that "The contract price... of the work shall include full compensation for all costs incurred" (9(b)). More specifically, with respect to the subject of taxes, section 4(h) of the contract stated,

> Except as otherwise provided in the Special Provisions, the contract prices shall include full compensation for all taxes which the Contractor is required to pay, whether imposed by federal, state, or local government, and no tax exemption certificate or any other document designed to exempt the Contractor from payment of tax will be furnished to the Contractor by the Department.

Section 7 of the contract authorized the engineer to order "changes in the contract as are required for the proper completion of the work" (7(b)). Such changes may result in additional compensation whenever they cannot "be fairly and reasonably paid for at contract prices" (7(e)).

The contract also contained a changed conditions clause, which made allowances for cost adjustments if costs were materially increased or decreased (7(h)). Changed conditions were defined as (1) "Subsurface or latent physical conditions at the site of the work differing materially from those represented in this contract" and (2)"Unknown physical conditions at the site of the work of an unusual nature differing materially from those ordinarily encountered."

After the contract was awarded, the state legislature increased the sales and use tax rate by 1%. The additional 1% tax burden was applied to material purchases, the rental or purchase of heavy equipment (such as earthmovers and cranes), and tires, fuel, oil, grease, and repair parts. Western paid the additional sales and use tax, which amounted to approximately $102,000 and now makes a claim to the department for additional compensation, alleging a changed condition based in a type I DSC.

Does the phrase "subsurface or latent physical conditions at the site" have any bearing on Western's claim? Is Western entitled to its DSC claim?

Exercise 9-7: P. J. Maffei Building & Wrecking Corp. and U.S. Government

In June 1976, the U.S. government issued an invitation for bids (IFB) for the demolition and removal of the United States pavilion and restoration of the grounds in Flushing Meadow Park in New York City. The pavilion had been built for the 1964 World's Fair. The IFB advised prospective contractors that the salvage value of the construction materials to be removed from the project site should be reflected in their bids because those materials would become the property of the contractor. The special conditions section of the IFB stated,

> 1.2 Some drawings of some of the existing conditions are available for examination at the New York City Parks Department's Administration Building at Flushing Meadow Park, Flushing, New York. These drawings are for information only and will not be part of the contract documents. The quantity, quality, completeness, accuracy and availability of these drawings are not guaranteed. Prospective bidders shall telephone Mr. S. Dubowy or Mr. S. Adler of the New York City Parks Department, at 212-699-4288, for an appointment to examine drawings of the existing conditions.

P. J. Maffei Building and Wrecking Corp.'s estimator visited the site, reviewed drawings obtained from a man named "Charlie" at the New York City Parks Department, consulted a "steel book," and subsequently arrived at a bid based on its estimate of the amount of salvageable steel in the pavilion. On October 7, 1976, the government awarded Maffei the demolition contract.

The contract contained a differing site conditions clause, which provides,

> The Contractor shall promptly, and before such conditions are disturbed, notify the Contracting Officer in writing of (1) Subsurface or latest [sic for latent] physical conditions at the site differing materially from those indicated in this contract, or (2) unknown physical conditions at the site, of an unusual nature, differing materially from those ordinarily encountered and generally recognized as inhering in work of the character provided for in this contract. The Contracting Officer shall promptly investigate the conditions, and if he finds that such conditions do materially so differ and cause an increase or decrease in the Contractor's cost of, or the time required for, performance of any part of the work under this contract, whether or not changed as a result of such conditions, an equitable adjustment shall be made and the contract modified in writing accordingly.

Maffei recovered 1,075 tons of steel, approximately 20% less than it had estimated it would salvage from the project. Maffei now requests an equitable adjustment on the basis that the parks department drawings indicated more steel than was salvaged.

What does the contract likely indicate? Should Maffei receive money for its claim?

Exercise 9-8: Carlos Teodori and Penn Hills School District Authority

On April 10, 1958, Carlos Teodori entered into a written contract with the Penn Hills School District Authority for excavation work to be done on a site to be used as an athletic field. The contract was awarded to Teodori after competitive bidding, for a total price of $134,485.

The contract provided that changes and alterations should be ordered in writing by the owner and also provided for changed conditions as follows:

> *"Conditions Differing From Those Shown on Plans Or Indicated In Specifications."*
>
> Should the Contractor encounter subsurface and/or latent conditions at the site materially differing from those shown on the Plans or indicated in the Specifications, he shall immediately give notice to the Architect of such conditions, before they are disturbed. The Architect shall thereupon promptly investigate the conditions and if he finds that they materially differ from those shown on the plans or indicated in the Specifications, he shall at once, make such changes in the Plans and/or Specifications as he may find necessary. Any increase or decrease of cost resulting from such changes shall be adjusted in the manner provided herein for adjustments as 'Extra Work'.

The method of computing payment for extra work was set forth in the contract by the following language:

> "By such applicable unit prices, if any, as are set forth in the Contract;" or "If no such unit prices are so set forth, then by a lump sum mutually agreed upon by the Owner and the Contractor"; or
>
> "If no such unit prices are so set forth, and if the parties cannot agree upon a lump sum then by the actual net cost in money to the Contractor of the materials and of the wages of applied labor (including premiums for Workmen's Compensation Insurance, Social Security and Unemployment Compensation) required by law, plus such rental for plant and equipment (other than small tools) required and approved for such Changes and Alterations, plus fifteen percent (15%) as compensation for all other items of profit,..."

A further pertinent provision of the contract provides

> *"All Work Subject to Control of Architect."*
>
> In the performance of the work, the Contractor shall abide by all orders, directions, and requirements of the Architect and shall

perform all work to the satisfaction of the Architect, and at such time and places, by such methods and in such manner and sequence as he may require.

Upon request, the Architect shall confirm in writing an oral order, direction, requirement or determination.

Teodori commenced work under the contract and was to have completed the job by June 14, 1958. A few days after the preliminary work had begun, Teodori was advised by a representative of the Sun Pipe Line Co. that a 6-in. high-pressure gasoline transmission line ran beneath the surface of the property at a depth of about 3 ft in the area where the excavating and grading work was to be done. The contract documents did not disclose the existence of the line.

The existence of the gasoline line was discovered by the architect, who called a meeting at which Teodori, the district authority's job inspector, and a representative of Sun Pipe Line Co. were present. At this meeting, Teodori was instructed to change his sequence of operations to avoid working in the area where the gasoline line was located until the gasoline line was relocated. No written directive as required by the contract was ever issued. The pipeline relocation was completed on July 2, 1958, and it was only then that Teodori was able to resume some semblance of normality in his earthmoving operation.

Teodori now claims damages in the sum of $98,613.95 over and above the original contract price, basing this amount on the increased cost of the work brought about by the existence of the high-pressure gasoline line, the delay in its relocation, and consequential changes in Teodori's plan of operation for the removal of earth. These factors caused a delay in completion of the work from June 14, 1958, to September 15, 1958.

Is Teodori entitled to extra compensation?

References

Al Johnson Construction Co. v. Missouri Pacific Railroad Co., 426 F.Supp. 639, 647 (8th Cir. 1976), aff'd 553 F.2d 103 (8th Cir 1976).

Arundel Corp. v. United States, 103 Ct.Cl. 688, 712 (1945), cert. denied, 326 U.S. 752, rehearing denied, 326 U.S. 808 (1945).

Bergman Construction Corp., ASBCA 9000, 1964 BCA 4426.

Blauner Construction Co. v. United States, 94 Ct. Cl. 503, 511 (1941).

Blount Bros. Construction Co., ASBCA 4780, 1 G.C. 686, 59–2 BCA 2316.

Caribbean Construction Corp., IBCA 90, 57–1, BCA 1315.

Currie, O. A., Ansley, R. B., Smith, K. P., and Abernathy, T. E. (1971). "Differing Site (Changed) Conditions." *Briefing Papers* No. 71-5, Federal Publications, Inc., Washington, DC.

Dravo Corp. v. Commonwealth of Kentucky, 546 S.W.2d 16 (1977).

Farnsworth & Chambers Co. v. United States, 171 Ct.Cl. 30, 376 F.2d 577 (1965).

Foster Construction C.A. & Williams Bros. Co. v. United States, 193 Ct. Cl 587, 602, 603, 604, 624, 435 F.2d 873 (1970).

Guy F. Atkinson Co., IBCA 385, 65–1 BCA 4642.

Hoffman v. United States, 166 Ct.Cl. 39, 340 F.2d 645 (1964).

Kaiser Industries Corp. v. United States, 340 F.2d 322, 325, 169 Ct.Cl. 310 (1965).

Leal v. United States, 276 F.2d 378, 383/384, 149 Ct.Cl. 451 (1960).

Metropolitan Sewerage Comm. v. R.W. Construction, Inc., 72 Wis.2d 365, 241 N.W.2d 371 (1976).

Mojave Enterprises v. United States, 3 Cl.Ct. 353, 358 (1983).

Morrison-Knudsen Co. v. United States, 345 F.2d 535, 170 Ct.Cl. 712 (1965).

Neale Construction Co., *ASBCA*, 2753, 58-1, 1710.

North Slope Technical Ltd., Inc., v. United States, 14 Ct.Cl. 242, 252 (1988).

Pacon, Inc., ASBCA 7643, 5 G.C. 76, 1962 BCA 3546.

Parvin, C. M., and Araps, F. T. (1982). "Highway Construction Claims—A Comparison of Rights, Remedies, and Procedures in New Jersey, New York, Pennsylvania, and the Southeastern States." *Public Contract Law Journal*, 12(2).

Ragonese v. United States, 120 F.Supp 768, 769, 128 Ct.Cl 156 (1954).

John K. Ruff v. United States, 96 Ct.Cl. 148, 163 (1942).

Sornsin Construction Co. v. State of Montana, 590 P.2d 125, 129, 180 Mont. 248 (1978).

Stock & Grove, Inc., v. United States, 493 F.2d 629, 631, 204 Ct.Cl. 103 (1974).

Stokes, M., and Finuf, J. L. (1986). *Construction Law for Owners and Builders*, McGraw-Hill, New York.

Thomas, H. R., Smith, G. R., and Ponderlick, R. M. (1992a). "Resolving Contract Disputes Based on Misrepresentations." *J. Constr. Engrg. and Mgmt.*, 118(3), 472–487.

———— (1992b). "Resolving Contract Disputes Based on the Differing Site Conditions Clause." *J. Constr. Engrg. and Mgmt.*, 118(4),767–779.

Town of Longboat Key v. Carl E. Widell & Son, Fla.App., 362 So.2d 719, 722 (1978).

United Contractors v. United States, 368 F.2d 585, 594, 595, 596, 597, 598, 177 Ct.Cl. 151 (1966).

Vann v. United States, 420 F.2d 968, 986, 190 Ct.Cl. 546 (1970).

Western Well Drilling Co. v. United States, 96 F.Supp 377, 379 (9 Cir 1951).

William F. Wilke, Inc., ASBCA Nos. 33,233, 33,748, 88–3 BCA 21,134.

Woodcrest Construction Co. v. United States, 408 F.2d 406, 410, 187 Ct.Cl. 249 (1969), cert. denied, 398 U.S. 958, 90 S.Ct. 2164, 26 L.Ed.2d 542 (1970).

Additional Cases

The following are additional cases related to issues associated with differing site conditions. The reader is invited to review the facts of the case, apply the decision criteria in the flowchart, reach a decision, compare it with the judicial decision, and determine the rationale behind the judicial decision.

Basin Paving Co. v. Mike M. Johnson, Inc., 107 Wn. App. 61 (2001).

Foundation Intern v. E. T. Ige Construction 81 P.3d 1216, Supreme Court of Hawaii, 2003.

SMC Corp. v. New Jersey Water Supply Auth. 759 A.2d 1223, Superior Court of New Jersey, Appellate Division, 2000.

Chapter 10

Misrepresentations

Contractor claims based on a theory of misrepresentation require that

- the owner made an affirmative representation,
- the contractor was reasonably entitled to rely on that representation,
- the contractor did rely on that representation,
- the representation was incorrect, and
- the contractor suffered damages resulting from its reliance on the misrepresentation.

Misrepresentation claims frequently involve subsurface or differing site conditions where the roles of the soil report, disclaimer clauses, and site visit requirements are often misunderstood, but claims based on misrepresentations are not limited to geotechnical issues.

Contract Language

Where the contract does not contain a differing site condition (DSC) or concealed conditions clause, the only recourse for a contractor seeking an equitable adjustment is to rely on an assertion of misrepresentation. The disclaimer and site visit clause detailed in the previous chapter may be relevant.

> Article 1.2.12 Subsurface Conditions: It is the obligation of the Bidder to make his own investigations of subsurface conditions prior

to submitting his Proposal. Borings, test excavations and other sub-surface investigations, if any, made by the Engineer prior to the construction of the project, the records of which may be available to bidders, are made for use as a guide for design. Said borings, test excavations and other subsurface investigations are not warranted to show the actual subsurface conditions. The Contractor agrees that he will make no claims against the State, if in carrying out the Project he finds that the actual conditions encountered do not conform to those indicated by said borings, test excavations and other subsurface investigations. Any estimate or estimates of quantities shown on the Plans or in the form of proposal, based on said borings, test excavations and other subsurface investigations, are in no way warranted to indicate the true quantities. The Contractor agrees that he will make no claims against the State, if the actual quantity or quantities do not conform to the estimated quantity or quantities, except in accordance with the provisions of Art. 1.8.4 (AIA 1987).

Background

Many of the principles of misrepresentation were developed in case-law decisions in the early part of the 20th century. There is always a danger that these older decisions have been overturned or are no longer applicable. Each case cited in the analysis was researched to ensure that the decision had not been reversed. Furthermore, other investigators have determined that the rules established in the earlier cases have not been modified (Vance and Jones 1978, p. 1478).

In the absence of a concealed conditions or differing site conditions (DSC) clause, the owner assigns the risk for unknown subsurface conditions to the contractor (Jervis and Levin 1988). The owner expects the contractor to include sufficient bid contingency to cover the risk of encountering latent physical conditions that may affect performance. Without a DSC clause, the contractor assumes the risk if the material is different than expected, and normally cannot recover additional related costs from the owner. In W. H. Lyman Construction Co. v. Village of Gurnee, the court stated,

It is well settled that a contractor cannot claim it is entitled to additional compensation simply because the task it has undertaken turns out to be more difficult due to weather conditions, the subsidence of the soil, etc.

Without a DSC clause, the only way a contractor is entitled to additional costs arising from latent subsurface conditions is to prove that the owner provided incorrect or misleading information that the contractor was entitled to rely on or the owner failed to disclose relevant information (Sweet 1989).

Rules of Application

Primary Issues Governing Misrepresentations

To be entitled to additional compensation resulting from inaccurate information, a contractor must show

- a positive representation of fact,
- that the facts provided proved to be inaccurate or nonexistent,
- that the claimant reasonably relied on the representation, or
- that the claimant suffered damages as a result of the claimant's reliance on the information given. (J. A. Johnson & Son v. State of Hawaii)

These requirements can be quite formidable, as described below and in Fig. 10-1.

Was There a Positive Representation?

"A misrepresentation is an assertion that is not in accord with the facts" (Restatement of the Law 1979). For a misrepresentation to exist, there first must be a positive, material statement purporting the conditions that the contractor can expect to encounter.

In the 1922 decision in MacArthur Brothers Co. v. United States, the U.S. Supreme Court set forth fundamental principles regarding misrepresentation disputes. The contractor was to build a canal at Sault Sainte Marie, Michigan. The alleged misrepresentation occurred from contractual assurances that part of the work could be done "in the dry." The fa-

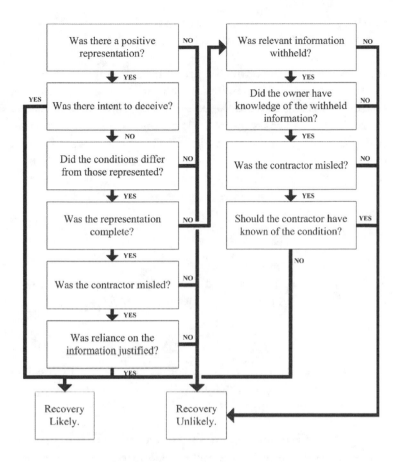

Figure 10-1. Decision Tree for Disputes Involving Misrepresentations.
SOURCE: Thomas et al. 1992, ASCE

cilities for keeping the work area dry were constructed by another contractor and proved to be inadequate. The court felt that the work was
advertised with

> ...no knowledge of impediments to performance, no misrepre-
> sentation of the conditions, exaggeration of them nor conceal-
> ment of them, nor, indeed, knowledge of them. To hold the
> government liable under such circumstances, would make it in-
> surer of the uniformity of all work and cast upon it responsibil-
> ity for all conditions which a contractor might encounter...
> (MacArthur Brothers Co. v. U.S.)

According to Vance, the MacArthur decision is sound law and has never been modified or overturned (1978, p. 1477).

> It is essential that a contractor prove that the representation in question was a positive, material statement about the nature of the concealed conditions anticipated during construction. A careful review of the cases indicates that this requirement must always be satisfied, and many claims fail because this requirement cannot be met. Various state and federal courts have stated: "It (the statement of conditions) was a positive and material representation as to a condition within the knowledge of the Government..."; (Hollerbach v. U.S.)

and

> ... the specifications spoke with certainty as to a part of the conditions to be encountered ... (Hollerbach v. U.S.)

and

> The plaintiff (contractor) ... had a right to rely upon the positive representations that were made by the defendant regarding the subsurface conditions ... (Morrison-Knudsen Co. v. U.S.)

The alleged positive statement must more than merely suggest the condition. The case of Wunderlich v. State of California is a watershed decision that clarified much of the confusion surrounding earlier misrepresentation cases. The dispute centered around the contractor's reliance on an internal state of California memo stating that a testing program indicated that a particular borrow pit was a possible source of material for the project. The memo went on to say that "Tests indicate that after processing, to meet gradation requirements, the material is suitable for imported base material, ..." The contractor knew that the test reports relied on to prepare the memo were available, but the test report was not consulted. When the supply of material at the borrow pit was insufficient, the contractor brought in new equipment and later used materials from more distant sites. The contractor claimed that the memo misrepresented the volume of suitable material. The Supreme Court of California denied the contractor's claim, stating that the memo did not represent that there would be adequate material to complete the project. The court stated,

If the agency makes a "positive and material statement as to a condition presumably within the knowledge of the government, and upon which... the plaintiff had a right to rely" then the agency is deemed to have warranted such facts, despite a general provision requiring an on-site inspection by the contractor. But if statements "honestly made" may be considered as "suggestive only," expenses caused by unforeseen conditions will be placed on the contractor, ..." (Wunderlich v. State of California)

In both the MacArthur and Wunderlich cases, a positive, material statement of the conditions to be encountered was lacking. Rather, it appears that in both cases, the work turned out to be harder than the contractor anticipated. Although these decisions may seem harsh, they are consistent with many other cases and highlight the degree of certainty with which the representation must be proven.

A number of later cases have followed the Wunderlich decision. In one Pennsylvania case, no misrepresentation was found when the court determined that the claimed representations were intended to be suggestive and were not to be relied on with exactness (Dravo Corp. v. Commonwealth of Kentucky Department of Highways). In an Arizona dispute over quantity estimates made for a borrow pit that proved to contain an insufficient volume of material, the contractor's claim was denied when the court said,

The representation relied upon by Ashton (the contractor) is not cast in the form of a positive assertion of fact, but is given as an estimate, and there was full disclosure of the basis for the estimate. (Ashton Co. v. State of Arizona)

The above cases highlight the fact that a contractor can only reasonably rely on affirmative representations. Opinions, suggestions, estimates, and interpretations are not positive representations.

Sweet, differentiating between facts and opinions, says that test results are clearly factual representations, whereas professional judgments that seek to draw inferences from this information are opinions (Sweet 1989). Thus, boring logs shown in the plans are positive factual representations, but interpretations of the results are not.

Positive representations must occur within the contract documents. In Raymond International, Inc. v. Baltimore County, the record showed that the specifications were materially wrong and inaccurate. The contract was for the repair of a bridge, and the plans represented an average surface deterioration in the existing piers of 6 in. despite the engineer's own reports that showed an average deterioration of only 2 in. (Raymond International v. Baltimore County).

In Coatsville Contractors and Engineers, Inc., v. Borough of Ridley Park, the contractor relied on a statement in the specifications that "The lake has been drained and shall remain in the drawdown condition until all silt debris removal work has been completed." The contract also contained an exculpatory clause stating that the contractor "will make no claim against the (Borough) because of any ... misrepresentation of any kind .... all losses or damages arising ... from unforeseen obstructions or difficulties ... shall be sustained by the contractor" (Loulakis 1986). Unfortunately, the lake was not drained throughout the period of performance. Despite the disclaimer, the contractor was allowed to recover his additional cost because the statement in the specifications imposed an affirmative obligation on the borough to maintain the lake in a drained condition (Coatsville Contractors & Engineers v. Borough of Ridley Park).

When the contract does not contain a DSC clause, borings and test pits positively represent conditions at the boring or test pit location and no more. As stated by one court,

> The borings were merely indications, at certain places and to certain depths, from which deductions might be drawn as to actual conditions along the line and to the depths of such borings. Both parties knew that deductions so drawn might prove untrue when necessary excavations were made. (Elkan v. Sebastian Bridge District)

The contractor could base a claim on an interpolation between borings. This position has been affirmed in other decisions (Morrison-Knudsen Co. v. U.S.). If such interpolations are included in the contract documents, however, they will probably be viewed as a positive representation. The boring logs may be entirely accurate and complete, but they do not represent all the conditions that will be found in the subsurface. In the words of one court,

> The bare statement that the boring sheet may be relied upon as accurate is entirely different from saying that the subsoil along the bridge line is as shown by the boring sheet. (Elkan v. Sebastian Bridge District)

Typically, contracts contain clauses that alert the contractor that what is provided may not be correct and is provided for information only. Depending on the jurisdiction, general disclaimers are seldom enforced, and it is clear from the case-law review that general exculpatory clauses will be given little, if any, weight in offsetting positive material statements. For a disclaimer to be enforced, it must be expressed, unqualified, and specific. The state of New Jersey has effectively used a subsurface conditions clause to disclaim subsurface conditions. As stated in Sasso Contracting Co. v. State relative to this clause,

> While we might agree ... that general exculpatory clauses will not relieve the State from responsibility for its express representations, it is otherwise where the relevant language of the contract is so straightforward, unambiguous and categorical as this is in placing responsibility for subsurface investigations on the contractor.

However, most disclaimers are not specific, and the rulings are usually more in line with E. H. Morrill v. State of California:

> The responsibility of a governmental agency for positive representations ... is not overcome by the general clauses requiring the contractor to examine the site, to check up on the plans, and to assume responsibility for the work ...

Case Study

Representations and Disclaimer

Miami-Dade Water and Sewer Authority v. Inman, Inc.
402 S.2d 1277 (1981)

Inman, Inc., entered into a contract with Miami-Dade Water and Sewer Authority to install wastewater force mains. The plans showed the location of underground utilities. The contract contained the following provision:

> Information shown on the Drawings as to the location of existing utilities has been prepared from the most reliable data available to the Engineer. This information is not guaranteed, however, and it shall be this Contractor's responsibility to determine the location, character and depth of existing utilities. He

shall assist the utility companies, by every means possible to de-termine said locations and the locations of recent additions to the systems not shown. Extreme caution shall be exercised to eliminate any possibility of any damage to utilities resulting from his activities. The location of all overhead utilities shall be verified and the Engineer notified of any conflict which might occur. The Contractor shall be responsible for determining which poles will need shoring during excavation and shall pro-vide shoring and support as required.

The contract contained no differing site condition clause.

During construction, the utilities were found at varying locations other than shown on the plans. Inman now seeks additional compen-sation. Is the disclaimer specific or general, and what features make it so? Was Inman misled and will the disclaimer language protect the sewer authority?

The Third District Court of Appeals of Florida agreed with the sewer authority on two counts. First, the court said that Inman was not misled:

> The quoted provision of the contract between Inman and the Au-thority does not represent the location of existing utilities is as shown on the plans. It simply represents that information per-taining to location "has been prepared from the most reliable data available to the Engineer." The fact, then, that the location is not as shown does not result in the Authority's liability unless the disclaimer is one that is, as the trial court implicitly found, inoperable as a matter of law.

In regard to the disclaimer, the court further stated that

> The holdings . . . exculpatory or disclaimer clauses which require the contractor, e.g., to examine the site, to check the plans, and to assume responsibility for the work, will not be allowed to defeat or overcome a contractor's justifiable reliance on an express or implied warranty or representation by the contracting authority. Thus, the rule . . . that the contractor is relieved from the impact of such a clause applies only when it is first misled. If, however, there is no misleading, a disclaimer or like clause . . . may be in-terposed to negate the liability of the contracting authority. It is neither alleged nor made to appear that that representation was

false or misled Inman, or that other representations were made, or withheld, which might preclude the Authority from further reliance on the disclaimer.

Thus, the position of the court is that there was no misleading of Inman by the sewer authority. It should be noted that the disclaimer is quite specific and may have been sufficiently specific to preclude Inman from recovering. Another important aspect is how the utility locations were represented. Had these been represented as precise locations, say, with dimensions and elevations recorded to the nearest hundredth, then the outcome might have been different. Thus, it should be obvious how seemingly minor facts can change the outcome even though the same rules are applied.

Was There Intent to Deceive?

Most misrepresentations occur where inaccurate information is provided without bad faith. However, several cases were identified where data were purposely changed to induce lower bids. In City of Salinas v. Souza and McCue Construction Co., the city was installing a sewer line in an area known by the city to be extremely wet and often having quicksand-like conditions. The city engineer directed the geotechnical firm to take borings at specific locations along the proposed center line to avoid the wettest areas. The court ruled for the contractor, finding that the city's actions were fraudulent.

Contracts sometimes contain clauses designed to provide immunity against liability for fraudulent misrepresentation, but these clauses have little effect. As stated in O'Neill Construction Co. v. City of Philadelphia, "...no one can escape liability for his own fraudulent statement by inserting in a contract, a clause that the other party shall not rely upon (the fraudulent statements)."

Did the Conditions Differ from Those Represented?

An essential requirement to recover additional costs is the need to prove that the actual conditions were materially different from those represented. Although this requirement appears trivial and obvious, it is discussed at length in most cases. If the contractor cannot prove that the conditions were materially different, then no equitable adjustment can be made.

Was the Representation Complete?

Although the representation may have been correct, relevant information may have been withheld. If relevant information is not provided, the contractor may be able to recover additional costs.

The case of United States v. Atlantic Dredging Co. arose from a contract for dredging a portion of the Delaware River. The government made test borings using the probe method, and the field logs showed that some borings had to be stopped because of obstructions. A map showing the results of the boring program was included in the contract, but it only showed the successful borings. The map made no mention of how the borings were made or the facts that obstructions were encountered and that a field log had been prepared. When the contractor found different material from that shown by the maps, it continued with the project thinking its interpretation was incorrect.

> It (the contractor) did not know at the time (of bidding or a subsequent change order) of the manner in which the test borings had been made. Upon learning that they had been made with the probe method, it then elected to go no further with the work, that is, upon discovering that the belief expressed was not justified and was in fact a deception. (U.S. v. Atlantic Dredging Co.)

The Atlantic Dredging case illustrates the concept of *withholding of information*. One important way for a withholding to occur is when "the defendant (owner) makes representations, but does not disclose facts which materially qualify the facts disclosed, or which render his disclosure likely to mislead" (Wiechmann Engineers v. California State Department of Public Works). The rules for withholding are given later in this chapter.

Was the Contractor Misled?

When information is misrepresented, the contractor must also show that it was misled. In Morrison-Knudsen Co. v. United States, the court stated that

> ... mere proof of the defendant's (owner) misrepresentations is not sufficient to justify a judgment in favor of the plaintiff. A further prerequisite to recovery by the plaintiff is proof that the plaintiff was misled by such misrepresentations.

Proof of being misled is often found in the contractor's estimate calculations. If the contractor did not rely on the misrepresentation in preparing the bid, then there is no damage and therefore no owner liability. If the contractor used the misrepresentation to prepare the bid, then it must also prove that the bid would have been different had the information been provided correctly.

Was Reliance on the Information Justified?

The contractor has to prove not only that it was misled but also that it was reasonably misled. The willingness of courts to examine this issue highlights the difficulties for a contractor in recovering additional costs where there is no differing site condition clause. Normally, contractors are not reasonably misled if other readily available information exists that would have given a more complete and contrary understanding of the alleged misrepresentation. This information may come from many sources, such as other contract clauses, test reports that were not included in the contract documents, site visits, and the contractor's own experience. If this other information acts to modify the representation, then the contractor's reliance may not have been justified.

Other Contract Clauses

Other contract clauses may modify or clarify data and other information furnished to bidders. All contract provisions must be read together to determine what is required.

In Morrison-Knudsen Co. v. United States, the court stated that the contractor was not justified in relying on borings showing no permafrost in the excavation area. The court said,

> ...the contract contained a provision that specifically informed the plaintiff (contractor) of the likelihood of encountering permafrost.... the incorrect data (soil borings) which the defendant furnished to the plaintiff...represented only a portion of the material which the defendant furnished to the plaintiff and other prospective bidders regarding subsurface conditions.... (Morrison-Knudsen Co. v. U.S.)

In this case, the court ruled that a misrepresentation existed at the two borings in dispute. However, the contractor was only reasonably misled by the misrepresentation in the area directly around the boreholes. The court awarded the contractor damages for permafrost found within 10 ft of the disputed borings.

Other Readily Available Information

Usually soil reports are not part of the contract documents but are made available to the contractor for review. The contractor must consult the soil report or assume the risk of not knowing its content.

In C. W. Blakslee & Sons, Inc., v. United States, the soil report was available for review in the resident engineer's office. The contractor only reviewed the wash boring map in the contract documents, which showed the stratification of the soil. It did not show boulders or indicate that the boring contractor had used explosives to get through boulders to continue some of the wash borings. The contractor did not review the boring logs. The contractor filed a claim when numerous boulders were found in the work area. The court, in denying the claim, stated,

> The method of making the borings and the fact that dynamite was used and similar information is recorded in the log book. Plaintiff (contractor) knew this but made no effort to consult the log book, which was available to them. Plaintiff therefore have no one but themselves to blame for the fact that at the time they submitted their bid they did not know that dynamite had been used by the defendant in making the borings and can not be heard to complain that they were misled or damaged by the defendant because of that fact. (C. W. Blakslee & Sons v. U.S.)

The need to review and consider all relevant information is further illustrated by Flippin Materials v. United States. The dispute involved a quarry operation for producing concrete aggregate. The boring profiles provided in the contract documents showed cavities in the rock without stating what was in those cavities. However, the field logs, which were not in the contract documents, showed that the cavities were filled with clay. This fact made the quarry material unsuitable for concrete aggregate. The court ruled that the boring profiles were correct as far as they went, and the contractor had been directed to review the field logs for the complete story. The contractor failed to do so. In ruling against the contractor, the court stated,

> ...but we think the fair residue of the opinions is that a contractor cannot call himself misled unless he has consulted the relevant Government information to which he is directed by the contract, specifications and invitations to bid. As we read them, the decisions of the Supreme Court and of this court do not permit the contractor to rest content with the materials physically furnished to him. (Flippin Materials v. U.S.)

Information outside the Contract Documents

Although a contractor is responsible for reviewing all readily available sources, information that is not part of the contract documents cannot be considered a positive representation. Only information provided in the contract documents is considered a positive, material statement.

In Foundation Co. v. State of New York, the state issued a contract requiring that caissons be sunk to bedrock. The plans did not show the expected depth of the caissons but did give an estimated quantity of material to be removed to reach bedrock. The contract did not state that borings had been made when, in fact, wash borings were taken and showed bedrock at about elevation 148 feet above sea level. Before bidding, the successful contractor learned of the borings and requested copies, which the state supplied. When the work began, bedrock was found much deeper than elevation 148. The contractor filed a claim requesting additional monies, arguing that the borings misrepresented the bedrock to be at elevation 148. The court, in denying the contractor's claim, stated,

> ...although damages might be recovered from the State for misrepresentations, upon which the bidder might rely, the boring sheet was not such a representation. It formed no part of the plans upon which the contract was based. It was not prepared or used for that purpose. It was an independent bit of information or supposed information in the possession of the State, to which the bidder resorted in making the investigations which it was required to make. If it relied upon this paper, it did so at its own risk. The most it could ask for in regard to this information was good faith. (Foundation Co. v. State of New York)

Site Visits

A common defense of owners is the site visitation clause that requires the contractor to become familiar with the site and local conditions. When a site visitation clause exists, courts require contractors to perform the visits in a reasonable manner. However, this requirement does not extend to making an independent subsurface investigation unless specifically directed to do so by specific contract clauses. Hollerbach v. United States is a landmark Supreme Court decision with respect to site visits. The court, referring to the site visitation clause, said,

> We think it would be going quite too far to interpret the general language of the other paragraphs as requiring independent investigation of facts which the specifications furnished by the government as a basis of the contract left in no doubt. (Hollerbach v. U.S.)

The Supreme Court stated in this case that contractors are not required to perform independent subsurface investigations. If the owner desires that the contractors make independent investigations, the contract should direct them to do so in clear, unambiguous terms.

Nevertheless, site visits must be performed in a reasonable and prudent manner. In Warren Brothers Co. v. New York State Thruway Authority, the contractor should have observed 12- to 18-in. rocks along a shoulder of a highway to be repaired. The contractor bid the job expecting rocks no larger than 6 in. In denying recovery, the court stated,

> Furthermore, it appears that an appropriate inspection of the job site by claimant (contractor), a requirement imposed by the proposal and contract, would have revealed the actual condition had not such an inspection been confined to driving along the highway in an automobile.

Case Study

Intent and Other Available Information

Suppose a contractor is contracted to construct a seawall along a tidal coastline. The plans shows on page 1 a diagrammatic detail of the seawall as shown in Fig. 10-2A. There was no differing site condition clause.

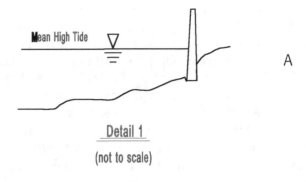

Detail 1

(not to scale)

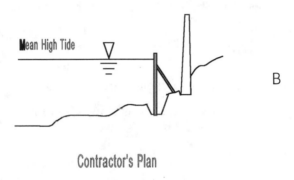

Contractor's Plan

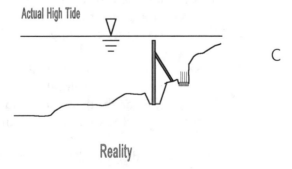

Reality

Figure 10-2. Diagrammatic Detail of a Seawall.

The contractor planned to construct a timber barrier as shown in Fig. 10-2B. The work proceeded as planned until one weekend, the tides were higher than expected, and the water overtopped the barrier and flooded the work site. This is shown in Fig. 10-2C. The contractor sought payment for the remedial work he had to perform. The basis of his argument was that the contract documents (Fig. 10-2A) represented a certain mean high tide, which he had scaled from Detail 1. The height of the barrier had been determined from this drawing.

What is the intent of Detail 1, and does it constitute a representation that the tide will not exceed a certain height? What features of Detail 1 are important in expressing its intent? Are there dangers in adding too much information on the plans to a drawing like Detail 1? What are the sources of other readily available information that would reduce reliance on Detail 1? Should the contractor recover his added costs?

Case Study

Site Investigation and Disclaimer

Green Construction Co. v. Kansas Power and Light Co.
1 F.3d 1005 (1993)

The Kansas Power and Light Co. (KPL) solicited bids for the construction of an earthen dam to create a reservoir at one of its power plants in Kansas. The dam was to be built out of clay found at the project site. KPL provided bidders with a geotechnical report of the subsurface conditions at the site but instructed the bidders to make their own investigation because there would be no future adjustment in price for unforeseen conditions. The contract contained the following:

> A.7 LOCAL CONDITIONS. Each bidder shall visit the site of the work and thoroughly inform himself relative to construction hazards and procedure, labor, and all other conditions and factors, local and otherwise, which would affect the prosecution and completion of the work and the cost thereof...
>
> It must be understood and agreed that all such factors have been properly investigated and considered in the preparation of every proposal submitted, as there will be no subsequent financial adjustment, to any contract awarded thereunder, which is based on the lack of such prior information or its effect on the cost of the work.

A.8 SUBSURFACE CONDITIONS. The determination of the character of subsurface materials ... shall be each bidder's responsibility. Borings, field testing, and laboratory tests have been performed for the project design. This information has been bound separately and is issued with these specifications. Rock cores from this site are available for inspection upon request at the Jeffrey Energy Center. Logs of test borings may not be indicative of all subsurface conditions that may be encountered.

Green Construction Co. was the successful bidder, but it did not conduct an investigation. The contract required Green to construct the dam with a moisture content that would yield the greatest strength. The allowable moisture range was +3% to -2% from optimum.

Green began work in the spring of 1985 and found that the soil in the borrow area contained a higher moisture content than indicated in the geotechnical report. The dam was finished in June 1986, and within a few weeks, cracks developed in the dam. Green tried unsuccessfully to fix the cracks. KPL refused to accept the dam and withheld $420,000 from the contract price.

Is Green entitled to the $420,000 because KPL misrepresented the moisture content?

The U.S. Court of Appeals addressed a number of arguments presented by Green. Regarding the misrepresentation, the court agreed with the district court, which found that when the contract expressly instructed bidders to conduct their own investigation of the site, the contractor bears the risk of excessive moisture content. This view was reinforced where the contract stated that the logs of test borings may not be indicative of all subsurface conditions that may be encountered. This case illustrates the importance of carefully reading the contract because the first sentence of paragraph A.8 is not especially direct, and the reader could easily overlook the language that makes this clause specific rather than general.

Contractor Experience

Courts have sometimes considered whether it was reasonable for the contractor to rely on the information, i.e., the contractor's experience. However, this criterion seems to be a "rule of last resort." An example is Morrison-Knudsen Co. v. United States. In this case, the court ruled that the contractor was not reasonable when it relied on two borings showing no permafrost to indicate that the entire site would not have permafrost. The court, in addition to other reasons, stated that the contractor should have known that some permafrost would be encountered because the contractor was "experienced in the area and had a general knowledge of the widespread, though discontinuous, existence of permafrost" (Morrison-Knudsen Co. v. U.S.).

Insufficient Time to Investigate

Several cases were identified in which the owner failed to provide adequate time for the contractor to verify the information provided (Kiely Construction Co. v. State of Montana; Alpert v. Commonwealth of Massachusetts; Haggard Construction Co. v. Montana State Highway Commission). In Haggard Construction Co. v. Montana State Highway Commission, the testimony revealed that the state expected the contractor to rely on the information in spite of a disclaimer that required the contractor to verify the conditions for itself. By allowing the contractor only 14 days to prepare its bid, the state induced a lower bid. In all the cases examined where there was inadequate time to verify the information provided, the owner was found to be liable.

Relevant Information Withheld

Another form of misrepresentation occurs when relevant information is withheld from the contractor.

Was Relevant Information Withheld?

In Warner Construction Corp. v. City of Los Angeles, the court stated,

> It is the general rule that by failing to impart its knowledge of difficulties to be encountered in a project, the owner will be liable for misrepresentation if the contractor is unable to perform according to the contract provision.

The court further stated three instances that may lead to a misrepresentation.

In transactions which do not involve fiduciary or confidential relations, a course of action for nondisclosure of material facts may
arise in at least three instances: (1) the defendant makes representations but does not disclose facts which materially qualify the
facts disclosed, or which render his disclosure likely to mislead;
(2) the facts are known or accessible only to the defendant, and defendant knows they are not known to or reasonably discoverable
by the plaintiff; (3) the defendant actively conceals discovery from
the plaintiff. (Warner Construction Corp. v. City of Los Angeles)

Failure to disclose all relevant facts can occur when the soil-boring
data are given in the contract documents, but particular information
about the borings is not provided or made available to the contractor. In
Christie v. United States, the U.S. government did not disclose in the soil
report that buried logs had been found during the boring operation.
When the drilling rig hit an obstruction, the crew moved the rig to another location where the boring could be completed. This new boring
location was subsequently recorded as if it were the planned position.
No mention was made of the unsuccessful borings. During construction,
the contractor discovered numerous buried logs and cemented sand and
gravel that greatly increased the cost of doing the work. Although the
boring logs were correct in what was presented, the court ruled in favor
of the contractor because vital information had been withheld.

Sometimes soil reports are purposely concealed from the contractor.
This situation is likely to lead to recovery of damages. In Valentini v.
City of Adrian, the city took borings that revealed quicksand along the
route of a proposed sewer. This information was never given to the con-
tractor, nor was the contractor informed that the borings had been made.
In awarding the contractor damages, the court stated,

> ... the city, through its consulting engineers had knowledge of
> the unfavorable subsurface conditions; that these conditions
> were not made known to the plaintiff (contractor); that as a result
> of encountering these unfavorable subsurface conditions of
> quicksand and excessive water, plaintiff's construction of the
> sewer was delayed and resulted in the greatly increased costs of
> construction.... (Valentini v. City of Adrian)

For a withholding to occur, the information has to be within the
knowledge of the owner or agent. The owner cannot be held liable for

information that it does not know exists. Owners are not required to make exhaustive searches of old records and interview all of their employees; rather, the information withheld must be known to the people actually involved in the construction. "The law puts no affirmative duty on public officers to search through old files for plans of existing structures before contracting..." (Annotation 1978).

Was the Contractor Misled?

The contractor must prove that the bid would have been significantly different if the information had been furnished. Although no dollar amount can be cited, courts use phrases like "greatly increased costs of construction" (Valentini v. City of Adrian) and "far more difficult and expensive to penetrate and excavate" (Christie v. U.S.). The difference must be substantial and not inconsequential.

Should the Contractor Have Known of the Condition?

The same steps that make a contractor not justified in relying on a representation make it aware of an unstated condition. These steps are (1) interpretation of the contract as a whole, (2) site visits, (3) other readily available information outside the contract, and (4) contractor experience.

An example of an inadequate site visit is found in Wiechmann Engineers v. California State Department of Public Works, in which a boulderous condition was found at the job site. A soil report that showed the boulderous condition was not included in the contract documents but was available for review on request. The contractor did not request a copy of the report even though the contractor knew it existed. Also, the boulders were readily apparent from a visual inspection of the site. The court ruled that the contractor should have known of the condition. The court, in ruling against the contractor, stated,

> ... knowledge of the boulderous condition was not known or accessible only to the State, nor did the State have such facts as were not known or reasonably discoverable by plaintiff (contractor), if plaintiff had made what would have been admittedly a reasonable and prudent inquiry. (Wiechmann Engineers v. California State Department of Public Works)

Case Study

Representations and Withholding

Jacksonville Port Authority v. Parkhill-Goodloe Co. Inc.
362 So.2d 1009 (1978)

The Jacksonville Port Authority issued plans to dredge the St. Johns River to a depth of 38 ft below the mean low tide. The contract provided for compensation to a depth of 40 ft. During the bidding period, the port authority issued Addendum No. 1, which contained 13 pages consisting of seven core borings made by Law Engineering and Testing Co. These borings indicated that the materials likely to be encountered within the 38-ft depth were sand, silt, and limestone fragments. One boring showed rock at 37.1 ft, and two others showed rock at 40 ft. Parkhill made 10 less sophisticated probings in the area, and these appeared to confirm the boring information supplied by the port authority.

The contract stated that the boring information furnished *"is not guaranteed to be more than a general indication of the materials likely to be found* adjacent to holes bored at the site of work approximately at the location indicated." In fact, the boring locations did not show that the underwater rock structure was in peaks and valleys with peaks extending into the area to be dredged.

Parkhill had prior experience in an area immediately east of the work area. There it had scraped over rock in dredging to a depth of 36 ft. There was no difficulty in reaching the 36-ft depth. Another company had also dredged to a depth of 37 ft in the same area east of the project. That company had to dredge an average of 7 or 8 ft of rock. That contractor had filed a claim after the work was complete. This information was not provided to any of the bidders on the present project.

Was the subsurface information misrepresented, and did the port authority withhold relevant information from Parkhill?

The Second District Court of Appeals of Florida provided detailed discussion on the representations in the boring information. On this point, the court stated,

> The negative language of the contract quoted above (referring to the disclaimer) constitutes a guarantee that the boring information furnished by (the port authority) gave a general indication of the materials likely to be found adjacent to holes bored at the site of work approximately at the locations indicated. The information

failed to give such general indication. We, therefore, must consider the legal effect of the giving by (the port authority) of this misleading information to bidders.

With regard to the presence of rock, there may possibly be some legitimacy in the argument that Parkhill should have known that rock might be encountered. However, the court demonstrated a widely held view that withholding of any information will often defeat the owner's position.

In furnishing bidders with information to the nature of the materials likely to be encountered in dredging the required depth, (the port authority) has a duty to furnish information which would not mislead prospective bidders and to not withhold from prospective bidders information that another contractor, in adjacent area, had encountered extensive rock in dredging to the (same) depth required by (Parkhill).

Courts seem to favor a contractor's position whenever there are appearances that the owner failed to impart all its knowledge, even in situations where other information may have alerted the contractor to the presence of the hidden conditions. In this case, the court ruled in favor of Parkhill. Although not addressed by the court, it appears from the review of numerous similar cases that the duty of the owner to provide all known information is greater when there is no DSC clause compared to situations where the contract contains a DSC clause.

Illustrative Example

The following example illustrates the use of the decision criteria shown in Fig. 10-1. The example is based on the decision in Con-Plex v. Louisiana Department of Highways.

Statement of the Facts

In March 1973, the Louisiana Department of Transportation (DOT) advertised bids to construct a new bridge across the Intracoastal Waterway in Calcasieu Parish, Louisiana, and to remove an existing pontoon bridge. Con-Plex (the contractor) obtained a set of plans and specifications and conducted a contractually required on-site inspection. Con-Plex prepared and submitted its bid in mid-May and, being the lowest qualified bidder, was awarded the job. A formal contract was executed on June 8, 1973.

The contract specifications stated the following:

> 102.05 EXAMINATION OF PLANS, SPECIFICATIONS. SPECIAL PROVISIONS, AND SITE OF WORK. The Department will prepare full, complete and accurate plans and specifications giving such direction as will enable any competent contractor to carry them out. The bidder is expected to examine carefully the site of the proposed work, the proposal, plans, specifications, supplemental specifications, special provisions and contract forms before submitting a proposal. The submission of a bid shall be considered prima facie evidence that the bidder has made such examination and is satisfied as to the requirements of the . . . contract.
>
> 105.04 COORDINATION OF PLANS, SPECIFICATIONS, SUPPLEMENTAL SPECIFICATIONS, AND SPECIAL PROVISIONS. (The contract documents) are intended to be complementary and to describe and provide for a complete project.

Additionally, Plan Sheet 207, showing details of the pontoon bridge, provided the following language:

> (1) For information purposes only
>
> (2) Additional details of existing pontoon bridge may be obtained from the Bridge Design Section
>
> C. Special Provisions:
>   The contractor shall familiarize himself with the conditions at the site with regard to all pilings to be removed. . . .

Over the years, the pontoon bridge had been damaged on several occasions, and all or almost all of the original pilings had been replaced without removing the damaged pilings. Many of the damaged piles were below the water surface. The additional details from the bridge design section gave no indication that the existing piles were still in place. The contract did not contain a DSC clause. On commencing removal of the existing pontoon bridge, it became apparent to Con-Plex that the plans and specifications did not accurately reflect the number or size of pilings to be removed. Because of the extra removal work required, Con-Plex requested additional compensation from DOT.

Analysis

The relevant questions from Fig. 10-1 are addressed below.

Was There a Positive Representation?

If the contract documents made no mention of the number and condition of the pilings but merely ordered removal of all deteriorated pilings, there would be no misrepresentation. Plan Sheet 207, however, contained a positive factual representation as to the number and character of the piles that Con-Plex contracted to remove.

Was There Intent to Deceive?

No intent to deceive is apparent.

Did the Conditions Differ from Those Represented?

There is little doubt that the number and character of piles that the contractor encountered were materially different from what was stated in the contract documents. Thus, the owner's position is substantially weakened, and the owner must rely on other issues to avoid liability.

Was the Representation Complete?

Yes, no information was withheld.

Was the Contractor Misled?

The contractor should have little difficulty in establishing that it was misled, especially because the contract was probably a unit-price contract.

Was Reliance on the Information Justified?

The pivotal factor in this dispute is whether the state had within its possession readily available information that showed the deteriorated piles. Based on the decision in Warner Construction Corp. v. City of Los Angeles, facts that "are known or accessible only to the defendant (owner), and the fact that the defendant (owner) knows they are not known to or reasonably discoverable by the plaintiff (contractor)" establish grounds for the contractor to recover his or her additional expenses. If the additional information from the bridge design section showed that the deteriorated piles were still in place, then the contractor's position would be substantially weakened to a point where he or she probably would not recover the added costs. If this element could not be shown, the contractor should prevail. The site visitation and disclaimer clauses cannot negate the obvious misrepresentation and the owner's superior knowledge of the latent conditions, unless it can be established that the contractor could have discovered the conditions during a routine site visit. The contractor is not expected to do extensive or underwater investigations unless specifically required to do so by the contract documents.

Suppose that the owner did not know about the deteriorated piles. This unlikely scenario assumes that the owner's preconstruction investigations did not reveal their existence. In this instance, the owner would probably prevail because the owner cannot be liable for facts or conditions that it does not know exists (Annotation 1978).

This outcome is typical of many misrepresentation disputes involving geological and other subsurface phenomena and illustrates the difficulty contractors face in seeking added expenses. It also illustrates that whereas the rules of law are consistent, the outcome is sensitive to the facts.

Other Issues

Role of the Soil Report

The soil report, which is seldom included in the contract documents, is often the center of controversy. However, when the contract does not contain a DSC clause, the role of the soil report is quite limited. Only information included in the contract documents provides a basis for a positive factual representation. Thus, the soil report can be used only to help clarify facts given in the contract documents.

The boring logs completed by an owner or architect, although not part of the contract, can sometimes be construed to be an affirmative representation.

Role of Site Visit Clauses and Disclaimers

The review of cases shows that disclaimers receive considerable discussion in most misrepresentation decisions. However, in no case was the disclaimer allowed to prevail where a positive representation existed. Instead, most cases where the disclaimer was discussed in depth were similar to the MacArthur (MacArthur Brothers Co. v. U.S.) and Wunderlich (Wunderlich v. State of California) cases in that the assertion of a positive representation was quite weak or did not exist. An analysis of these and other cases shows that, despite the rhetoric, disclaimers are narrowly construed.

The contractor must be allowed sufficient time to verify the information provided (Kiely Construction Co. v. State of Montana; Alpert v. Commonwealth of Massachusetts). In Raymond International, Inc., v. Baltimore County, the court noted that the contractor could not verify four years of periodic underwater inspections by Baltimore County and thus was justified in relying on the information provided by the owner. The court felt that expecting the contractor to conduct diving tests to verify the information in the plans and specifications was too burdensome.

The analysis also uncovered several cases where there were ambiguities that were supposedly ruled against the drafter (owner). The case of Haggard Construction Co. v. Montana State Highway Commission is instructive. At issue were statements in the contract that borrow material from a certain source would be adequate in terms of quantity and quality. The state unsuccessfully argued that the disclaimer specifically exonerated the state from responsibility for the accuracy of the information. A synopsis by Vance summarized the view of the court (Vance 1978, p. 1485).

> Statements concerning materials available were deemed to be only suggestive or "merely indications" in the California case (Wunderlich), whereas definite assertions as to matters of fact were found to have been made in Haggard. (Haggard Construction Co. v. Montana State Highway Commission)

As further stated by Vance, the issues were narrowed to a consideration of whether exculpatory language can be a valid defense against positive assertions in the contract. The court rejected the state's defense and awarded damages to the contractor. This view seems to reflect the prevailing judicial attitude, depending on the jurisdiction, that owners cannot rely on disclaimers to avoid liability for positive assertions as to matters of fact.

Exercise 10-1: J. A. Thompson & Son and State of Hawaii

J. A. Thompson and Son, Inc., a California corporation, was awarded a contract by the state of Hawaii to construct a four-lane divided highway over a portion of Kalanianaole Highway beginning at the Kailua junction and extending toward Waimanalo for a distance of approximately 7,000 ft. The contract price for the job was $564,989.45. Of that sum, the bid price for excavation was 54.6 cents per cubic yard for 287,000 yard^3, or a total of $156,702. Thompson visited the site as required by the bid documents. Outcroppings of rock could be observed near where Thompson later found hard rock.

During construction, Thompson encountered solid rock in the area designated on the plans at about Station 49 + 50, where test hole No. 6 had been drilled. The log for hole No. 6 disclosed the presence of basalt boulders, commonly called "blue rock." The term "blue rock" is used to denote very hard rock.

The contractor provided the following comparison of the actual boring logs and the logs shown on the plans for Hole No. 6:

MEASURE	ACTUAL BORING LOG	LOG AS SHOWN ON PLANS
0–15 ft	Hard, dry red clay with decomposed lava rock	Red clay with decomposed lava rock
15–30 ft	Hard, red and yellow-gray-black clay, slightly plastic, with decomposed lava rock	Slightly plastic, red and yellow-gray-black clay with decomposed lava rock
30–32 ft	Brown, yellow, slightly plastic, damp clay with medium-hard decomposed lava rock and little red and black clay	Slightly plastic, brown-yellow clay with decomposed lava rock and red and black clay
32–40 ft	Little, hard, red clay with hard basalt boulders	Red clay with basalt boulders

MEASURE, CONT.	ACTUAL BORING LOG, CONT.	LOG AS SHOWN ON PLANS, CONT.
40–80 ft	Firm, brown clay with hard basalt boulders or cracked basalt clay	Brown clay with basalt boulders or cracked basalt strata with red strata mixed with very red clay

The contract contained the following provision:

> 2.4 Mass Diagram—If a mass diagram has been prepared for a project, it will be available to the bidders upon the following conditions.

The swell or shrinkage of excavated material and the direction and quantities of overhaul as shown on the mass diagram were stated for the purpose of design only, and in like manner as specified in Article 2.3 above, concerning the furnishing information resulting from its subsurface investigation, the department assumed no responsibility whatever for the interpretation or exactness of any of the information shown on the mass diagram and did not, either expressly or by implication, make any guarantee of the same; the department reserved the right to change the direction and quantities of overhaul and the swell or shrinkage factors shown on the mass diagram, and no additional compensation was allowed by reason of such changes, except as otherwise provided in Article 4.3 of these specifications.

The mass diagram was made available to Thompson. The calculations thereon were based on a shrinkage factor of 23%. Thompson claims that the contract represented that the excavated material would shrink by 23%. In fact, the material swelled by 48%. This differential amounted to 118,000 yard^3 of excess material.

The contract contained no DSC clause.

Is Thompson entitled to an equitable adjustment?

Exercise 10-2: Williams-McWilliams Co., Michigan Wisconsin Pipeline Co., and U.S. Government

This dispute arose when Williams-McWilliams Co., dredging under government contract in Atchafalaya Bay, encountered and damaged a natural gas pipeline belonging to Michigan Wisconsin Pipeline Co.

Michigan Wisconsin and Williams-McWilliams now seek damages from the U.S. government for injury to the pipeline. Both contend that liability should lie with the United States for furnishing faulty specifications that failed to show the presence of the pipeline and on which Williams-McWilliams relied.

Congress provides that no structure shall be built in navigable waters "except on plans recommended by the Chief of Engineers and authorized by the Secretary of the Army" (Rivers and Harbors Act of 1899 Sec. 10, 33 U.S. Code Sec. 403). The authority to permit construction is delegated to division and district engineers (33 Code of Federal Regulations Sec. 209.120 et seq.). This authority is exercised by the New Orleans District of the U.S. Army Corps of Engineers. The Permits and Statistics Branch of the New Orleans District processes applications to do construction in navigable waters within its jurisdiction. It gives public notice describing prospective construction and inviting public comment or protest, issues permits for approved construction, and retains file copies of construction permits. In 1965, the Placid Oil Co. applied to the engineers for permission to construct a 20-in. natural gas pipeline extending from the Eugene Island area offshore, inland at St. Mary Parish to Patterson, Louisiana. After routine processing, the application was approved and the pipeline was constructed.

Copies of permits for construction projects are filed at the permit section of the U.S. Army Corps of Engineers' New Orleans offices. The public may consult the files to learn of the existence and location of underwater structures. The permits are filed according to waterway names; most permits involve only one waterway and are filed under that name. Some permits, however, often including those concerning pipelines, involve several waterways. According to Charles W. Decker, Chief of the Permits and Statistics Branch of the New Orleans District, it is the policy of the engineers to discourage filing applications for separate permits for each waterway crossed by such a structure. Instead, the engineers encourage applicants to include all navigable waterways crossed by a proposed pipeline in one application. A single permit for all crossings is issued and filed under the name of one major waterway crossed by the pipeline. This was done in the case of the Placid Oil 20-in. line. Although the permit section's file on that line indicates that it crosses at least 12 waterways, a single permit was issued for all of them. That permit was filed under the heading "LTAV"—LT for the Louisiana–Texas section of the Intracoastal Waterway, and AV for the portion of the Waterway between the Atchafalaya and Vermillion Rivers. Only this single copy of the permit was filed by the engineers. According to Decker, no cross-reference index of the copies of the permits is maintained by the permit section as to any other waterways.

One waterway crossed by the Placid Oil line was not clearly shown on the application drawings and was not listed among the waterways to be crossed in the public notice of proposed construction. This was the Atchafalaya Fairway, containing the Atchafalaya River channel, which extends into Atchafalaya Bay to the south of St. Mary Parish. Nothing either in the 20-in. pipeline's permit file or in the public notice explicitly indicates that the pipeline crossed the channel. This omission was particularly misleading, according to Decker, because of the distance between the Intracoastal Waterway and the Atchafalaya Fairway in the bay. Decker states that someone familiar with the permit section's single-permit filing system who found no permits under a particular heading might cross-check for permits under the headings of nearby waterways, but that in the case of waterways as far apart as the Intracoastal Waterway and the Atchafalaya Fairway, such cross-checking was unlikely.

In 1967, ownership of the 20-in. pipeline was transferred from Placid Oil Co. to the Michigan Wisconsin Pipeline Company. Soon afterward, Michigan Wisconsin applied for a permit to construct a 30-in. natural gas pipeline parallel to the 20-in. pipeline already in place. The public notice that was issued concerning the 30-in. pipeline did contain a reference to the Atchafalaya Fairway; however, no new permit was issued for the 30-in. line. Instead, the existing permit for the 20-in. line was amended to include the 30-in. line. No other filing concerning the new 30-in. line was made at the permit section.

The permit section received a "completion letter" giving notice of completion of the 30-in. line on May 7, 1971. It was the practice of the permit section after such construction had been completed to wait until the construction had been inspected and then to forward the completion notice to the Engineering Division. The Engineering Division includes the Service Branch and is the division in which specifications and drawings are prepared for construction projects led by the engineers. When the Service Branch receives such completion notices, its employees chart the new construction on "base maps" that are used when specification drawings are made up. At the time of the accident, the completion letter for the 30-in. Michigan Wisconsin pipeline was still in the hands of the inspector. The Survey Section did not receive the letter until October 18, 1971. Therefore, construction of the additional pipeline was not yet shown on the base maps at the time of the accident.

In 1971, the New Orleans District of the U.S. Army Corps of Engineers, the same district at whose offices the Michigan Wisconsin pipeline construction permit was on file, let a contract for the biyearly maintenance dredging of the Atchafalaya River channel in Atchafalaya Bay. The engineers prepared specifications and distributed them to prospective bidders. Included among the specification documents was a drawing of the area to be dredged. This drawing was prepared by George Meyn, an employee of the Waterways Section of the New Orleans office of the engineers. Meyn visited the permit section to see if the permit section files contained any permits for construction crossing the channel to be dredged. He told the permit section employees that he wished "to see all of the permits on the project" and that they showed him the file headed "Atchafalaya River—Morgan City to the Gulf." Because Meyn was familiar with the permit section files, he received no further assistance from permit section employees. He spent "two to three hours" examining the files and found no pipeline construction permits therein. The reason Meyn did not find the permit involved here was that it had been filed by the engineers under an Intracoastal Waterway heading (LTAV) rather than an Atchafalaya River heading. Accordingly, the drawing that he prepared and included in the specifications did not show the Michigan Wisconsin pipelines, which lay across the Atchafalaya River channel.

Williams-McWilliams submitted the low bid and was awarded the contract to dredge the channel. Attached to the contract and made a part thereof was a copy of the specifications prepared by the U.S. Army Corps of Engineers, including the drawing prepared by Meyn. The dredging contract provided in part that Williams-McWilliams was to position its dredge in the channel by electronic means. Offshore Raydist, Inc., was engaged by Williams-McWilliams to provide the electronic positioning. The contract also contained two "site inspection" clauses, placing on the contractor the responsibility "for having taken steps reasonably necessary to ascertain the nature and location of the work and the general and local conditions which can affect the work and the cost thereof" and requiring the contractor to acknowledge that he "has investigated and satisfied himself as to the conditions affecting the work."

A Williams-McWilliams engineer, accompanied by two U.S. Army Corps of Engineers field engineers and an Offshore Raydist technician, made a site inspection using a fathometer, which revealed no structures or other impediments to dredging on the bottom of the channel. This result conformed with Williams-McWilliams' experience of two years earlier, when the company dredged the same channel without mishap.

Williams-McWilliams began dredging on July 12, 1971, using the dredge *Arkansas*. The *Arkansas* was positioned by Offshore Raydist, as provided by the contract. Offshore apparently misinterpreted its half-scale chart, and as a result positioned the *Arkansas* 250 ft too far to the west. The *Arkansas* began dredging on a line 250 ft away from, and parallel to, the intended dredge line. On July 13, the *Arkansas* port spud encountered the Michigan Wisconsin 30-in. natural gas pipeline. The line was ruptured, and loss was incurred by Michigan Wisconsin.

Is the government liable for damages to the pipeline?

Exercise 10-3: P. T. & L. Construction Co. and State of New Jersey

This dispute involved a substantial contract for a small section of Interstate Route 78 as it passes through heavily built-up areas of Union County in Springfield, New Jersey. It involved 1.4 mi of construction where a multilane superhighway was cut under two heavily traveled local roads, Vaux Hall Road and Burnett Avenue (Fig. 10-3). P. T. and L. Construction Co., Inc., was to do site clearance and underground and roadway work; another contractor, Ell-Dorer Contracting Co., was to do the bridge construction.

At the work site, the south side of the road was bordered by single-family dwellings, the north side by mixed uses, primarily residential but including a large commercial development with paved parking areas. The roadway was an east–west cut within this built-up area with an existing downward slope to the west. The plan was to move extra soil from the east end of the project to the west end, to provide drainage both along and across the roadway, to bridge the superhighway for the two local roads, to finish the grade, and to pave the divided superhighway.

The contract was awarded on October 31, 1972, for a bid price of $9,337,584.45. P. T. and L. and Ell-Dorer commenced work on November 8, 1972. The contract called for completion by November 15, 1974. The contract was not completed until June 11, 1976.

Several details pertaining to the East Rahway River are germane to this dispute. A tributary of the East Rahway River cut across the site. Downstream, where the tributary met the East Rahway River at the Morris Avenue bridge, there was an obstruction that caused general water backup into the proposed construction site. The area frequently flooded and had

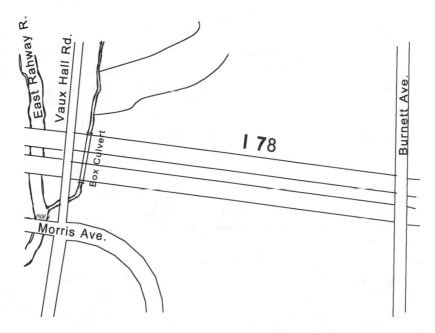

Figure 10-3. Diagram of roads in Springfield, New Jersey.

been the subject of frequent public meetings. In 1964, during initial planning for the project, a letter was written to the state by a local engineering firm that disclosed to the state that working conditions at the site would impose unusual difficulties for a construction contractor and that the "extent of the removal of the wet excavation . . . will depend on climatic conditions and the time of year . . ." This letter was not made available to the bidders. The U.S. Army Corps of Engineers was to have alleviated the backup problem by removing the obstruction and rerouting the streambed, but this project was canceled before design on I-78 was complete. The state had designed the project based on the corps's project to relieve the flooding. The bidders were not informed of this fact.

The contract contained several excavation items, including general excavation because of wet conditions, 6-in. stripping and 18-in. stripping to remove vegetation, and ditch excavation. The contract also called for Zone 3 fill material to be excavated from the project site and to be used in the roadway embankments, even though the DOT's standard specifications and accepted engineering practice are to require the use of more porous Zone 2 fill material in wet conditions or under water. The contract (P-1, Sheet 18) indicated by way of both pictorial description (i.e., an arrow) and the use of the word "flow" that a branch of the East Rahway River that traversed the project would furnish the drainage for the west end of the project. But, the contract did not specify either the construction of a cofferdam around the box culvert to be built through the west end, or the replacement of a stone base under the box culvert, both of which are required when a box culvert is to be built under wet conditions.

The contract did not contain a DSC clause. It did require that the contractor "make his own investigations of subsurface conditions prior to submitting his Proposal." Additionally, Article 1.2.12 relating to ground conditions required the contractor to agree to "make no claim for additional payment or extension of time for completion of work...because of any misinterpretation or misunderstanding of the contract, on his part, or of any failure to fully acquaint himself with all conditions relating to the work."

P. T. and L. made the required site visit. It was done at the end of an extraordinarily dry summer, and the tributary of the East Rahway River was nothing more than a dry riverbed. P. T. and L. saw low areas covered with vegetation and grass dry as a bone. P. T. and L. did not observe the downstream obstruction at the Morris Avenue bridge that caused the water to back up into the site.

The job was plagued from the start by poor working conditions. After the first heavy rain, water collected on the site, sometimes leaving the west end fill site 3–4 ft under water. This condition made the roadway excavation material unsuitable as fill material for the bridge and roadbed embankments, thereby delaying completion of the fills. In an attempt to drain the area, P. T. and L. built a temporary ditch before beginning construction of the box culvert. This work was largely unsuccessful. Sheeting was required to contain the porous fill behind the south wing wall of the Vaux Hall Road bridge; this construction delayed completion of the bridge for three and a half months from the time it took to recognize the problem until the remedy was complete. Change orders were approved by the state for the ditch and sheeting.

P. T. and L. stripped 9.87 acres an average depth of 2 ft. This work was almost 10 times the amount of stripping called for in the contract, and it took 171 days to complete rather than the 3 days originally allocated to the job. A major cost overrun occurred in the 18-in. stripping item of the contract. P. T. and L. now seeks reimbursement for its added costs.

Will the disclaimer protect the state from liability? Is P. T. and L. entitled to recover their extra costs?

Exercise 10-4: Charlotte, N.C., and Ruby-Collins

In June 1986, the city of Charlotte, North Carolina, solicited bids for the construction of a water main. The proposed water main would begin at the intersection of Tuckaseegee Road and Vanizer Street in northwestern Charlotte and end at the intersection of Sharon Amity and Providence Roads in southeastern Charlotte, a distance of approximately 6 mi. The city divided the water main project into three distinct segments. The city eventually awarded the entire water main project to Ruby-Collins, Inc., which submitted the lowest bid, $11.5 million. Ruby-Collins constructed the water main and received more than $12.5 million from the city for work performed, including change order work.

The city retained HDR Infrastructure, Inc., of North Carolina (HDR) to serve as the engineer for the project. As the engineer, HDR designed the water main and administered the contract between Ruby-Collins and the city. HDR in turn retained Soil and Material Engineers, Inc. (SME), to evaluate the subsurface conditions along the route of the pipeline, including assessing soil and groundwater conditions, providing design and construction guidance concerning backfill placement and compaction requirements within roadway areas, and preparing a report of subsurface explorations (hereafter "the SME report"). The city provided the SME report to potential bidders.

The design of the water main project required Ruby-Collins to install the water main mostly under existing, paved streets. The design, therefore, required Ruby-Collins to dig a trench, install the pipeline, and backfill the trench. Under the design's specifications, the backfill needed to be free of rocks, cobbles, roots, sod or other organic matter, and frozen material. It also needed to meet moisture-density requirements established by HDR. HDR also retained SME to determine if Ruby-Collins completed the backfill process in accordance with the specifications.

In computing its bid, Ruby-Collins reviewed the pipeline route, made sample borings during its own prebid site investigation, and studied the SME report. During its prebid investigation, Ruby-Collins did not evaluate the condition of the soil to be excavated regarding its suitability for use as backfill. Ruby-Collins assumed that the soil that it intended to excavate in digging the trench and to use as backfill would be sufficiently dry to compact immediately without processing. Ruby-

Collins based this assumption on the fact that the soil already supported the existing streets.

In computing its bid, Ruby-Collins did not include costs for hauling, storing, drying, or replacing most of the trench backfill. Ruby-Collins included in its anticipated costs, instead, the use of a conveyor belt that would allow the excavated soil from the trench to be loaded onto the conveyor belt, transported back along the pipeline route approximately 100 yard, and deposited immediately into the trench to cover the newly laid pipe. The process is called the "cut and cover" method.

Ruby-Collins thereafter submitted its bid, which the city subsequently accepted. After executing the contract, Ruby-Collins began construction of the water main. During the course of its performance, Ruby-Collins discovered that approximately 20% of the soil under the paved street had a moisture content that made the soil unsuitable for immediate use as backfill. Ruby-Collins regularly discussed with the city the problem of meeting the specifications regarding the backfill compaction; however, Ruby-Collins needed either to dry the soil or to haul in other backfill. Ruby-Collins eventually abandoned its "cut and cover" method and hauled other backfill material to the site. This change in method added considerably to Ruby-Collins' cost of performance.

Because of the moisture encountered during the construction of the water main, Ruby-Collins in October 1986 requested a change order encompassing the backfill problems. HDR reviewed Ruby-Collins' request for a change order and determined that the work was within the scope of the contract. By a letter dated December 15, 1986, HDR recommended to the city not to issue a change order.

The contract did not contain a DSC clause. Furthermore, there is no indication that any information in the SME report was incorrect. Paragraph 4 of "Information for Bidders" provides as follows:

> 4. GEOTECHNICAL INVESTIGATIONS
>   It shall be the Contractor's obligation to satisfy himself as to the nature, character, quality, and quantity of subsurface conditions likely to be encountered. Any reliance upon the geotechnical information made available by the Owner or the Engineer shall be at the Contractor's risk. The Contractor agrees that he shall neither have nor assert against the Owner or the Engineer any claim for damages for extra work or otherwise, or for relief from any obligations of this Contract based upon the failure by

the Owner or the Engineer to obtain or to furnish additional subsurface information in the Owner's or Engineer's possession or based upon any inadequacy or inaccuracy of the information furnished.

Certain subsurface information may be shown on separate sheets or otherwise made available by the Owner or Engineer to Bidders, Contractors, and other interested parties. Neither such information nor the documents on which it may be shown shall be considered a part of the Contract Documents or Contract Drawings, it being understood that such information is made available only as convenience, without expressed or implied representation, assurance, or guarantee that the information is adequate, complete, or correct, or that it represents a true picture of the subsurface conditions to be encountered, or that all pertinent subsurface information in the possession of the Owner or Engineer has been furnished.

Any holder of Contract Documents will be permitted to make test borings, test pits, soundings, etc., on the site of the work if he so desires subject to his first obtaining approval from the Engineer and the N.C. Department of Transportation. It is understood that the party or parties receiving such approval must assume all risks and liabilities contingent thereto.

It shall be the obligation of the Contractor to inquire of the Owner and Engineer whether pertinent subsurface information has been obtained by the Owner with respect to the Work.

Paragraph 7 of "Information for Bidders" provides as follows:

7. BIDDER'S RESPONSIBILITY

Each Bidder is responsible for inspecting the site and for reading and being thoroughly familiar with the Contract Documents. The failure or omission of any Bidder to do any of the foregoing shall in no way relieve any Bidder from any obligation in respect to his Bid.

Bidders must satisfy themselves of the accuracy of the estimated quantities in the Bid Schedule by examination of the site and a review of the drawings and specifications including Addenda. After Bids have been submitted, the Bidder shall not assert that there was a misunderstanding concerning the quantities of Work or the nature of the Work to be done.

Exercise 10-5: Pennsylvania Turnpike Commission and York Engineering & Construction Co.

The Pennsylvania Turnpike Commission, on April 17, 1939, advertised for bids for the construction of a certain section of the Pennsylvania Turnpike. The bids were due in the office of the Turnpike Commission on April 28, 1939. The York Engineering and Construction Co. obtained a copy of the plans and specifications and made a surface examination of that section of the turnpike, which was as complete an examination as could be made in the short period of time granted in which to file bids (11 days). The information contained in the plans and specifications indicated that the material to be excavated would consist principally of loose earth and approximately 50,000 yard^3 of limestone, visible at the east end of the section. The York Engineering and Construction Co. was awarded the contract and entered into a formal agreement to perform the work.

On the plans were set forth the estimated amounts of grading quantities. In the contract, it is set forth under "Instructions to Bidders":

> Each Bidder shall familiarize himself with all of the attached forms, Instructions, General Conditions, Specifications, Drawings, etc., as he will be held responsible to fully comply therewith. Each bidder must visit the site and acquaint himself with conditions affecting the work.

Section 36 of the contract provided

> The bidder's attention is called to the fact that the estimate of quantities of work to be done and materials to be furnished under these specifications, as shown on the proposal form, is approximate and is given only as a basis of calculation upon which the award of the contract is to be made. The Commission does not assume any responsibility that the quantities shall obtain strictly in the construction of the project nor shall the contractor plead misunderstanding or deception because of such estimate of quantities or of the character of the work, location or other conditions pertaining thereto. The Commission re-

serves the right to increase or diminish any or all of the above mentioned quantities of work or to omit any of them, as it may deem necessary.

Section 37 provided

> Wherever subsurface materials information is indicated on the drawings it is understood that it was obtained in the usual manner and with reasonable care, and the location, depths, and the character of the material have been recorded in good faith. There is no expressed or implied agreement that the depths or the character of the material have been correctly indicated and the bidders should take into account the possibility that conditions affecting the cost or quantities of work to be done may differ from those indicated.

There was no DSC clause.

Considering the magnitude of the project, four to eight weeks would have been required for a complete subsurface investigation, and York was, therefore, compelled to rely on the plans as to subsurface conditions. These plans showed a shrinkage factor for the fill that indicated that the character of the material to be excavated was largely loose earth because only loose earth shrinks when mechanically moved from one place to another. The engineers for the Turnpike Commission had investigated the subsurface conditions of this section and found it to be predominantly rock, which should have been shown on the plans because the fill had a swell factor instead of a shrinkage factor.

After the contractor commenced excavating, it was discovered that the subsurface, at the point of excavation, was limestone rock containing clay seams, and this condition prevailed at each subsequent cut into the subsurface. The additional expense of working this unanticipated type of subsurface material was much greater and, on account of the swell in the waste excavation, also increased its cost of removal.

Is the shrinkage factor sufficient to be a positive material representation? Will the disclaimers protect the turnpike commission from liability? Does the fact that rock was visible have any bearing on the outcome? Is York entitled to extra compensation?

Exercise 10-6: Utah DOT and Thorn Construction Co.

On March 27, 1973, the Utah Department of Transportation (DOT) contracted with Thorn Construction Company, Inc., for the construction of an access road at Rockport State Park, near Wanship, Utah. Before submitting its bid, several representatives of Thorn and a low-level Utah DOT representative, Virgil Mitchell, toured the site.

The standard specifications in Sec. 105.17 contained the following:

> If, in any case, where the contractor deems that additional compensation is due him for work or material not clearly covered in the contract or not ordered by the Engineer as extra work as defined herein, the contractor shall notify the Engineer in writing of his intention to make a claim for such additional compensation before he begins the work on which he bases the claim.

Before Thorn submitted its bid, several representatives of Thorn and a representative of the DOT toured the work site to inspect conditions and potential sources of borrow material. The DOT representative escorted Mr. Thorn and the others to the "Utelite property," which was close to the construction site. The DOT representative stated that the Utelite property was available and could be used for borrow. Several other sources of borrow material were discussed, but these sites were not visited. Thorn then computed his bid based on the cost of using the Utelite property.

After bids were received, the DOT conducted tests on the material, and it was found that the Utelite pit could not be used. Thorn then obtained borrow material from the "Crandall pit," which was 1.7 mi farther from the construction site than the Utelite pit. Thorn submitted a claim the basis of which is that the extra distance and certain other conditions at the "Crandall pit" increased its costs.

The standard specifications contained the following site visitation clause in Sec. 102.05:

> ... The bidder is required to examine the site of the proposed work, the proposal, plans, specifications, supplemental specifications, special provisions, and contract forms before submitting a proposal. The submission of a bid shall be considered prima facie evidence that the bidder has made such examination and is satisfied as to the conditions to be encountered in performing the

work and as to the requirements of the plans, specifications, supplemental specifications, special provisions, and contract.

Does the site visitation clause relieve the state from liability? What kind of representation was made by the DOT representative? Is Thorn entitled to more money?

Exercise 10-7: Indianapolis and Twin Lakes Enterprises

In 1988, the city of Indianapolis, Indiana, hosted the Pan American Games. In preparation for the games, the city sought bids for the construction of a rowing course on the Eagle Creek Reservoir. Twin Lakes Enterprises, Inc., was the successful bidder. The contract called for Twin Lakes to dredge the site of the rowing course of silt and sand by means of a suction hose to produce a flat, mud bottom in the bed of the reservoir. Thus, the contract documents defined the job as a sand and silt dredging operation.

In years past, there had been dumping of materials on land adjacent to the reservoir. The streambed for the creek near the dredge site was moved after these activities had begun. Thus, it came about that large obstructions had been dumped in the reservoir with the city's knowledge at the location which eventually became the dredge site, and the city had known of the large obstructions both before and after it entered into the contract with Twin Lakes.

A site visit was part of Twin Lakes's obligations, which it made. One of the city employees present during the site visit had knowledge of the obstructions, but nothing was said. This employee had told city officials that the suction method contemplated in the contract could not be used because of obstructions on the floor of the lake. None of this information was conveyed to Twin Lakes.

The site visitation clause stated,

> Bidders shall examine the site and thoroughly familiarize themselves with the site and all conditions in connection therewith. Lack of familiarity with the site and present conditions will not be considered as justification for changes or extra charges of any kind, since any contract, in whole or in part will be based on the assumption the bidder knows, understands and accepts these conditions.

The contract did not contain a differing site conditions clause.

As fate would have it, Twin Lakes could not use the suction method contemplated in the contract. It incurred considerable cost overruns. It was not until after Twin Lakes was terminated from the project that Twin Lakes learned that the city had knowledge of the obstructions.

What role does the site visitation clause play in this dispute? Should Twin Lakes be paid for its added costs?

Exercise 10-8: Post & Front Properties and Roanoke Construction Co.

In August 1988, Samuel Ashford of Post and Front Properties, Ltd. (P&F), met with the president of Roanoke Construction Company, Inc., to discuss the possibility of completing renovations to a building owned by Post and Front. At this meeting, the president of Roanoke asked Mr. Ashford how much money was left in his construction loan fund. Ashford told Roanoke he had $180,000 remaining. Roanoke determined that $180,000 was sufficient to complete the renovation and subsequently entered into an oral contract where Roanoke would act as general contractor. He was to be paid cost plus 10%.

Roanoke began work in September 1988 and soon thereafter learned that only $12,000 was left in P&F's construction loan account and that the bank was not going to allow P&F access to those funds. Roanoke also learned that the bank, in July 1988, had authorized foreclosure proceedings on the property because of the delinquent status of the construction loan. Roanoke soon thereafter terminated its work and invoiced P&F for $110,000. P&F refused to make payment.

Is there any intent to deceive? Is P&F liable to Roanoke?

Exercise 10-9: Ideker Corp. and Missouri State Highway Commission

Ideker Corp., Inc., was the successful bidder on a construction project let by the Missouri State Highway Commission on I-35 in Harrison County, Missouri. Plans and specifications for the project were prepared by the highway commission and relied on by Ideker in calculating and submitting its bid.

The contract contained three boilerplate provisions found in the contract, bid proposal, and standard specifications. The contract contained a provision which provided that the contractor was fully informed regarding all conditions affecting the work to be done by reason of its own investigation and not from any estimates of the highway commission. The bid proposal stated that the bidder "declares . . . that he has carefully examined the location of the proposed work. . . ." Section 102.5 of the standard specifications provides, in part, that "submission of a bid shall be considered proof that the bidder has made his own examination and is satisfied as to the conditions to be encountered in performing the work." The contract did not contain a differing site condition clause.

The plans prepared by the highway commission stated that the project was a "balanced" job in the sense that the profile of the grade of the highway to be constructed was designed so that the cuts were equal to the fills. The plans reflected a "shrinkage" factor of 1.28. On the plans, cuts and fills were segregated into twenty-one (21) designated areas, each known and referred to as a "balanced zone."

When excavated material removed from cuts exceeds that which can be accommodated by the fills in a balanced zone, the excess material, or waste, must be removed and disposed of, or "wasted," at some site other than the fills in the balance zone. The plans did not provide for any waste disposal areas. According to the highway commission, the project was "designed" and "intended" to be a "balanced" job on the basis of the shrinkage factor of 1.28. Ideker submitted its bid on the basis of a "balanced" job.

Shortly after commencing work on the project, it became apparent that the excavated material exceeded the fills and that considerable waste was regularly encountered. The highway commission made several changes in grade to help alleviate the waste encountered by Ideker. Unfortunately, voluminous amounts of waste continued to plague the project, and at completion, 355,937 yard[3] of waste had been disposed of by Ideker. In retrospect, the shrinkage factor was 1.13, instead of 1.28.

Ideker was paid at the contract unit price (40.5 cents per cubic yard) for all material excavated from the cuts. Ideker claimed $287,701 for damages for additional costs incurred in disposing of the waste.

Were the plans correct as shown? Is Ideker entitled to its claim?

Exercise 10-10: Waldinger Corp. and Daniel Hamm Drayage Co.

In 1976, the Caterpillar Tractor Co. entered into a general construction contract with the W. E. O'Neil Construction Co. for the construction of foundry buildings and a pollution control facility in Mapleton, Illinois. In early 1976, O'Neil entered into a subcontract with the Waldinger Corp. for the mechanical portion of the general contract. Waldinger was to design, construct, and erect certain pollution control equipment.

In April 1976, Waldinger hired FMC Corp. to design and construct pollution control equipment principally consisting of large clarifier tanks and operating mechanisms for use inside the tanks. Waldinger later hired the Daniel Hamm Drayage Co. to rig and assemble the FMC equipment.

Waldinger's contract with FMC provided for delivery of various items within a specified number of weeks after final approval of the design drawings by Waldinger and Caterpillar. For example, the internal mechanisms for the clarifier tanks were to be delivered 46 weeks after design approval. Waldinger approved FMC's design drawings in October 1976; delivery was therefore projected for September 1977.

By March 1977, Waldinger was behind schedule on its portion of the Mapleton project. O'Neil urged Waldinger to begin erection of the pollution control equipment, some of which was already at the job site.

Gary Nicholas was the Waldinger executive responsible for the Mapleton project. Greg Roth was Waldinger's manager of purchasing. At Nicholas' direction, Roth contacted Dave Garrett, then vice president and chief estimator for Hamm, to negotiate a bid for the erection of the pollution control equipment. Roth and Garrett met in Des Moines, Iowa, in March 1977. Roth indicated to Garrett that some of the equipment to be erected had already arrived at the job site, and that Waldinger was eager for erection of the equipment to begin as soon as possible. Although Roth provided Garrett with Caterpillar's specifications, Roth did not provide Garrett with a copy of Waldinger's contract with FMC. The FMC–Waldinger contract contained modifications to the Caterpillar specifications. For example, Waldinger's contract with FMC excluded the services of an FMC erection supervisor, although an erection supervisor was required by the Caterpillar specifications.

Although Roth had previously been advised by FMC that all equipment would not be on site until September 1977, Roth failed to so advise Garrett at the March 1977 conference. Instead, Roth advised Garrett that all equipment provided by FMC would be delivered no later than July 25, 1977.

Relying on Roth's representations as to the equipment delivery schedule, the availability of an FMC erection supervisor, and the match-marking of FMC's equipment, Garrett estimated that Hamm could complete work on its portion of the project within 18 consecutive work weeks. Roth and Nicholas concurred in this estimate, and Garrett prepared a bid accordingly.

In April 1977, after negotiations, Waldinger accepted Hamm's bid in the sum of $174,850. Garrett advised Roth that Hamm could begin work immediately. Based on Roth's representations as to the delivery schedule, Garrett anticipated that the remainder of the FMC equipment would be delivered to the job site by the time Hamm completed work on equipment already at the site. At the request of Jim Tippery, Waldinger's chief field representative for the Mapleton project, Hamm manned the project within a few days of Waldinger's acceptance of Hamm's bid. Therefore, Hamm expected to complete work on the project in mid-August of 1977.

Unfortunately, the equipment deliveries were not made as anticipated by Hamm. On June 9, 1977, Nicholas wrote a letter to FMC, confirming a telephone conversation wherein FMC had advised Waldinger that the FMC equipment would not be delivered to the job site until late October of 1977. Garrett received a copy of Nicholas' letter. In response to Garrett's subsequent complaints regarding the delays in deliveries, Tippery advised Garrett that Waldinger was insisting on, and expected, delivery earlier than October.

Final deliveries of the FMC equipment were not completed until late December 1977. Hamm had left the job site on December 7, 1977, because of extreme weather conditions and returned to the site in early January 1978. The midwinter deliveries resulted in further delays; Hamm did not substantially complete work on the project until April 1978.

Because of the five-month delay in the delivery of the FMC equipment, Hamm incurred expenses in the amount of $269,294. Hamm sought compensation, arguing that it would not have bid the project at the agreed-on price had it known of the delivery schedule.

Is there a positive material statement of fact? If the contract contained a DSC clause, would it apply in this instance? Should Hamm be paid?

Exercise 10-11: Pinkerton & Laws Co. and Roadway Express

The Pinkerton and Laws Co. ("P&L"), contracted in November 1981 to construct a motor freight terminal in Ringgold, Georgia, for Roadway Express, Inc., in accordance with plans and specifications prepared by Roadway's architect. P&L subcontracted the excavation, grading, and fill portion of the work to Jerome Bradford Construction Co.

Relevant portions of the contract documents provide as follows:

> Changes in Work: Changes in work may be ordered only upon written order from Owner. Cost or credit to Owner resulting therefrom shall be calculated in accordance with the method chosen by Owner pursuant to ARTICLE No. 1-18 of the General Conditions.
>
> 1-18. *Changes in the Work:*
>
> The Owner, without invalidating the contract, may order extra work, or make changes by altering, adding to, or deducting from the work, the contract sum being adjusted accordingly. All such work shall be executed under the conditions of the original contract, except that the claim for extension of time caused thereby shall be adjusted at the time of ordering such change.
>
> No extra work or change shall be made without a written order from the Owner, in which event the Contractor shall proceed with such extra work or change, and no claim for an addition to the contract sum shall be valid unless so ordered.
>
> 1.020 Special Conditions
>
> 2.2 *Examination of Site:*
>
> The Contractor shall, before submitting his proposal, examine the site, inform himself of the conditions and make his own estimates of the facilities and difficulties attending the execution of the work.
>
> Specifications
>
> 2.010 Soil Boring Logs
>
> 1.01 *General:*
>
> Included in this Section of the Specifications are copies of forty-four soil boring logs taken at the site. These logs are part of

subsurface investigations performed between February 9 and April 2, 1980, under separate contract by Law Engineering Testing Company, Atlanta, Georgia.

2.200 Earthwork

PART 1—GENERAL

1.01 *Scope of Work:*

A. Under this Section of the Specifications the Contractor shall furnish all labor, plant and materials required to complete the following general items of work:

1. Preparation of areas to receive fills and the constructions of such fills.
2. General excavation and site grading.
3. Finish grading including all slopes, ditches and subgrade preparation.
4. Excavations and backfilling.
5. Spreading of stockpiled top-soil.

B. All general site grading shall be completed before work is started on the building foundations.

The Contractor shall visit the site, inform himself of the conditions and make his own estimates of the facilities and difficulties attending the execution of the work.

3.03 *Fill Construction:*

A. Immediately prior to the placing of fill materials, the subgrade shall be compacted to a minimum density of 95% of maximum laboratory dry weight as determined by AASHTO Test Designation T-180 (Modified Proctor).

B. Fill materials shall be spread in uniform layers having a maximum thickness, measured loose, of eight inches. Each layer shall be compacted to the density herein specified before the next layer is placed.

C. Fill material shall be compacted at the optimum moisture content plus or minus three percent. The moisture content of each lift shall be adjusted by either aeration with a disk harrow or sprinkling as may be necessary to facilitate proper compaction.

D. The fill material shall be mechanically compacted to a minimum density of ninety-five percent (95%) of maximum laboratory dry weight as determined by AASHTO Test Designation T-180 (Modified Proctor).

The contract between P&L and Roadway did not contain a differing site conditions clause.

Before P&L submitted its bid, a company representative visited the site to observe the conditions. He walked approximately one-half of the site but did not walk through the areas that were covered with dense undergrowth. Before the contract was signed, P&L did not ask to see the other reports made by Law Engineering regarding its subsurface investigations that were not included in the contract documents. P&L did not conduct any soil tests at the site before signing the contract. However, P&L's initial bid letter contained the following qualifications:

1. Progress Payments—Owner shall be invoiced by the first of the month and shall make payment by the Tenth (10th) of the month.

2. Retainage shall not exceed 5% of the contract amount. This may be accomplished in various ways, for instance 10% for the first half of the project and none thereafter.

3. Modified proctor (95%) compaction on fill construction is contingent on residual soils moisture content of which there is no information given. Our quotation anticipates that this material falls within the acceptable range prior to the start of the construction.

Representatives of P&L and Roadway subsequently discussed these bid qualifications because Roadway refused to accept a qualified bid. P&L increased its price by $30,000 because Roadway would not change the payment terms to which P&L had objected. After Roadway's representative stated that there were no unusual soil conditions at the site, P&L removed the remaining qualifications.

When the work began, P&L began experiencing difficulty in achieving the required soil compaction because of excess moisture in the soil. As a result, the earthwork portion of the contract took longer to complete than P&L had anticipated. Because of these difficulties in achieving the proper soil compaction, Bradford abandoned the job in July 1982. P&L then completed the earthwork portion of the project. P&L never presented a written request for change order or an extension of time to complete the work.

In December 1983, after encountering problems with the construction of one wall of the terminal building, P&L asked Roadway to provide all information that Law Engineering had supplied to Roadway. Roadway furnished the remaining Law Engineering reports and documents to P&L some eight months later, in July 1984.

The additional Law Engineering documents included the following information:

> The laboratory results indicate the soils at the site are generally wetter than the optimum moisture content. Therefore, it appears that the site will require some drying out to achieve 95 percent compaction.
>
> During this investigation, extremely soft surface conditions were noted, particularly in the western half of the site. Extensive dozer assistance was required for our drill rig to obtain access to boring locations. This condition can be attributed to very heavy rainfall during March and the tightness of the residual clay which does not allow rapid drainage. Therefore, water tends to sit on the surface and soften a zone 12 to 18 inches deep. This surface zone consists of topsoil and probably a loose plow zone in many areas of the site. It is our experience that soft conditions can extend up to 3 feet in isolated areas where previous clearing has left stump holes which were eventually filled. We have indicated on the boring records where the water softened soils were deeper than about 18 inches.
>
> *Grading:* Prior to initiating fill placement in any areas, the organic surface soils should be stripped. Your plans for stripping the upper 12 inches appear to be adequate. However, after stripping, 6 to 12 inches of water softened soils are likely to remain in most areas, unless substantial site drainage has occurred. Such soft soils should be scarified in place, dried and recompacted. Scarification and drying can best be accomplished by a disc harrow which turns the soils numerous times during a period of several hours on a warm, sunny day. Turning the soils once will not suffice to air dry the wet clays. Therefore, it is important that site grading be accomplished during a dry season. Some undercutting may be necessary where wet soils extend too deep to scarify and recompact.
>
> Structural fill should be placed in maximum 8-inch lifts and compacted to at least 95 percent of the soil's maximum dry density as determined by the Standard Proctor compaction test. The

upper 12 inches of subgrade beneath pavements should be compacted to at least 98 percent. Soil moisture during placement
should be within 3 percent of the optimum moisture content.
Comparison of in-place moisture to optimum values of Proctors
#1 and #2 indicate that drying of the borrow soils will be neces-
sary prior to placement as structural fill. Provisions should be
made to scarify and dry sections of the borrow area prior to moving the soils into the fill areas. Effective use of grading equipment will be required to coordinate drying of one area while
hauling already dry soils from another area.

There is no dispute that P&L encountered excessive moisture in the
soil at the project site and that its performance was delayed and made
more difficult as a result.

How do you reconcile the fact that Roadway did not provide the Law
Engineering report but P&L never asked to see it? Is P&L entitled to recover its added cost as outlined in its claim?

Exercise 10-12: Flippin Materials Co. and U.S. Government

Flippin Materials Co. was part of a joint venture of nine large construction enterprises that built the Bull Shoals Dam in Arkansas for the U.S.
government. Flippin's role was to manufacture sand and crushed rock
from limestone found in a government-owned mountain (Lee Mountain) near the Bull Shoals area and to deliver this aggregate to the dam
site for use in the concrete required for the Bull Shoals Dam.

For coarse aggregate, Flippin was to be paid $2.41/ton, and the esti-
mated quantity was 2,700,000 tons. For fine aggregate, the price was
$2.54/ton and the estimate was 1,100,000 tons. Flippin was also to be
paid $0.30 per cubic yard for waste material stripped from above and in
and around the rock formations.

A major consideration in bidding on quarrying work of this kind is
the nature of the material underlying the surface of the quarrying area.
Limestone is a good source of rock for aggregate, and it was known to be
there. If the cavities in the limestone are empty or filled (partially or
wholly) with sand, the job is a normal one because sand does not hinder
the breaking up of limestone for the aggregate; but if the limestone cav-

ities are filled with clay, unusable for aggregate and contaminating the rock, there is much additional material to be stripped and wasted.

Before the contract was executed, the government drilled 33 holes in the mountain to ascertain the character of the underlying material. The government then supplied to Flippin the drawings of the profiles (or logs) of these borings, which showed cavities in the limestone formation by solid black markings; there was no statement, however, whether the cavities were void or wholly or partially filled with sand or clay. The actual cores of the borings were retained and were available to, and inspected by, Flippin. On the basis of the contract drawings and core-borings, as well as the other facts of which it was aware, Flippin had no reason to know that a large part of the subsurface material in Lee Mountain was contaminated with clay.

The plaintiff found much more clay-contaminated rock than (it now says) it expected or was led to expect and was therefore required to remove much more waste material. It also says that because of the clay it was forced to change from an operation at the side of the mountain to a more difficult and expensive one at the crest. Its claim is that it did not know and was not told of the probable extent and area of the clay in Lee Mountain but that the defendant did have the information (from its precontract test borings), which it failed to reveal to the plaintiff, as it should have.

The government also had field logs, recording the actual findings of the inspectors and geologist, which contained considerably more information than was shown on the contract drawings. From these field logs, it could readily be determined that a great many of the cavities shown in the contract drawings were clay-filled. The field logs were kept at the U.S. Army Corps of Engineers' field office, where the actual boring-cores were on display and inspected by Flippin's representatives; Flippin failed to make such a request.

At a prebid conference, Flippin was informed in writing that "results of explorations and tests, including cores from some of the borings, are available at the Bull Shoals suboffice, Mountain Home, Arkansas." Section SC-12 of the contract specifications similarly provided that the "results of all borings and tests, including samples of core, which have been made by the Government, of the materials contemplated for use under these specifications are available at the Bull Shoals suboffice, Mountain Home, Arkansas, for examination by the bidders."

How do you balance the failure of the government to impart all the relevant information it had in its possession with failure of the contractor to ask for this information? Should Flippin's claim be honored?

Exercise 10-13: Public Constructors and State of New York Department of Public Works

Public Constructors, Inc., and the State of New York Department of Public Works (NYDPW) entered into a highway construction contract dated November 10, 1965, for the construction of a portion of the new Route 17 in Delaware County. The contract provided for the construction of 5.68 mi of main road, 7.72 mi of access road, and four bridge structures. The plans were formally approved and accepted by appropriate NYDPW officers on September 21, 1965. The contract was advertised for bids in early October 1965, and final bids were received during the latter part of that same month, approximately three and a half weeks later. Public, as the low bidder, was awarded the contract on November 10, 1965, and began work on or about December 20, 1965. The amount of Public's bid was $11,856,562.50. The project was completed on January 15, 1969.

In 1952 and 1953, many years before the letting of this contract, NYDPW began subsurface explorations along the contract site. The results were transcribed into boring logs and forwarded to the state's Soils Bureau in Albany. The Albany Bureau also made its own on-site investigations in 1952, and in 1956, seismic tests were conducted at the site. In 1962, the state through its Binghamton field office conducted further on-site soils explorations, mainly the taking of 145 test borings. Analyses of these test borings were prepared and compiled by the unskilled field workers. (This compilation is referred to as the "Binghamton borings.") In 1963, the samples were inspected by skilled technicians at the Albany Soils Bureau, whose findings were written up and became known as the Albany laboratory logs. The 1952–1953 borings and the Albany laboratory logs were not released to Public Constructors. The 1952–1953 borings and the Albany laboratory logs presented a different picture of the contract site than did the Binghamton field borings, which, in addition to presenting information tending to indicate far fewer problems, were prepared by untrained staff members who failed to comply with the state's own specifications for recording the information.

When work began, Public encountered wet subsurface conditions that disrupted its excavation and embankment operations. When deep cuts were made, fill areas were not sufficiently extensive to permit the placing of excavated material in the required thickness of layers to dry out, and therefore when excavated material was wet, Public was compelled to cease operations, resulting in extensive delays. Public claims that the state provided incorrect and misleading information indicating that subsurface soil consisted primarily of coarse-grained material that had the capacity to shed moisture, whereas in fact the subsurface soil consisted primarily of fine-grained materials lacking the capacity to permit compaction in moist conditions, and that conditions of excessive moisture were present that the state failed to disclose. The effect of encountering these unanticipated conditions was to require Public to increase the scope and number of drainage structures and to change its earthwork operations in light of the embankment problems that it encountered.

The NYDPW relied on the position that adequate information was provided in the material furnished to bidders and that Public itself must bear the responsibility for not being aware of the conditions because of its failure to conduct adequate prebid, on-site investigations. The state further claims that it is excused from liability by virtue of a provision in the contract advising bidders that they were not entitled to rely on the accuracy of descriptions of subsurface conditions contained in bidding documents.

Is Public entitled to its claim?

Exercise 10-14: E. H. Morrill Co. and California Department of Public Works

The E. H. Morrill Co. and the Department of Public Works of California contracted in 1962 for the construction of the MonoInyo Conservation Facility in accordance with plans, specifications, and special conditions attached to the written contract. Special Condition 1A-12 provided

> SPECIAL SITE CONDITIONS. The site is situated on a terminal moraine. The soil is composed of granite boulders, cobbles, pebbles, and granite sand. Boulders which may be encountered in the site grading and other excavation work on the site vary in size from one foot to four feet in diameter. The dispersion of boulders varies from approximately six feet to twelve feet in all directions, including the vertical.

Section 4 of the General Conditions read, in part

> Examination of Plans, Specifications and Site of Work: The bidder shall examine carefully the site of the work and the plans and specifications therefor, and shall satisfy himself as to the character, quality, and quantity of surface and subsurface materials or obstacles to be encountered. He shall receive no additional compensation for any obstacles or difficulties due to surface or subsurface conditions actually encountered.
>
> If discrepancies . . . are found in the plans and specifications prior to the date of bid opening, bidders shall submit a written request for a clarification.
>
> Where investigations of subsurface conditions have been made by the State in respect to foundation or other structural design, and that information is shown in the plans, said information represents only the statement by the State as to the character of material which has been actually encountered by it in its investigation, and is only included for the convenience of bidders.
>
> Investigations of subsurface conditions are made for the purpose of design, and the State assumes no responsibility whatever in respect to the sufficiency or accuracy of boring or of the log of test borings or other preliminary investigations, or of the interpretation thereof, and there is no guaranty, either expressed or implied, that the conditions indicated are representative of those existing throughout the work. . . . Making such information available to bidders is not to be construed in any way as a waiver of the provisions of the first paragraph of this article and bidders must satisfy themselves through their own investigations as to conditions to be encountered.

Morrill contends that the special condition was false in that it misrepresented the true character of the site and that the boulders found were substantially larger and more concentrated than represented. It is further alleged that the state knew or should have known that the representation was false, because of the state's superior knowledge of the site. Although Morrill inspected the site, it alleges that it was unable to discover facts to contradict the representations, and that it relied on those representations in submitting the bid that became the basis for the contract.

Did the state misrepresent the site? Is Morrill entitled to its claim?

References

Alpert v. Commonwealth, 357 Mass. 306, 258 N.E.2d 755 (1970).

American Institute of Architects (AIA). (1987). *General Conditions of the Contract for Construction*, AIA Document A201. American Institute of Architects, Washington, DC.

Annotation. (1978). "Public Contracts: Duty of Public Authority to Disclose to Contractor Information Allegedly in Its Possession, Affecting Cost or Feasibility of Project," 86 ALR.3d 182, 246.

Ashton Co. v. State, 9 Ariz. App. 564, 454 P.2d 1004 (1969).

Christie v. United States, 237 U.S. 232, 239, 59 L.Ed. 933, 35 S.Ct. 565 (1915).

City of Salinas v. Souza & McCue Construction Co., 424 P.2d 921, 57 Cal.Rptr. 337, 66 Cal.2d 217 (1967).

Coatsville Contractors & Engineers, Inc., v. Borough of Ridley Park, 506 A.2d 862 (1986).

Con-Plex v. Louisiana Department of Highways, 439 So.2d 567, La.App 1st Cir. (1983).

C. W. Blakslee & Sons, Inc., et al. v. United States, 89 Ct.Cl.226, 246 (1939), cert. denied, 309 U.S. 659, 60 S.Ct. 512, 84 L.Ed. 1007 (1940).

Dravo Corp. v. Commonwealth Department of Highways, 564 S.W.2d 16, 19 (Ky.App., 1977).

E. H. Morrill Co. v. State, 423 P.2d 551, 554, 56 Cal.Rptr. 479, 65 Cal.2d 787 (1967).

Elkan v. Sebastian Bridge District, 291 F. 532, 538 (8 Cir. 1923).

Flippin Materials v. United States, 312 F.2d 408, 414, 160 Ct.Cl. 357 (1963).

Foundation Co. v. State of New York, 135 N.E. 236, 238/239, 233 N.Y. 177 (1922).

Haggard Construction Co. v. Montana State Highway Commission, 149 Mont. 422, 427 P.2d 686 (1967).

Hollerbach v. United States, 233 U.S. 165, 169, 43 S.Ct. 533, 58 L.Ed. 898 (1914).

J. A. Johnson & Son, Inc., v. State, 51 Hawaii 529, 465 P.2d 148 (1970).

Jervis, B. M., and Levin, P. (1988). *Construction Law Principles and Practice*, McGraw-Hill Book Company, New York, p. 139.

Kiely Construction Co. v. State, 154 Mont. 363, 463 P.2d 888 (1970).

Loulakis, M. (1986). "Disclaimers of Liability." *Civ. Eng.*, 56(10), 32.

MacArthur Brothers Co. v. United States, 258 U.S. 6, 66 L.Ed. 433, 42 S.Ct. 255 (1922).

Morrison-Knudsen Co. v. United States, 345 F.2d 535, 539, 540, 170 Ct.Cl. 712 (1965).

O'Neill Construction Co. v. City of Philadelphia, 6 A.2d 525, 528, 335 Pa. 359 (1939).

Raymond International, Inc., v. Baltimore County, 45 Md.App. 247, 412 A.2d 1296 (1980).

Restatement of the Law. (1979). "Contracts," 159 ALR.2d 424.

Sasso Contracting Co. v. State, 414 A.2d 603, 605, 173 N.J.Super. 486 (App.Div. 1980), cert. denied 85 N.J. 101, 425 A.2d 265 (1980).

Sweet, J. (1989). *Legal Aspects of Architecture, Engineering, and the Construction Process*, 4th Ed., West Publishing Company, New York.

Thomas, H. R., Smith, G. R., and Ponderlick, R. M. (1992). "Resolving Contract Disputes Based on Misrepresentations." *J. Constr. Engrg. and Mgmt.* 118(3), 472–487.

United States v. Atlantic Dredging Co., 253 U.S. 1, 12, 40 S.Ct. 423, 64 L.Ed. 735 (1919).

Valentini v. City of Adrian, 79 N.W.2d 885, 888, 347 Mich. 530 (1956).

Vance, J. C., and Jones, A. A. (1978). "Legal Effect of Representations as to Subsurface Conditions." *Selected Studies in Highway Law*, Vol. 3, Transportation Research Board, Washington, DC, 1471–1494.

Warner Construction Corp. v. City of Los Angeles, 466 P.2d 966, 1001, 2 Cal.3d 285, 85 Cal.Rptr. 444 (1970).

Warren Brothers Co. v. New York State Thruway Authority, 309 N.Y.S.2d 139, 139, 34 N.Y.2d 770 (1974).

W. H. Lyman Construction Co. v. Village of Gurnee, 403 N.E.2d 1325, 1328, 84 Ill.App.3d 28, 38 Ill.Dec. 721 (1980).

Wiechmann Engineers v. California State Department of Public Works, 107 Cal.Rptr. 529, 535, 31 Cal.App.3d 741 (1973).

Wunderlich v. State of California, 423 P.2d 545, 548, 56 Cal.Rptr. 473, 65 Cal.2d 777 (1967).

Additional Case

The following is an additional case related to issues associated with negligent misrepresentation. The reader is invited to review the facts of the case, apply the decision criteria in the flowchart, reach a decision, compare it with the judicial decision, and determine the rationale behind the judicial decision.

Bilt-Rite Contractors, Inc., v. The Architectural Studio (Pennsylvania).

Chapter 11

Defective Specifications

This chapter focuses on the theory of *implied warranty* as it applies to defective plans and specifications and the rules used to resolve defective specification disputes. Because the directive may originate from the specifications or plans, the term "defective specifications" refers to both.

Contract Language

There is no specific language in the contract dealing with defective specifications. The contract does not contemplate defects. AIA A201 (1987), Art. 4.2.1 says,

> The Contractor shall carefully study and compare the Contract Documents and shall at once report to the Architect any error, inconsistency or omission he may discover. The Contractor shall not be liable to the Owner or Architect for any damage resulting from any such errors, inconsistencies or omissions in the Contract Documents.

The warranty clause may also be relevant. A typical material and workmanship warranty clause can be found in AIA A201, Art. 4.5 (1976), which states in part,

> The Contractor warrants to the Owner and Architect that all equipment and materials furnished under this Contract will be new unless otherwise specified and that all work will be of good

quality, free from faults and defects and in conformance with the
Contract Documents.

Under this clause, the contractor's obligations are limited to the quality of equipment, materials, and workmanship. The exact wording of the
clause is important because the contractor may be obligated to do more
or the risk allocation may have been changed.

Background

Defective Specifications vs. Differing Site Conditions

It is not always clear if a disagreement should be resolved as a differing
site condition (DSC) or as a defective specification. Each situation has
different criteria that must be met, and recoveries are based on different
principles.

Information or Instruction

The primary difference between defective specifications and differing site
conditions is the type of data at issue. Differing site conditions deal with
information about the conditions to be encountered, whereas defective
specifications deal with instructions or details of construction.

Reliance

With a DSC claim, a contractor shows that he relied on the information the
owner provided. With defective specifications, it is not necessary to show
that the contractor relied on the instructions because he had no choice and
was required to follow the instructions provided by the owner.

Legal Theories — Implied Warranty

Defective specification claims are premised on the implied warranty of
the adequacy of the plans and specifications supplied by the owner. Impossibility and commercial impracticability are closely related alternative theories for presenting defective specification issues, with particular

application in the case of performance specifications. DSC claims are premised on the DSC clause in the contract, or in the absence of such a clause, under a misrepresentation theory.

Implied warranty is a broad legal theory that applies to many aspects of construction. For example, there are implied warranties relating to the quality of workmanship provided by contractors and the suitability of products furnished by manufacturers. This book covers only implied warranties applied to the construction plans and specifications.

In most situations, by providing a method specification, the owner implies a warranty that the specifications are adequate to achieve the desired purpose. As explained by the U.S. Supreme Court, this concept rests on the presumed expertise of an owner where it sees fit to prescribe detailed specifications (J. D. Hedin Construction Co. v. U.S.). In the United States v. George B. Spearin decision, the Supreme Court concluded that if a contractor is bound to build according to detailed specifications, the contractor is not responsible for the consequences of the defects.

The argument for an implied warranty rule is particularly convincing when the detailed design in question is complex, state-of-the-art, or relies on engineering data not readily available to the contractor. Foundation designs, for example, often fit this last case. The contractor is not obligated to evaluate soil conditions and anticipated loading conditions to ascertain if the foundation will function properly.

Rules of Application

Fig. 11-1 is a flowchart of the rules for contractors and administrators to apply to defective specification disputes.

Primary Issues Governing Disputes over Defective Specifications

In dealing with disputes over defective specifications, several overriding questions arise:

- What caused the failure?
- Which party had control over the feature that led to the failure?
- Did the contractor follow the specifications? (method directive)

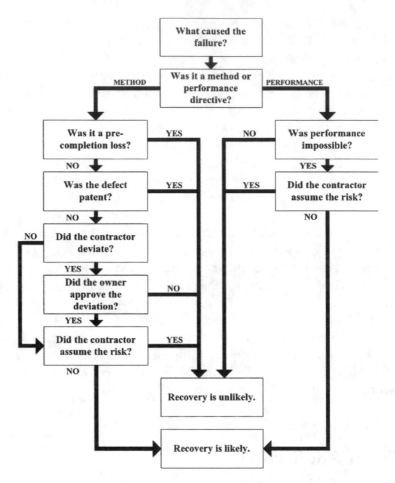

Figure 11-1. Decision Tree for Disputes Involving Defective Specifications.
SOURCE: Thomas et al. 1995, ASCE

- Could the work have been successfully performed by any method? (performance directive)
- Did the contractor warrant the outcome?

What Caused the Failure?

The implied warranty concept applies only to method specifications (Stuyvesant Dredging Co. v. U.S.). Because most contract clauses are a mixture of method and performance requirements, it is imperative to identify what caused the failure. The following steps should be helpful:

1. Determine the nature and extent of the failure.
2. Identify the apparent cause of the failure.
3. Identify the root cause of the failure.

An example is worthwhile to illustrate how to determine the cause of failure. Suppose a building fails because cracks are found in a footer. There are many potential causes. The failure may be a workmanship problem caused by the contractor adding water to the mix or not protecting the plastic concrete from freezing. Under AIA A201, Art. 4.5 (1987), this cause is within the contractor's control. However, if the failure was because the footer was undersized, the bearing capacity of the soil was incorrectly determined, or the service loads were underestimated, the design-related issues pertain to the part of the specification that the owner controls. The process of identifying the cause of the failure is illustrated in Fig. 11-2.

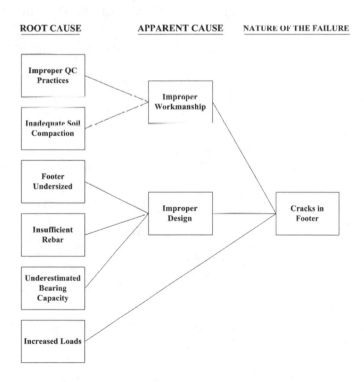

Figure 11-2. Flow Diagram Illustrating Cause of Failure.
SOURCE: Thomas et al. 1995, ASCE

There are two defective situations: when the item cannot be constructed, and when the end product fails to meet the end requirements. These two situations are discussed below.

In the first situation, it may be impossible or commercially impractical to construct an item. Frequently, change orders and design changes are an acknowledgment that the design was defective.

The second type of failure, which is more common than the first, results when the end product does not measure up to the owner's expectations as expressed in the contract documents (Haehn Management Co. v. U.S.). Sometimes, this problem arises when the contract contains both a method that the contractor must follow and an outcome that the method must achieve. Several examples illustrate this situation.

The specification was considered defective in McCree and Company v. State of Maine, where the contractor was unable to achieve the compaction required. The contract also included a detailed method that the contractor was required to follow.

Similarly, an Illinois court ruled that the specification was defective in W. H. Lyman Construction Co. v. Village of Gurnee, where the required method for sealing manhole bases did not result in the manholes meeting the infiltration limits required by the contract.

Another common source of dispute is when the final outcome does not satisfy the owner. Typically, disputes occur when the owner requires the contractor to take corrective action because the outcome is unacceptable.

In Puget Sound National Bank of Tacoma v. C. B. Lauch Construction Co., the owner required the contractor to apply a third coat of paint to an apartment complex when the required two coats showed excessive fading. In awarding compensation to the contractor, the court noted,

> The contract called for a two coat paint job, not three, and whether the job was sufficient or not, it was the specification under which [the contractor] did the painting.... He was to apply two coats of paint of a specified kind, and whether or not it was sufficient was a matter over which [the contractor] had no control. (Puget Sound National Bank of Tacoma v. C. B. Lauch Construction Co.)

It is not sufficient for the contractor simply to show that he or she followed the plans and specifications. This issue was explored in Mayville-

Portland School School District No. 10 v. C. L. Linfoot Co., when the contractor refused to repair or replace a tank he installed, which was damaged and was unfit for its intended purpose. Linfoot claimed that he installed the tank according to the plans and specifications, and therefore, was not responsible for the damage. The court stated,

> These North Dakota cases and the cases cited from other jurisdictions, therefore do not automatically relieve the contractor of liability for defects when he has followed plans and specifications furnished by the other party. The contractor, however, may be relieved of liability if the plans or specifications furnished by the other party were defective or insufficient, and such defects or insufficiency caused the damage complained of. (Mayville-Portland School District No. 10 v. C. L. Linfoot Co.)

The Mayville-Portland decision discussed several cases that imply that the contractor simply has to show that he followed the plans and specifications and does not have to show that the plans were defective, but the court noted that the plans were, in fact, defective in all these cases (Mayville-Portland School District No. 10 v. C. L. Linfoot Co.).

Which Party Had Control over the Feature That Led to the Failure?

Specifications can be a method or performance (end result) type. The central question here is whether the contractor had any latitude or control over the aspect that caused the failure. Thus, the importance of isolating the cause of the failure is apparent. The operative parts of the contract may be a simple directive to the contractor to follow a certain sequence of construction or it may be a requirement to construct a footer of a certain size, with so much reinforcement. In both instances, the contractor may have no latitude. As indicated in these two examples, the directive aspect is not limited to specifying procedures. The central issue is the latitude the contractor had in how the work was done. The directive may be found in the plans or specifications.

Alternatively, the specifications might say that the contractor is supposed to secure and protect the work, perform the work in a workmanlike manner, and produce a product that performs in a certain way. This part of the contract documents is performance-related because it gives the contractor a choice in how to perform the task.

Method Directive

Was It a Precompletion Loss?

The concept of implied warranty generally does not apply if the failure occurs before completion of the structure or component. The contractor is normally responsible for protecting the work during construction, and the theory of implied warranty does not extend to a precompletion loss (Utility Contractors, Inc., v. U.S.). The owner does not warrant that a partially completed structure will not be damaged, and he or she does not have an obligation to design a structure such that it is impermeable to any damage during the course of construction. As stated in Utility Contractors, Inc., v. United States,

> ... absent any contract provision to the contrary, the government implicitly warrants that satisfactory performance will result. This is not a warranty against pre-completion losses.... (The owner was not required to) provide protective measures so the project could be built under all situations, including heavy rainstorms.

An exception to the precompletion loss rule is when the component would have failed anyway. Blue Bell, Inc., v. Cassidy illustrates this exception. During the course of construction of an industrial building, two building columns failed due to excessive soil settlement. This settlement also caused a portion of the roof to collapse. There was no evidence that the contractor was negligent in protecting his uncompleted work; rather, the evidence indicated that the columns would have failed even if the structure was complete. In this situation, the contractor was not liable for the building failure (Blue Bell v. Cassidy).

A similar conclusion was reached in Miller v. Guy H. James Construction Co., where the contractor was awarded his repair costs when heavy rains washed out his partially completed ditch liner. The court allowed recovery because the final design slope was too steep and allowed the runoff to flow at excessive velocity. The drainage system design was defective, and the finished project as designed would probably have been damaged even if completed (Miller v. Guy H. James Construction Co.).

Was the Defect Patent?

When the contractor knew, or should have known, that the contract documents were defective, the contract usually imposes on the contractor a duty to call the defect to the attention of the owner (Beacon Con-

struction Co. of Massachusetts v. U.S.). This duty is based on the principle that a contractor cannot knowingly produce something useless and
then charge the customer for fixing it (R. M. Hollingshed Corp. v. U.S.).
To be considered obvious (patent), the error must be glaring and significant. Few situations are such that the error is so glaring and significant
as to be considered a patent error.

A patent error was discussed in Allied Contractors, Inc., v. United
States. Allied was constructing a Nike launching area for the U.S. Army
Corps of Engineers. The contract required the construction of two un-
supported 4-in. thick masonry walls against an earthen embankment.
The contract did not require any support for the walls, but the walls
were to be waterproofed by another contractor and backed by a 2-ft-
thick concrete wall. Heavy rains caused the masonry walls to collapse.
The contractor claimed that it built the walls exactly as called for by the
plans and that improper design was the cause of the failure. Rejecting
the contractor's argument, the court found,

> It is not true that [Allied] was justified in blithely proceeding
> with its work in the face of obvious and recognized errors. The
> obligation was cast upon [the contractor] to do something about
> it." (Allied Contractors, Inc., v. U.S.)

Patent errors do not protect the owner if the contractor discovers a
significant error and calls it to the owner's attention. In Ridley Investment Co. v. Croll, the owner was responsible where the contractor notified the owner of a defective design because of unsuitable soil under a
floor slab. The owner directed the contractor to continue work without
making provisions for additional support, but excessive settlement
caused damage to the facility.

Case Study

Patent Defect in the Contract Requirements

Enid Corp. v. H. L. Mills Construction Co.
101 So.2d 906 (1958)

Enid Corp. and H. L. Mills Construction Co. (Mills) entered into an oral
contract whereby Mills was to build certain roads in Biscayne Key Estates. The roads were to be built to grade stakes that were set by Enid's

engineer. Enid continually inspected the work from day to day and at all times had control of the elevation to which the finished roads were to be built. The stakes were set for a 5-ft elevation of the completed roads. Both parties knew the character of the subsoil at the time of entering into the work and at the time the work was being done; moreover, each was fully aware of the possibility that roads built on such soil might settle. The roads were inspected during the progress of the work by a county inspector who was principally concerned with the depth of rock and with the surfacing. The roads were completed according to the county specifications and at the elevation set and required by Enid.

During the course of the work, Mills was fearful that the roads might settle and wanted to build them higher than the elevation set by Enid. Mills advised Enid that the roads might settle, but Enid would not permit Mills to build the roads higher because a higher road would have required additional fill on the lots.

Soon after the work was complete, there was a slight settlement of some portions of the road so that the roads were below the 5-ft elevation as required by Enid. Can Enid require Mills to raise the road elevations at Mills's expense?

The Third District Court of Appeals of Florida said no. The court said,

> ... the court had a right to believe the roads were constructed in accordance with (Enid's) direction and against (Mills') admonitions that they might subside. In such a situation (Enid) assumed the risk of subsidence . . . . If the Court had not found that (Enid) was fully aware of the condition of the subgrade both before the oral contract and when poor conditions were encountered, we would have been presented with a different situation.

Did the Contractor Deviate?

The contractor may not deviate from the specifications, even if the specification is defective. This exception to the implied warranty rule is based on the general rule that when an owner specifies a particular method or design, the contractor has no right to depart from those plans and specifications. If a contractor deviates, he or she becomes the guarantor of the strength and safety of the structure (Clark v. Pope).

This rule was used in Valley Construction Co. v. Lake Hills Sewer District, when the contractor deviated from the specifications while installing a sewer line. The contract called for the contractor to hand-shape the trench bottom so that the pipe would rest uniformly on the bottom. During excavation, the contractor encountered hardpan material and determined that hand-shaping was impossible. He orally requested permission to use bedding gravel, which was a unit-price pay item in the contract requiring permission from the engineer. When the engineer disapproved the contractor's request, the contractor chose to deviate from the contract by using a cushion-course method for installing the pipe. When 48 sections of the pipe that the contractor installed using this method broke after a heavy rain, the contractor refused to replace the broken sections. The court acknowledged that expert witnesses agreed that bedding material was required and that hand-shaping the trench was not an adequate method, but stated,

> Be that as it may, respondents (the contractor) agreed to follow the specifications provided by appellant (the owner); as long as they did so, they would not be liable for disastrous consequences.... [The contractor] would only be discharged from nonperformance or poor workmanship by following the specifications. (Valley Construction Co. v. Lake Hills Sewer District)

In Robert C. Regan v. Fiocchi, a masonry subcontractor installing brick veneer walls did not install wall ties at the spacing required by the specification. When several walls bulged, the deviation was discovered, and the subcontractor was ordered to correct the condition. The subcontractor refused to perform the corrections. The court ruled in favor of the owner, given that the contractor chose to depart from the specifications (Regan v. Fiocchi).

The contractor may be able to overcome the deviation rule if he or she can show that the deviation was minor and had nothing to do with the failure. For example, in Burke City Public School v. Juno Construction, the contractor was able to recover when he proved that the damage to a roof was not caused by his slight deviations from the specifications but was caused solely by the defective design. Clearly, contractors place themselves at great risk when they deviate from the plans and specifications which they have agreed to follow.

Did the Owner Approve the Deviation?

The contractor is likely to recover his or her costs if the owner approves the deviation. W. H. Lyman Construction Co. v. Village of Gurnee is a recent case that illustrates this point. The village contended that the contractor was liable for additional costs because he deviated from the specifications to seal the manhole bases. The court rejected this argument because the deviation was approved by the engineer, W. H. Lyman Construction Co. v. Village of Gurnee).

The contractor may be able to recover if the owner knows of a deviation but accepts the completed structure anyway. The rule states that

> Where the owner accepts a structure without complaining, within a reasonable time, of defects or contract deviations which are known to him or which are open, obvious and apparent, he is precluded from seeking damages for those defects or deviations. (Havens Steel Co. v. Randolph Engineering Co.)

Did the Contractor Assume the Risk?

Assumption of risk is a matter of how other clauses in the contract are worded, and the determination may be based on certain legal interpretations. Therefore, if the dispute comes down to this issue, legal guidance is advised.

Owners sometimes include exculpatory and specifically worded warranty clauses in the contract in an effort to shift the responsibility for the adequacy of the plans and specifications to the contractor. However, disclaimers and exculpatory clauses are not often successful if the owner has provided a detailed design that the contractor is required to follow. Conversely, warranty clauses are occasionally so specific that the risk of the defective specifications shifts to the contractor. Exculpatory clauses are not favored by courts and are strictly construed against the party seeking to benefit from them.

It has long been held that general disclaimers, such as those requiring the contractor to visit the construction site and to check the plans and specifications, have limited effect on the implied warranty of the sufficiency of the plans and specifications if the defect is patent. As stated in Unites States v. George B. Spearin,

> The obligation to examine the site did not impose upon him (the contractor) the duty of making a diligent inquiry into the history of the locality, with a view to determining, at his peril, whether the sewer specifically prescribed by the government would prove adequate. The duty to check plans did not impose the obligation to pass upon their adequacy to accomplish the purpose

in view. And the provision concerning contractor's responsibility cannot be construed as abridging rights arising under specific provisions of the contract.

The defect in the Spearin case was not readily apparent from a site visit. Specifically worded disclaimer clauses are sometimes determined to be valid. A clause included in the contract in Philadelphia Housing Authority v. Turner Construction Co. serves as an example. The method specification for interior painting for a housing development was defective, and the contractor was forced to use a different and more expensive paint than specified in the contract. However, the contract also included the following exculpatory clause:

> By submitting a bid the bidder agrees that he has examined the site and the specification and drawings, and where the specification requires in any part of the work a given result to be produced, that the specifications and drawings are adequate and the required result can be produced under the specification and drawings. No claim for any extra work will be allowed because of alleged impossibilities in the production of the results specified or because of inadequate or improper plans and specifications and wherever a result is required, the successful bidder shall furnish any and all extras and make any changes needed to produce, to the satisfaction of the local authority, the required result. (Philadelphia Housing Authority v. Turner Construction Co.)

The court found that the clause had shifted the risk to the contractor, and the contractor was not allowed to recover the additional costs.

However, specific disclaimers are not often valid. The outcome in W. H. Lyman Construction Co. v. Village of Gurnee is more typical. In addition to the detailed design for the submerged manhole bases, the specification included a clause holding the contractor solely responsible for meeting the infiltration limits set in the contract. The provision required the contractor to indicate in writing, with his proposal, if he could not comply with the infiltration requirements. The method required by the specifications proved defective, and the contractor eventually received permission to seal the manhole bases using a method originally prohibited by the plans and specifications. The court found the disclaimer to be an attempt to shift responsibility for the adequacy of the specifications without providing the contractor the opportunity to choose the method of sealing the bases. This provision was determined by the court to be against public policy that an owner should imply a warranty of the sufficiency of method specifications (W. H. Lyman Construction Co. v. Village of Gurnee).

The differences in Philadelphia Housing and Lyman would appear to be the latitude granted the contractor. In Philadelphia Housing, the contractor could perform the painting in the manner he chose. In Lyman, the contractor was instructed on how to seal the manholes. The attitude of the courts seems to be that the owner cannot describe what he wants and how the contractor is to accomplish the work and then disclaim responsibility when the method does not work.

If the language of the warranty clause is clear and specific, the clause is likely to be upheld. When the clause is less explicit, the court examines other contract clauses and the actions of the parties to determine the intent of the warranty clause. The language of the other clauses and surrounding facts are also relevant. Generally, courts are hesitant to shift the risk of the adequacy of the design unless the contract has clearly spelled out the intent that the risk was shifted to the contractor.

Warranty clauses like the standard 1987 American Institute of Architects Guaranty Clause, Art. 3.5, require the contractor to remedy any defects caused by faulty materials and workmanship that appear within a specified time period, which is usually one year. Although occasionally challenged, courts have consistently held that, when a general warranty clause extends only to materials and workmanship, the contractor is not responsible for a failure due to a defective design (Teufel v. Wiernir). Occasionally, however, warranty clauses are written such that the contractor guarantees the performance of the finished product regardless of the reason it fails or is defective. In Shopping Center Management Co. v. Rupp, the contractor provided two submersible sewage pumps that met the contract requirements, were approved by the architect, and were installed according to the plans. Shortly after installation, the pumps failed because the pumps were not built to operate under the conditions at the site. The warranty clause read,

> The contractor shall guarantee the satisfactory operation of all materials and equipment installed under this contract, and shall repair or replace, to the satisfaction of the owner or architect, any defective material, equipment, or workmanship which may show itself within one year after the date of final acceptance. (Shopping Center Management Co. v. Rupp)

The court held that under the language of the guarantee, the contractor assumed the risk that the equipment would operate satisfactorily. The wording of this clause should be compared to the wording of the AIA Art 4.5 (1976).

Another court came to a similar conclusion: that the warranty clause required the contractor to guarantee the installation of a heating system that failed. The warranty clause held the contractor responsible "for anything that goes wrong a year from the date of completion." The contract required the work to be done in strict accordance to plans and specifications, and there was no evidence that the contractor did not conform to the requirements, but the court concluded that the contractor made an express and comprehensive warranty that the heating system would give reasonably satisfactory performance for a year after its installation (Shuster v. Sion).

Occasionally, other clauses have been found to establish an implied warranty. For example, in Emerald Forest Utility District v. Simonsen Construction, a combination of clauses amounted to an express guarantee to provide a working sewer, free from defects. Although the owner failed to provide sufficient plans and specifications, the court determined that a clause requiring the contractor "to complete the structure according to the contract and to prepare the site and structure in a workable condition for final acceptance" combined with a provision that all work be "able to pass any inspection, tests or approvals provided for in the contract" was an express warranty that the sewer line would be acceptable (Emerald Forest Utility District v. Simonsen Construction).

A similar decision was reached by the Louisiana Supreme Court in Brasher v. City of Alexandria. The court concluded that a clause requiring the contractor to "correct any deficiencies existing in the sewers, manholes or other appurtenances, and put the entire system in working condition" was an express requirement to provide a complete working sewer system.

Performance Specifications

Where directions are of the performance type, the contractor bears a considerably greater risk in producing the desired end result. The applicable legal rule generally is that the contractor must prove that performance is impossible or commercially impractical to be excused from compliance with a performance specification.

The primary feature of a performance directive is that it will "set forth an objective or standard to be achieved, and the successful bidder is expected to exercise his ingenuity in achieving that objective or standard of performance, selecting the means and assuming a corresponding responsibility for that selection" (J. L. Simmons Co. v. U.S.).

Was Performance Impossible or Commercially Impractical?

Degree of Impossibility

A performance specification is defective only if the requirement set forth is impossible or commercially impractical. Absolute impossibility implies that the work is physically impossible or beyond the state of the art. Commercial impracticality exists if the work is physically possible, but at great cost. As stated by Williston (Jaeger 1961),

> The true distinction is not between difficulty and impossibility. A man may contract to do what is impossible as well as what is difficult, and be liable for failure to perform. The important question is whether an unanticipated circumstance has made performance of the promise vitally different from what should reasonably have been within the contemplation of both parties when they entered into the contract. If so, the risk should not be thrown on the (contractor).

Thus, to prevail on a claim based on commercial impossibility, a contractor must show that the difficulty is far beyond what the parties contemplated when the contract was made. In determining whether something is beyond the contemplation of the parties, contractors normally prove this precisely by showing how much harder the work was than anticipated. As stated by one court,

> The doctrine ultimately represents the ever-shifting line, drawn by courts hopefully responsive to commercial practice and mores, at which the community's interest in having contracts enforced according to their terms is outweighed by the commercial senselessness of requiring performance. (Natus Corp. v. U.S.)

To determine what the parties contemplated, courts examine the entire contract plus the actions of the parties. For example, if the work required the contractor to use a special piece of equipment or method that was clearly not envisioned by the contract, this fact might be an indication that the difficulty was beyond what was contemplated. Another indicator might be if the difficulty was so great that the contractor could not possibly complete the project within the scheduled completion date.

The decision in Tombigbee Constructors, Inc., v. United States is an illustration of a commercially impractical specification. Although the contractor was able to achieve the required 95% compaction, the court was influenced by testimony that "compaction was achieved slowly, with difficulty and great cost, with the use of a variety of equipment, and without possibility of meeting the construction schedule." The court also found it significant that, approximately halfway through the job, the owner consented to a change order to allow the contractor to add Portland cement to the soil for the remainder of the project. The court treated the change order as an admission that the compaction could not be achieved within the time set forth by the contract.

The Tombigbee decision is contrasted by Baton Rouge Contracting Co. v. West Hatchie Drainage District, where the contractor was unable to recover additional costs when he encountered difficulty maintaining a required 1:1 slope on the bank of a channel he was dredging. The contractor argued that the 1:1 slope was commercially impractical and that a flatter slope was more desirable. The court ruled against the contractor, where the contractor was able to achieve the slope as specified. The project was also completed within the specified time. The significant difference between the two cases was not the degree of difficulty encountered, but the difficulty appearing to be beyond the contemplation of the parties.

Performance Must Be Impossible by Any Method

Because the contractor is not limited to one method, he or she has the added burden to show that the outcome could not be met by any reasonable method, and not just the one chosen. In Koppers Co. v. United States, Koppers chose to abandon efforts to produce runway matting for an airport when their first attempt to meet strength requirements was unsuccessful rather than use a different core material or alter their fabrication procedures. Based on their initial attempt, Koppers concluded that the specification was commercially impractical because of the abandoned performance. The court found that Koppers did not show that "a competent contractor either could not have performed the contract or that performance involved unreasonable and excessive costs" (Koppers Co. v. U.S.).

Subjective or Objective Impossibility?

Impossibility can be either objective or subjective. The difference can be thought of as the difference between "it cannot be done" (objective) and "I cannot do it" (subjective). A contractor is not excused from performing the contract due to subjective performance (Restatement of the Law 1979). Two cases illustrate this point.

In Ballou v. Basic Construction Co., the contractor failed to produce 200 precast concrete columns satisfactorily. The contractor argued that the columns, as designed, were extremely difficult to construct and that the tight construction schedule required by the contract made performance commercially impractical. The court found that the columns were possible to manufacture because the contractor had already manufactured 45 columns and that the failure to manufacture the columns was purely subjective. No objective impossibility was shown. Simply because the contractor could not manufacture 200 acceptable columns within the allotted time did not excuse performance (Ballou v. Basic Construction Co.).

In B's Co., Inc., v. B. P. Barber and Associates, Inc., a subcontractor claimed that the installation of two water mains under a river was impossible when two attempts to install the lines were unsuccessful. The court rejected the subcontractor's claim because another contractor was able to complete the project by using an alternative method of installation. The court found that,

> ... the evidence shows this to be a most difficult job requiring an experienced crew and proper equipment, but the trial judge found that it was not impossible to perform. It appeared most difficult or perhaps impossible for the B's Co. but apparently a routine operation for an experienced operator in the field. (B's Co. v. B. P. Barber and Associates)

Did the Contractor Assume the Risk?

Where a specification is shown to be impossible or commercially impractical, the contractor may not recover if it can be shown that he assumed the risk of impossibility. The assumption of risk generally occurs when a contract clause specifically places the risk on the contractor. The clause must be specific, and it must be clear that the risk of impossibility has been placed on the contractor.

A contractor may also assume the risk of impossibility where it can be shown that the owner or designer relied on the contractor's knowledge or expertise. Although this situation is uncommon with most ordinary construction operations, it may occur if a project uses a new or state-of-the-art construction technique or product. Under these circumstances, a contractor may possess information or expertise superior to that of the designer. If the owner relies on the expertise of the contractor, risk of failure may shift to the contractor. One must answer two inquiries: (1) which party had the greatest expertise in the subject matter and (2) which party took the initiative in drawing up the specifications and promoting a particular design.

Where these criteria have been met, the case most often cited is Bethlehem Corp. v. United States. There the U.S. Army contracted for the construction of an environmental test chamber. Because the designers had limited experience in the design of this type of structure, they consulted Bethlehem to determine which performance characteristics were achievable. Bethlehem advised the designers on the limits of possible performance, and this advice was used to develop the specification, which was later revised before the advertisement based on a review by Bethlehem. Bethlehem bid on the project and was subsequently awarded the contract. The chamber was constructed, but did not meet the performance requirements for control of relative humidity. The court determined that the specification was impossible to perform, but that Bethlehem had assumed the risk of nonperformance. Bethlehem was aware that it was being consulted as a leader in the field and that the army's designers did not have expert knowledge (Bethlehem Corp. v. U.S.).

Bethlehem is contrasted by City of Littleton v. Employers Fire Insurance Co., where the court refused to find that the contractor had assumed the risk of impossibility. During the course of construction, two 5-million-gal. water tanks collapsed. The parties entered into a supplemental agreement for reconstruction, but the contractor subsequently refused to attempt reconstruction when he determined that the revised design was impossible to construct without another collapse. Both the contract and the supplemental agreement were based on engineering data provided by the owner, and there was no showing that the contractor possessed any superior expert knowledge. Additionally, nothing in the contract could be construed as shifting the risk of impossibility to the contractor (City of Littleton v. Employers Fire Insurance Co.).

Illustrative Example

An example based on J. L. Simmons v. United States illustrates the rules of application.

Statement of Facts

In October 1949, the Veterans Administration awarded a contract for approximately $7 million to J. L. Simmons Company for the construction of a hospital and related facilities in Chicago, Illinois. During construction of the pile foundations, it was discovered by the inspector that the cast-in-place concrete piles would not support the design loads. To correct the problem, the owner substituted a composite pile. The revised specifications prescribed in detail the methods, sequence, and procedures for driving and forming the piles. For example, "The casing and a close fitting interior core were to be driven to a depth approximately equivalent to the length of the upper section of the pile. The core was then to be removed and a pipe section inserted. The core was then to be replaced and the pipe section driven to the required penetration and bearing" (J. L. Simmons Co. v. U.S.). The contractor proceeded with the installation of the pile foundations. After almost all 1,700 piles had been driven, the contractor detected movement in some of the pile clusters and notified the owner. After a complete evaluation of the piles, extensive restoration work was required to correct the movement problems, leading to the dispute. The relevant questions from Fig. 11-1 are discussed below.

Analysis

What Caused the Failure?

The key issue of this dispute was what had caused the pile foundations to move. Testimony by soil experts confirmed that mass movement and drifting of piles and pile groups was inevitable under the sequence of operations required by the specification. Thus, the cause of the failure was determined to be the sequence of operations specified in the contract.

Was It a Method or Performance Directive?

The owner contended that Simmons was contracted to produce the ulti-mate design objective by application of its own skills and by the con-struction methods of its own choice, subject only to minimum standards prescribed for quality and workmanship. In essence, the owner's posi-tion was that the work was done under a performance specification. However, every detail of the pile work was spelled out, including the sequence; thus, the part of the specification that related to the failure was clearly a method directive. Had the sequence not been specified, work would have been done under a performance directive.

Was It a Precompletion Loss?

Although the project had not been completed, the failure was not related to the contractor protecting the work. Would the component have failed anyway? The answer to this question is probably yes.

Was the Defect Patent?

This was not an issue in this dispute.

Did the Contractor Deviate?

This rule was not questioned in court, and there was no attempt to show that the contractor had not followed the specification. It is assumed the contractor complied with the contract.

Did the Contractor Assume the Risk?

The contract contained standard government disclaimers and warranty clauses. There was no express wording in the contract to shift the risk of defective specifications to the contractor.

Synopsis

Based on the rules of application, the contractor should recover the added cost. This analysis is consistent with the ruling of the court, which found that the contractor was entitled to the cost to restore the piles, plus the costs of delays to the overall project.

Exercise 11-1: Blount Brothers and U.S. Government

Blount Brothers Corp. and the U.S. government entered into a contract dated August 12, 1982, for additions, alterations, and improvements to the existing Air Force Hospital at Wright-Patterson Air Force Base in Ohio.

Provision 6.2.2 of the contract specified that the concrete for use in the hospital walls be made using exposed tan and brown aggregate. It stated,

> Aggregates for normal weight concrete shall conform to ASTM C 33. Maximum nominal aggregate size shall be 1" for slabs on grade and footings and ¾" for all other work. Coarse aggregate for exposed ribbed concrete walls shall be *tan and brown washed river gravel*. Adequate supply of approved gravel shall be required to provide uniform concrete for all exposed work. (emphasis added)

In addition, Provision 8 stated with relation to the concrete mix that the goal was "uniformity in texture, color and distribution of aggregate in [the] mix." Provision 19.2 explained how the ribbed concrete walls would be bush-hammered to expose the aggregate consistently and completely. Nothing in the contract specified that the overall visual impression of the concrete panels had to be a specific color or that a physical majority of the stones had to be tan or brown.

The contract required Blount to prepare sample concrete panels for examination and approval by the government. On December 20, 1982, Blount poured the required sample concrete panels. The government examined the sample panel and observed that the exposed aggregate was not tan and brown. The contract did not call for bush-hammered paneling that had a predominant visual impression of tan and brown. The aggregate was approximately 85% various shades of gray, white, and blue. At most, only 15% of the aggregate might be considered remotely approaching the color requirement of tan and brown.

Blount decided to pour a second sample concrete panel containing a greater concentration of aggregate in the ribs. This panel was poured in early February 1983, but the coloration problem remained unsolved. The aggregate still was largely gray and white. The government insisted that the aggregate did not meet the contract specifications.

Other aggregate sources were investigated, although no other sample panels were cast. Several aggregate samples from another stockpile at the subcontractor's place of business were examined, but with no satisfaction. Blount tried to locate proper gravel sources at locations in Ohio, Indiana, and Kentucky that were suggested by the government, but these too were rejected. They also investigated additional sites in Alabama and Georgia and offered these sources to the government. The government was still not satisfied.

The main question was whether Blount's failure to perform was caused by their own actions or by defective specifications.

Exercise 11-2: Marine Colloids and M. D. Hardy

On November 22, 1974, Marine Colloids, Inc., a seaweed processor, sent to three contractors a request for the submission of bids on a construction project. Among other aspects of the project, Marine Colloids sought estimates of the cost of building a large metal-clad building with an accompanying firewall, plus a slab foundation for the future expansion of an existing building known as the "pilot plant." The description of the firewall in the bid request read in its entirety,

> Firewall.
>
> A firewall is to be constructed at the north end adjacent to the slab area described under Item 1. The wall is to be constructed of 12" block construction for a two hour fire rating. The top edge of the structure is to be concrete capped, with a 22 GA galvanized iron flashing covering this concrete. The firewall will be flashed to the building at both side walls and roof.
>
> A sleeve of 12" channel shall be set in the wall to allow access of necessary service piping, etc. and will receive a plate assembly on each side of the block wall, filled with insulation. A design of the holes for the two plates will be provided the contractor at a later date.

Accompanying the bid request was a drawing of the entire proposed project. If a contractor's bid was accepted by Marine Colloids, that contractor was obliged under the terms of the bid request to "guarantee soundness of construction for a minimum period to be specified as one year from completion of the contract."

Malcolm Hardy, president of M. D. Hardy, Inc., and William Greet, the special projects engineer for Marine Colloids, met two days before the bids were due. At that meeting, Greet explained the scope and purpose of the proposed construction project and many specific dimensions of the works to be built. Greet told Hardy that the firewall must be built of 12-in. concrete blocks, extend 2 ft beyond the walls and eaves of the building, and be able to restrain a fire for a minimum of 2 hours; thus, Greet defined the height and width of the wall and the basic materials of which it would be constructed. Although Hardy inferred that further design of the firewall might be necessary, he assumed that Marine Colloids would make design alterations through subsequent change orders issued to the contractor. Furthermore, Hardy was not too concerned about the stability of the firewall, inasmuch as the project called for the ultimate erection of buildings abutting the wall on either side. As far as Hardy knew, the contractor was required to build only a free-standing curtain wall that would stand between two buildings without being bonded to them; he was not being asked to construct either a bearing wall or an end wall that would be exposed to the elements.

Bid requests were submitted by M. D. Hardy, Inc., and two other construction firms. Hardy was the low bidder, but the bids from the other two firms on the firewall were so similar to Hardy's as to indicate that all three firms had the same understanding of how the firewall was to be built. Marine Colloids accepted Hardy's bid and included in its acceptance the following language: "Above construction to be all as per W. E. Greet specifications and drawings by M. D. Hardy, Co." Hardy had drafted certain drawings of the projected construction, but those drawings only followed specifications and plans that had already been created by Greet.

When Greet attempted to secure a building permit, the building inspector informed him that the Rockland building code required firewalls of the proposed height to be at least 16 in. thick. At Marine Colloids' request, Hardy submitted an estimate of the increase in cost. Marine Colloids accepted the proposal and authorized Hardy to construct the wall at the revised thickness.

In accordance with Marine Colloids' specifications, Hardy erected the metal-clad building, constructed the firewall, and poured the foundation for the pilot plant expansion. Hardy completed the firewall on July

9, 1975. The next day, Marine Colloids told Hardy that the pilot plant expansion would be delayed indefinitely. In the fall of 1975, Greet communicated to Hardy his concern that the firewall might not be stable in the absence of the abutting pilot plant expansion. Hardy told Greet that he had no confidence in the firewall's ability to serve as an exposed end wall rather than as an interior curtain wall. Even after Hardy and Greet had expressed their misgivings about the stability of the wall, Marine Colloids made no effort to avoid the risk of the wall collapsing.

The firewall stood for almost seven months exposed on the northern end of the metal-clad building. On February 2, 1976, during a winter storm, the firewall fractured horizontally and fell to the north, damaging the existing pilot plant and other property. At the time the wall collapsed, winds had been sweeping across the roof of the metal-clad building from south to north at speeds of up to 70 knots. According to engineer Paul Atwood, the peculiar configuration of the buildings caused a wind tunnel to form between the metal-clad building and the pilot plant, in turn subjecting the firewall to intense forces of suction. Atwood testified that the wall was properly constructed for use as a firewall and that it collapsed only because of the effect of the wind tunnel, a phenomenon that would not have occurred had the pilot plant been expanded as originally planned.

Marine Colloids directed Hardy to rectify the damage at Hardy's expense, alleging faulty workmanship. Hardy refused.

What effect does the warranty clause that the contractor shall "guarantee soundness of construction for a minimum period to be specified as one year from completion of the contract" have on the outcome of this dispute? Should Hardy have to correct the damage at its own expense?

Exercise 11-3: Western Foundation Corp. and Veterans Administration

J. D. Hedin Construction Co. contracted with the Veterans Administration (VA) to construct hospital facilities at Ann Arbor, Michigan. The contract was entered into on August 24, 1949, and provided that construction was to be completed within 540 calendar days from the date of receipt of the notice to proceed, which was issued on September 22, 1949. The project consisted of a 500-bed hospital proper and numerous appurtenant buildings. The hospital was not completed until July 31,

1953, which was 1,408 days after receipt of the notice to proceed and 868 days after the scheduled completion date.

The plans and specifications for the foundations were prepared by the Structural Division of the Veterans Administration. These government-prepared specifications were based on subsurface investigations conducted for the VA by the Interstate Engineering Co. These explorations disclosed that the soils were a glacial moraine, consisting of an accumulation of earth, stone, sand, silt, clay, and gravel, with occasional boulders, and in some cases nested boulders. The boring logs showed a stratum of compact sand and gravel at varying elevations. They also confirmed the existence of highly resistant materials and boulders. On the basis of this information, the VA prepared the specifications for the foundation piles. The VA had not driven any test piles as part of its investigation of subsurface conditions. The idea was rejected because of the cost involved. The specifications required piles to be of cast-in-place concrete, and they were to be formed by one of three specified methods. Each of the three methods specified that piles were to be encased in a thin steel shell of approximately 0.05-in. thickness. Piles were called for and used in about 90% of the building foundation.

The invitations for bids, specifications, and drawings were issued on June 27, 1949, and provided for the receipt of bids and bid openings on August 9, 1949. Drawings included logs of all borings and test pits, the manner in which they had been made, and the number of hammer blows required to penetrate the subsurface materials at various test locations. In short, the information submitted to the bidders contained all pertinent engineering data and information as to the pile-driving conditions that would be encountered at the site.

Hedin, who was the low bidder, subcontracted the pile-driving operations to Western Foundation Corp., an experienced pile-driving firm. The subcontractor arrived at the site on October 17, 1949, and promptly prepared its equipment and materials for the actual driving operations. The initial test pile was driven on November 8, 1949, and proved unsuccessful. The difficulty with driving resulting from the extreme compaction of the dense subsoil. It became apparent that the piles could not be driven "dry" through the compact subsoil, and the subcontractor had to resort to "jetting" to loosen the soil so that the pile could be driven. On November 16, 1949, the preparations for "jetting" having been accomplished, three piles were driven in a satisfactory manner. On the next day, three additional piles were driven, but caused the three piles driven on the previous day to collapse or otherwise become unsatisfactory. On the third day, a final pile was driven, which caused the three piles driven on the second day to become unsatisfactory. Only the last pile of the seven was satisfactory. The difficulty was caused by the fact that the thin-shelled piles could not withstand the actual pressures exerted from the driving of subsequent piles. Thus, only the last pile driven would remain satisfactory. On the basis of these driving results, it became apparent that the thin-shelled piles were inadequate for the job, and Hedin's superintendent in charge of construction at the project site so advised the VA in Washington on November 18, 1949. Moreover, Hedin suspended the pile-driving operations pending the VA decision and requested that a representative from the VA's central office be sent to the site immediately. Thereafter, the subcontractor attempted to drive the specified thin-shelled piles in other areas to ascertain if the subsoil conditions encountered in the driving of the first series of piles prevailed throughout the area. These efforts were also unsuccessful.

As of November 30, 1949, the VA had taken no action to correct the piling operation. On that date, the subcontractor proposed the use of a heavy steel pipe pile with a ¼-in. wall rather than the 0.05-in. wall required in the specifications. On December 6–7, 1949, several of the heavy steel pipe piles were successfully driven, and a representative of the VA deemed the heavier pile satisfactory. It was not until January 17, 1950, that the VA issued Change Authorization No. 3, which authorized Hedin to proceed with the driving of the heavy steel piles. Western Foundation prepared its equipment for the driving of the heavier steel piles, shipped to the construction site the duly authorized pipe pile, and accomplished the necessary preparatory work. Pile-driving operations resumed on February 3, 1950, and were substantially completed by May 16, 1950. Western Foundation took 59 driving days, or 102 calendar days, to place the heavier piles.

Should Western be paid the added cost of the heavier pile or was this work done under a performance specification, as alleged by the VA?

Exercise 11-4: J. D. Hedin Construction Co. and Veterans Administration 1

J. D. Hedin Construction Co. contracted with the Veterans Administration (VA) to construct hospital facilities at Ann Arbor, Michigan. The contract was entered into by the parties on August 24, 1949, and provided that construction was to be completed within 540 calendar days from the date of receipt of the notice to proceed (September 22, 1949). The project consisted of a 500-bed hospital proper and numerous appurtenant buildings and was not completed until July 31, 1953, which was 1,408 days after receipt of the notice to proceed and 868 days after the scheduled completion date.

The plans and specifications for the foundations were prepared by the Structural Division of the VA. These government-prepared specifications were based on subsurface investigations conducted for the VA by the Interstate Engineering Co. These explorations disclosed that the soils were a glacial moraine, consisting of an accumulation of earth, stone, sand, silt, clay, and gravel, with occasional boulders, and in some cases nested boulders. The boring logs showed a stratum of compact sand and gravel at varying elevations. They also confirmed the existence of highly resistant materials and boulders. On the basis of this information, the VA prepared the specifications for the foundation. Piles were called for and used in about 90% of the building foundations. The remaining foundations were spread footings. The VA determined that spread footings were to be used in two wings of the hospital building. They were also to be used in several other structures.

The contract drawings set forth the elevations where each spread footing was to be placed. Generally, the footings were to rest 6 in. below the gravel strata found in the subsoil at the site. In preparing for the pouring of the footings, the ground was to be excavated to the top of the gravel strata and then the bed for the footings was dug by hand.

On December 8, 1949, while excavating the footings, Hedin did not hit the gravel strata at the elevations indicated on the drawings. The VA's project superintendent was made aware of this situation and on that date notified, by telegram, the central office of the VA in Washington of this discrepancy. On January 1, 1950, having received no response to his telegram, the project superintendent again advised the central office that the gravel stratum was at a deeper elevation than shown on the drawings and requested an immediate reply. During the interim, on December 13, 1949, Hedin also called this situation to the contracting officer's attention and asked for instructions, but received no response.

Finally, on January 9, 1950, the VA's project superintendent received a memorandum from the central office wherein detailed instructions were given with respect to the difficulties arising from the spread footings. The memorandum specified the footings that were to be lowered less than 2 ft and stated that for those footings that needed to be lowered more than 2 ft, the elevation of the gravel strata should be submitted first before a lowering would be authorized. The memorandum also directed the superintendent to obtain a proposal from the contractor covering the cost involved in lowering the spread footings. The request appeared to be extremely impractical, if not impossible, because in the VA project superintendent's words, "the number of footings and the amount that each must be lowered is indeterminate until such time as all footings have been installed." Moreover, the requirement that Hedin excavate down to the gravel level seemed extremely impractical. If Hedin were to do this, then the excavations would be left open awaiting a decision and would cause, as the VA's project superintendent recognized, "excessive caving and reexcavation." Accordingly, Hedin suggested a unit-price proposal, whereby Hedin would be paid the agreed price multiplied times the number of feet each footing would have to be lowered.

The VA persisted in requiring Hedin to provide the elevations where footings were to be lowered more than 2 ft below the contract elevations before a change was authorized. Hedin was also requested to submit a lump-sum proposal instead of the unit price per additional foot originally suggested by the VA's superintendent. Hedin was so advised by letters dated February 13, 1950, and February 28, 1950. The lump-sum proposal was submitted on March 7, 1950, and on March 17, 1950, Hedin submitted the elevations. Change Authorization No. 6, accepting Hedin's proposal of March 7, 1950, was not issued until April 25, 1950.

In some areas, the gravel stratum was located so much lower than shown on the contract drawings (in some instances 15–20 ft) that spread footings actually could not be used, and it became necessary to redesign for the use of pile foundations. The VA superintendent advised the central office of this fact on March 29, 1950, but the VA took no action. Again, on May 15, 1950, the VA superintendent advised the central office that if piles were going to be used in lieu of spread footings, such a determination should be made quickly because other pile-driving operations on site were to be finished the following day and there would be increased costs for moving the driving equipment back once it was removed. It was not until June 1, 1950, that the VA authorized the substitution of pile footings for the spread footings.

Is Hedin entitled to additional compensation?

Exercise 11-5: J. D. Hedin Construction Co. and Veterans Administration 2

J. D. Hedin Construction Co. contracted with the Veterans Administration to construct hospital facilities at Ann Arbor, Michigan. The contract was entered into by the parties on August 24, 1949, and provided that construction was to be completed within 540 calendar days from the date of receipt of the notice to proceed (September 22, 1949). The project consisted of a 500-bed hospital proper and numerous appurtenant buildings and was not completed until July 31, 1953, which was 1,408 days after receipt of the notice to proceed and 868 days after the scheduled completion date.

The specifications provided for permanent drainage of the buildings and site by sewers. On the east side of the site, the storm sewers were to run into a new 15-in. storm sewer. On the west side, the sewers were to drain into a 10-in. trunk line, which flowed into an outlet near the southwest corner, and thence the discharge was to be through a manhole and off-site culverts. A new 30-in. storm sewer approximately 230 ft long was also to be constructed on the west side of the site.

Hedin subcontracted the work for the sewer system. The work was scheduled to be completed during the early stages of the project so that fill from the excavations of the buildings could be placed over the sewer lines and the area could be graded for roads and parking areas. It was also necessary that the 15-in. and 30-in. storm sewers be constructed during the first months of construction so that other utility lines, which ran over or across the sewer system, could be installed. The soil at the east and west sides of the site where the 15-in. and 30-in. sewer lines were to be constructed was extremely "swampy," thus necessitating up to 15–18 ft of fill. Laying of the 15-in. storm sewer line at the east side of the site began on November 1, 1949. The instability of the soil in this area caused the subcontractor to experience extreme difficulties in the installation of the sewer lines because the subsoil in that area could not support the weight of the pipe and fill on it. Laying of the 15-in. line was completed on November 28, 1949, but it was still not stable.

On November 30, 1949, Hedin proposed to the VA contracting officer the removal of the unstable soil along the 15-in. sewer line and replacement of it with coarse gravel. The VA contracting officer failed to reply to Hedin's proposed solution.

On January 24, 1950, the VA superintendent advised Hedin that, in view of the unstable soil conditions in the vicinity of the 15-in. line, further installation at the east end of the site should be suspended. The letter advised Hedin that the 15-in. line as laid was not acceptable because it had dropped, heaved, and was out of line because of the unstable soil condition. He further indicated that it would be necessary to remove the line and put it on a firm foundation by a method to be determined by the VA.

Placing the 30-in. concrete sewer pipe on the west side of the site began on November 2, 1949, and continued until November 8, 1949, when the VA superintendent ordered that any further work on the line be suspended because of a sinkhole encountered along 75 ft of the trench. The VA superintendent described the condition as "a bog with a very soft muddy bottom which seems bottomless." The VA superintendent thought it would be necessary to put the pipe on a concrete mat or reroute the line. The condition encountered was at once communicated to the central office of the VA. On November 9 and 15, 1949, Hedin advised the contracting officer that in accordance with the directions of the VA superintendent, it had stopped work on the 30-in. line until a decision was made by the VA. Hedin also indicated that until a decision was rendered, this stoppage was holding up work of excavation, fill, grading, and installation of the other lines.

On November 14, 1949, the VA superintendent proposed to Hedin that the sinkhole in the area of the 30-in. sewer line be dug out and filled with coarse gravel. He also submitted a drawing specifying the manner of performing the work and requested a cost proposal from Hedin. Hedin submitted the cost proposal on December 5 and again on December 27, 1949. No action was taken thereon by the VA central office. Hedin, without waiting for approval, dug out the sinkhole and replaced the excavated materials with gravel. However, Hedin's actions were unsuccessful, and so Hedin advised the VA superintendent on January 24, 1950. At this point, the matter was referred to the central office for a solution.

On February 15, 1950, the VA superintendent requested that Hedin submit a proposal covering the amount of corrective work necessary to install the 15- and 30-in. lines and manholes in a stable position. Hedin replied on February 23, advising the VA that the proposal would be submitted as soon as the drawings, previously promised, showing where the sewer was to be changed from clay to concrete pipe, were supplied. On March 9, 1950, the VA superintendent wrote Hedin providing general information on methods to be used in correcting the sewer problems. The letter requested a proposal for a concrete slab to be placed on the gravel pile already in place beneath the 30-in. sewer line. As to the 15-in. line, alternatives were suggested of either (1) sheeting the tile underdrain, with the sewer line to be of cast-iron pipe, or (2) setting cast-iron sewer pipe or creosote wooden piles with the pipe itself anchored to a pile cap. On April 3, 1950, Hedin submitted its proposal in response to the VA request of March 9, 1950. It also advised the VA that all of the cast-iron pipe had to be ordered. No action was taken by the VA on Hedin's proposals.

In May and June of 1950, Hedin protested the VA's delay in responding. The grading and plumbing subcontractors were also protesting the delay. During the months of May, June, and July of 1950, the VA's superintendent reported to the central office that the failure of the central office to arrive at a final decision in regard to the sewer installation was causing the subcontractor additional expense. The VA superintendent observed that if the work had to be carried over until the spring and summer of 1951, it would interfere with roads, walks, grading, and drainage work and prolong the entire project.

Finally, on July 25, 1950, the VA issued Change Authorization No. 13, permitting a change, where required in the judgment of VA's superintendent, from clay or concrete pipe to cast iron for the storm and sanitary sewers. Supporting piles were authorized for the pipe and manholes in the eastern part of the work, and the 30-in. pipe on the western side was to be supported by a concrete slab. The obligation of the government for the changes was limited to $45,000 by the change authorization. By letter of the same date, the VA's superintendent gave to Hedin his decision on those locations where cast-iron sewer pipe should be used. On August 3, 1950, Hedin acknowledged receipt of the change authorization of July 25 and advised the VA that during the months intervening between its proposals and the change authorization (April 3 to July 25), the cost of performing the work had materially increased and that delivery of the cast-iron pipe could not be made by the supplier until November 15, 1950. Hedin also requested that the VA advise it as to "what means, what type of pipe you now desire a quotation on inasmuch as cast iron pipe is not available." Hedin also protested the change authorization. The VA did not reply to Hedin's letter.

Having received no reply or further instructions from the VA, on August 31, 1950, Hedin wrote the VA regarding the change authorization of July 25, 1950, and requested that it be permitted to proceed with the 30-in. line portion of the sewer, inasmuch as the lack of available cast-iron pipe only affected the 15-in. line. Again the VA did not reply to Hedin's request.

On September 1, 1950, Hedin wrote the VA once more and submitted correspondence from its sewer subcontractor and suppliers indicating that cast-iron pipe delivery was indefinite. Hedin again protested the continued delay and advised the VA that the job was "now in a de-plorable condition," requiring construction of temporary roads so work could proceed during the rainy season. Hedin noted its appeal of the change authorization, although expressing the view that a change authorization was not a change order under the contract and that it had never received any orders in respect to the sewer lines. The VA again did not reply to this letter.

On September 29, 1950, Hedin advised the VA that there were several thousand yards of excavated materials stockpiled, which now had to be rehandled for a second time. Hedin requested immediate action so that sewers and water lines could be installed without any further delay, but the VA did not reply to Hedin's requests.

Finally, the VA's contracting officer issued Change Order T on October 23, 1950, and for the first time directed that the corrective sewer work be done at a cost not to exceed $75,000. Hedin thereupon attempted to get a definite commitment on delivery of cast-iron pipe but had difficulty in doing so. On December 7, 1950, Hedin submitted to the VA for approval a description of the pipe it would be able to obtain, and the VA granted approval on December 15, 1950. Delivery of the cast-iron pipe commenced on February 16, 1951, and continued into March 1951.

Winter weather made ground conditions unfavorable for the laying of the sewer line. The VA's superintendent recognized that if the sewer system could not be installed before winter weather set in, the work would have to be carried over until the spring and summer of 1951. The sewer system was finally completed in the fall of 1951.

What was the cause of the failure? What liability does the VA have for defective plans and specifications?

Exercise 11-6: Weston Racquet Club and Republic Floors of New England

In the fall of 1978, Richard Trant, the president of Weston Racquet Club, Inc., sought to replace the playing surfaces on the club's tennis courts. The old surfaces had discolored, had bubbled in many places, and had separated from the concrete undersurface.

After examining several products, Trant negotiated with Stephen Weeks, the president of Republic Floors of New England, Inc., for the installation of ChemTurf, a polyurethane material manufactured by CPR.

Weeks made a site visit to Weston's courts and examined their condition. Before installing ChemTurf, he performed adhesion tests to determine if the material would be appropriate for the location. He also discussed the suitability of the installation with CPR's president, Anthony DiNatale. Although Weeks assured Trant that ChemTurf would bond to the concrete surface, Trant remained unconvinced. Weeks suggested that Trant discuss the matter with James Gilchrist, ChemTurf's developer and a chemist at CPR. Mr. Gilchrist told Trant that the primer he developed had a "bond that was greater than any vapor pressure that could be produced" and that there would be no problems with the product coming loose.

In April 1979, Trant wrote Weeks a letter explaining the requirements for the surface and that it was his understanding that both Republic and ChemTurf's manufacturer would guarantee that the surface would have "no debonding, no crazing, no bubbles, no discoloration." A check for $10,000 accompanied Trant's letter. Republic began installation in May 1979 and substantially completed the work in late June 1979.

On June 26, 1979, Weston and Republic signed a letter of agreement. The agreement contained the following provisions:

> **Condition A**. The manufacturer of ChemTurf, CPR Industries and the installer (Republic), agree to jointly and severably guarantee the installation of ChemTurf at (Weston) as follows:
>
> 1. CPR and Republic acknowledge that the surfaces are to be used both indoors and outdoors.
>
> 2. CPR (Republic) will guarantee the CHEMTURF installation at (Weston) for a period of two (2) years beginning on 6/30/79 and ending on 5/31/81, with respect to the stability of the performance characteristics of the material and with respect to all workmanship and performance of the surface both indoors and outdoors, and against such problems as debonding, delamination, separation, bubbles, crazing, cracks, discoloration, and dead spots. In the event [of] such occurrences, CPR/(Republic) will restore the surfaces or any portion thereof to original form, function, and performance characteristics.

The contract contained a number of comments, including one whereby Republic "agrees that all conditions regarding temperature and preparation of substrata to receive ChemTurf tennis surface have been met." The final relevant paragraph of the agreement stated, "Joint materials and installation guarantees will be furnished when final payment is received."

Even before the contract was signed, bubbles started to appear under the surface. Weeks told Trant that some bubbling was normal and that it would stop.

In early July, Trant refused to pay Republic the balance due on the contract because the bubbling persisted and worsened. Republic refused to make repairs. It was later determined that there was a preexisting condition with the concrete substrata that made it incompatible with ChemTurf. Thus, the ChemTurf surface was destined to fail.

Does the fact that a contract was signed after the work was complete have any bearing on this dispute? What can be said about Republic assuming the risk that ChemTurf would not bond? Is Republic responsible for the preexisting condition? Should Republic fix the tennis courts at its own expense?

Exercise 11-7: Rhone-Poulenc Rorer Pharmaceuticals and Newman Glass Works

Rhone-Poulenc Rorer Pharmaceuticals, Inc., contracted with Turner Construction Co. to install opaque spandrel glass in Rhone's headquarters and research facility near Philadelphia. In turn, Turner, as the general contractor, entered into a subcontract with Newman Glass Works to supply and to install the opaque spandrel glass that composed the curtain wall. The subcontract specified the type of glass Newman was to install and listed three manufacturers from whom glass could be purchased. The specification for the glass read,

> d. Spandrel Glass
>
> (a) Type 8: ¼ inch thick heat strengthened float glass coated on the face with opaque colored ceramic coating or black polyethylene opacifier on the rear surface.

For each manufacturer, the subcontract specified a product identification number, the color, and the type of glass that was acceptable. Section X of the subcontract stated,

The Subcontractor shall . . . take down all portions of the Work and remove from the premises all materials . . . which the Architect or Turner shall condemn as unsound, defective or improper or as in any way failing to conform to this Agreement or the Plans, Specifications or other Contract Documents, and the Subcontractor, at its own cost and expense, shall replace the same with proper and satisfactory work and materials . . .

Section 4.5.1 of the general conditions read,

4.5 Warranty

4.5.1 The Contractor warrants to the Owner and the Architect . . . that all Work will be of good quality, free from faults and defects and in conformance with the Contract Documents.

In compliance with the specifications, Newman began installing the spandrel glass that it had purchased from Spectrum Glass Products, Inc., one of the three listed manufacturers. Spectrum had attached the opacifier coating (a polyethylene film) to the glass with a glue that product literature stated normally could be expected to perform in temperatures exceeding 180 °F. The glass was exposed to such temperatures after installation.

Before Newman completed the installation, the opacifier coating began to delaminate from portions of the glass. Rhone and its architect noticed the delamination because portions of the installed glass exhibited a mottled appearance. Rhone demanded in writing that Newman replace the defective glass. Newman refused to do so. All indications are that Newman followed the specifications.

Does Newman have to replace the glass at its own expense?

Exercise 11-8: ALTL and The Architectural Group

In 1988, ALTL, Inc., decided to build additional warehouse space by placing a preengineered metal building adjacent to an existing, shorter building. The first set of plans drawn up for the project did not note that the disparity in height between the two buildings created a potential Canadian snow load problem. The term "Canadian snow load" refers to the fact the higher roof line of the new building would result in a heavier load of snow and ice accumulating on the lower roof, especially at the joint between the two buildings. One of the consultants hired by ALTL brought this issue to the attention of The Architectural Group (TAG), the firm drawing up the plans. The outcome of discussions was that a note was added to the plans indicating that this problem required that the lower roof of the existing building be strengthened. The notation read, "reinforce existing roof in this area w/additional purlins between existing roof to support extra snow load."

The plans, with the snow load notation, were sent to William Bolt, who bid on the project and contracted with ALTL to construct the building. Bolt claims that he requested and received further oral instructions about the installation of the purlins from TAG, but TAG denies that such instructions were given. Regardless, Bolt proceeded according to the oral instructions that he claims to have received. However, Bolt knew that the instructions dictated an improper manner of installing the additional purlins. The new purlins were placed between the original purlins and bolted to the existing frame of the building using a gusset plate, rather than being bolted to the roof deck, as the original purlins had been. Yet, the manner in which Bolt affixed the purlins complied with the snow load notation.

In early 1994, inclement weather resulted in a large accumulation of snow, ice, and slush on the roof of the lower building, culminating in its collapse. The collapse resulted in property damage in the amount of $210,980.89.

What was the cause of this failure? Does ALTL's knowledge of the faulty notation have any bearing on this dispute? Does ALTL have to repair the damage?

References

Allied Contractors, Inc., v. United States, 381 F.2d 995, 999, 180 Ct.Cl. 1057 (1967).

American Institute of Architects (AIA). (1976). *General Conditions of the Contract for Construction*, AIA Document A201. American Institute of Architects, Washington, DC.

American Institute of Architects (AIA). (1987). *General Conditions of the Contract for Construction*, AIA Document A201. American Institute of Architects, Washington, DC.

Ballou v. Basic Construction Co., 407 F.2d 1137, 1141 (1969).

Baton Rouge Contracting Co. v. West Hatchie Drainage District, 304 F.*Supp.* 580, 587 (1969).

Beacon Construction Co. of Massachusetts v. United States, 314 F.2d 501, 504, 161 Ct.Cl. 1 (1963).

Bethlehem Corp. v. United States, 462 f.2d 1400, 1404, 19 Ct.Cl. 247 (1972).

Blue Bell, Inc., v. Cassidy, 200 F.*Supp.*, 443, 447 (1961).

Brasher v. City of Alexandria, 41 So.2d 819, 832, 215 La. 827 (1949).

B's Co., Inc., v. B. P. Barber and Associates, Inc., 391 F.2d 130, 137 (1968).

Burke City Public School v. Juno Construction, 273 SE.2d 504, 507, 50 N.C.App. 238 (1981).

City of Littleton v. Employers Fire Insurance Co., 453 P.2d 810, 814, 169 Col. 104 (1969).

Clark v. Pope, 70 Ill. 128, 132 (1873).

Emerald Forest Utility District v. Simonsen Construction, 679 SW.2d 51, 53 (Tex. App. 14 Dist. 1984).

Haehn Management Co. v. United States, 15 Cl.Ct. 50, 56 (1988).

Havens Steel Co. v. Randolph Engineering Co., 613 F.Supp. 514, 527 (D.C. Mo. 1985).

Jaeger, W. (1961). *A Treatise on the Law of Contracts*, 3rd Ed., Vol. 18, Lawyers Co-Operative Pub. Co., Rochester, NY, 8.

J. D. Hedin Construction Company v. United States, 347 F.2d 235, 241, 171 Ct.Cl. 70 (1965).

J. L. Simmons Co. v. United States, 412 F.2d 1037/1360, 188 Ct.Cl. 684 (1969).

Koppers Co. v. United States, 405 F.2d 554, 565/566, 186 Ct.Cl. 142 (1968).

Mayville-Portland School District No. 10 v. C. L. Linfoot Co., 261 NW.2d 907, 911 (1978).

Miller v. Guy H. James Construction Co., 653 P.2d 221, 224 (Okl.App. 1982).

McCree and Co. v. State of Maine, 91 NW.2d 713, 725, 253 Minn. 295 (1958).

Natus Corp. v. United States, 371 F.2d 450, 456 178 Ct.Cl. 1(1967).

Philadelphia Housing Authority v. Turner Construction Co., 23 A.2d 426, 427, 343 Pa. 512 (1942).

Puget Sound National Bank of Tacoma v. C. B. Lauch Construction Co., 245 P.2d 800, 803, 73 Ida. 68 (1952).

Robert C. Regan v. Fiocchi, 194 NE.2d 665, 44 Ill.App. 2d, 336 (1963).

Restatement of the Law: "Contracts," 159 ALR.2d 424 (1979).

Ridley Investment Co. v. Croll, 192 A.2d 925, 926/927, 56 Del. 208 (1964).

R. M. Hollingshed Corp. v. United States, 111 F.Supp 285, 286, 124 Ct.Cl. 681 (1953).

Shopping Center Management Co. v. Rupp, 343 P.2d 877, 879/880, 54 Wsh. 624 (1959).

Shuster v. Sion, 136 A.2d 611, 612, 86 RI. 432 (1957).

Stuyvesant Dredging Co. v. United States, 11 Cl.Ct. 853, 860, 834 F.2d 1576 (Fed. Cir. 1987).

Teufel v. Wiernir, 411 P.2d 151, 154/155, 68 Wsh. 31 (1966).

Thomas, H. R., Smith, G. R., and Wirsching, S. M. (1995). "Understanding Defective Specifications." *J. Constr. Engrg. and Mgmt.* 121(1), 55–65.

Tombigbee Constructors, Inc., v. United States, 420 F.2d 1037, 1049, 190 Ct.Cl. 615 (1970).

United States v. George B. Spearin, 248 U.S. 132, 136, 63 L.Ed.166 (1918).

Utility Contractors, Inc., v. United States, 8 Cl.Ct. 42, 49 (1985).
Valley Construction Co. v. Lake Hills Sewer District, 410 P.2d 796, 801, 67 Wsh.2d 910 (1965).
W. H. Lyman Construction Co. v. Village of Gurnee, 403 NE.2d 1325, 1329/1330/1332, 84 Ill.App.3d 28, 38 Ill.Dec. 721 (1980).

Additional Cases

The following are additional cases related to issues associated with defective specifications. The reader is invited to review the facts of the case, apply the decision criteria in the flowchart, reach a decision, compare it with the judicial decision, and determine the rationale behind the judicial decision.

Centrex Construction Company v. United States (Texas).
Diamond "B" Constructors, Inc., v. Granite Falls School District.
Edsall Construction Co., Inc., ASBCA.
Interstate Contracting, Inc. v. City of Dallas, Texas, U.S. Court of Appeals for the Fifth Circuit, 4/22/05.
Soave v. National Velour Corp., Supreme Court of Rhode Island. 12/14/04.
State v. Arkell (Minnesota).
Vincent T. Preston v. Condon Construction and Realty, Inc., Wisconsin Court of Appeals.

Chapter 12

The No-Damages-for-Delay Clause

In recent years, there have been a number of construction disputes involving no-damages-for-delay clauses in the construction contract. These are contract clauses in which, in the event of a schedule delay caused by the owner, the drafter of the contract limits his or her liability to a time extension only. The contractor is not entitled to monetary damages caused by the delay. The clause creates a formidable barrier for the contractor who is harmed economically by the delay.

Contract Language

The no-damages-for-delay clause is not contained in the standard contract forms published by the American Institute of Architects (AIA 1987) Document A201, Engineers Joint Contract Documents Committee (EJCDC) Document 1910–8, and Standard Form 23A (Federal Contract). A typical no-damages-for-delay clause is as follows:

> The Contractor agrees to make no claim for damages for delay in the performance of this contract occasioned by any act or omission to act of the City or any of its representatives, and agrees that any such claim shall be fully compensated for by an extension of time to complete performance of the work as provided herein. (Kalisch-Jarcho v. City of New York)

However, the language of the clause is sometimes less direct. Clauses such as the following may escape the casual reading of the contractor.

> EXTENSIONS OF TIME. Should the Contractor be delayed in the final completion of the work by any act or neglect of the Owner or Engineer or of any employee of either, or by any other contractor employed by the owner, or by strike, fire, or other cause outside of the control of the Contractor and which, in the opinion of the Engineer, could have been neither anticipated or avoided, then an extension of time sufficient to compensate for the delay, as determined by the Engineer, will be granted by the Owner provided that the Contractor gives the Owner and the Engineer prompt notice in writing of the cause of the delay in each case and demonstrates that he has used all reasonable means to minimize the delay. Extensions of time will not be granted for delays caused by unfavorable weather, unsuitable ground conditions, inadequate construction force, or the failure of the contractor to place orders for equipment and materials sufficiently in advance to insure delivery when needed. (Peter Kiewit & Sons Co. v. Iowa Southern Utility Co.)

The above clause provides for an extension of time as the remedy for certain delays. In this instance, nowhere else in the contract were there allowances for an equitable adjustment of the contract sum resulting from delays. In the absence of such a provision, cost adjustments for delays were not contemplated by the parties, and a time extension was determined to be the sole remedy. This viewpoint is in contrast to Article 8.3.3 of AIA A201, Delays and Extensions of Time, which states, "This paragraph 8.3 does not preclude recovery for damages for delay by either party under other provisions of the Contract Documents" (AIA 1987).

The above clauses were all included in the time extension provisions of the contract; however, language related to monetary damages may be found in other parts of the general conditions or in the special conditions. It should be obvious that a careful reading of the entire contract is important.

Judicial Attitude

Courts frequently have been asked to enforce a no-damages-for-delay clause. Generally, the no-damages-for-delay provision is valid and enforceable if it meets the ordinary rules governing the validity of contracts. However, courts have used restraint in enforcing no-damages-for-delay clauses because of their harshness, and such clauses are strictly construed against the owner, who is usually the drafter of the exculpatory provision (Western Engineers v. Utah State Road Commission).

Western Engineers, Inc., v. Utah State Road Commission enumerated the applicable rules of application from 10 A.L.R.2d 803. To avoid application of the no-damages-for-delay clause, it must be shown that (1) the delay is the result of fraudulent, malicious, capricious or unreasonable acts on the part of the owner to delay or harass the contractor in performance of the work; (2) the delay was unreasonably long, such that the contractor would be justified in abandoning the contract; (3) the delay is not within the specifically enumerated delays to which the clause is to apply; and (4) the delay was not of the type contemplated by the parties at the time the contract was signed (10 A.L.R.2d 803). With the exception of No. 2, these issues are discussed in detail below.

Rules of Application

The interpretation process involves three primary inquiries. These are

- Is the delay enumerated in the contract?
- Was the delay of the type contemplated by the parties?
- Did the owner act responsibly?

The no-damages-for-delay clause seeks to place the entire economic risk of delays on the contractor, and the sole remedy for the contractor is a time extension. To recover monetary damages, the contractor must prove that the clause should not be enforced. A positive response to each of the above three inquiries will result in the enforcement of the no-damages-for-delay clause. However, different jurisdictions have applied different criteria to individual cases. In Massachusetts, because a no-damages-for-delay clause is required in public contracts, courts have been slow to recognize exceptions to enforcing no-damages-for-delay clauses like the ones enumerated in this chapter. In New York state, the standards for applying exceptions has been rather strict. In some states, legislation bars the use of no-damages-for-delay clauses in public construction. Many other states are thought to recognize exceptions to the no-damages-for-delay clause.

Is the Cause of the Delay Enumerated in the Contract?

No-damages-for-delay clauses can be broadly worded, covering many or all delays, or they can cover specific types of delay as enumerated in the contract. As in any contract interpretation issue, a careful reading of the contract is essential. When no-damages-for-delay clauses are broadly written and are unambiguous, the clause likely indemnifies the owner from a wide variety of unforeseen delays (Western Engineers, Inc., v. Utah State Road Commission). However, in some instances, no-damages-for-delay clauses are written to address specific aspects of the contractor's undertaking (74 A.L.R.3d 199 (1967)). Under these circumstances, the clause will exclude many other causes of delay.

Peckham Road Co. v. State of New York illustrates this situation. Peckham Road Co. contracted with the state of New York to construct a new road and reconstruct a part of the Ossining-Kitchawan State Highway. The new highway passed through the Hess and Seymore-Bradley properties. The contract contained a special note stating,

> The Contractor should understand that immediate possession of all buildings within the highway limits is not now available. Negotiations for possession are now in process and such buildings, with their surrounding premises, will be available to the Contractor as soon as they have been vacated. All bids submitted should be made on the basis that work can be performed only on vacant buildings, and that as to occupied buildings and surrounding premises, the State is taking all reasonable steps to gain early possession. The Contractor must realize, however, that the proceedings to obtain possession can become lengthy legal proceedings, and no claim shall be brought against the State for failure to obtain early possession.

The work began on March 8, 1961. The appropriation papers were not filed until May 24, 1961, and the last property was not vacated until September 25, 1961. Although the project was completed on time, Peckham submitted a claim for delay damages. The New York Supreme Court rejected Peckham's claim because it was precisely the type provided for in the contract and was therefore contemplated by the parties.

In another case, Ace Stone, Inc., v. Township of Wayne, the contractor, Ace Stone, was awarded a contract to construct a sewer line. At the preconstruction meeting, the engineer conveyed the importance of time and told Ace that it would be held to the 80-day completion date. The engineer also advised Ace that to expedite the work, it was to start work in three separate locations designated by the engineer. Ace was further advised that there were no easement problems and that the work could commence without interruption.

The time extension clause provided that the township could defer the beginning or suspend the whole or part of the work whenever the engineer considered it necessary or expedient to do so, and that if delayed, the contractor was entitled to a time extension only.

Unfortunately, Ace was delayed by easement problems, which meant that it could only work at one of the three locations identified by the engineer. Ace could not proceed in an orderly, continuous, and economical manner; work stoppages ensued; and work had to be carried out in the winter months. Ace's claim for delay damages was subsequently denied by the court on the basis of the time extension clause.

The Supreme Court of New Jersey spoke at length about whether the nature of the delay encountered by Ace was within the scope of the clause, citing numerous supportive cases where the contractor's work was delayed because of easement and site access problems. McGuire and Hester v. City and County of San Francisco and Sheehan v. City of Pittsburgh illustrate situations where broadly worded clauses were not determined to apply to delays in site access and recovery was allowed. Conversely, in Christhilf v. Mayor and City Council of Baltimore and A. Kaplen & Son, Ltd., v. Housing Authority of City of Passaic, recovery by the contractor was denied because of the precise wording of the no-damages-for-delay clause. Site access and lack of easements were the cause of the delays in both cases. The operative language of the clause in Christhilf stated that delays caused by "failure or inability of the city to obtain title to or possession of any land" would entitle the contractor to an extension of time only. The operative language in Kaplen spoke directly of the site not being released to the contractor. Thus, the contract must be carefully read.

Was the Delay of the Type Contemplated by the Parties?

Where broadly worded, the no-damages-for-delay clause is enforced, provided the delays are of the type contemplated by the parties at the time the contract was created. Thus, it is necessary to determine the contemplation or common intention of the parties. This intention is gathered from the language of the contract, read "in the light of the existing facts with reference to which it was framed" (30 A.L.R. 209 (1967)).

Kalisch-Jarcho, Inc., v. City of New York was a case involving the construction of the heating, ventilation, and air conditioning portion of the New York City police headquarters. The contract contained a general, broadly worded no-damages-for-delay clause absolving the city of liability for delays caused by any act or omission to act.

The 1,000-day completion schedule was extended by an additional 28 months because of the city's endless revisions of scores of plans and drawings and its alleged failure to coordinate the activities of the prime contractors (Kalisch-Jarcho was one of four prime contractors). In a complex decision, the Court of Appeals of New York determined that the delays were clearly, directly, and absolutely within the contemplation of the parties.

Peter Kiewit and Sons Co. contracted in 1966 as general contractor to construct a fossil fuel electrical power plant for Iowa Southern Utilities Co. Kiewit's work included pile driving for foundations, pouring concrete slabs, erecting walls and ceilings for various structures, and painting. Separate contracts were awarded to specialty contractors for other portions of the work. Kiewit's contract contained the following provision:

> If the work of the contractor is delayed ... the contractor shall have no claim against the owner on that account other than an extension of time. The contractor expressly agrees that the construction period named in the contract agreement includes allowances for all hindrances and delays incident to the work. (Peter Kiewit & Sons Co. v. Iowa Southern Utility Co.)

The engineer, Black and Veatch, prepared an overall construction schedule in the form of a bar chart. The schedule, titled "G-1," was made part of the general, mechanical, and electrical contracts. The G-1 schedule was not incorporated in the boiler, turbine, or steel contracts because these contracts were awarded before the issuance of the G-1 schedule. Because problems continually arose on the site, the new work schedules became obsolete almost as soon as they were issued. This obsolescence required Black and Veatch to make frequent schedule modifications on site.

The contract also contained a contract clause titled "Project Management," which stated,

> In the event conflicts arise between contractors concerning scheduling or coordination, the Engineer will make the final decision resolving the conflict. The Engineer's decision shall not be the cause for extra compensation or for extension of time. (Peter Kiewit & Sons Co. v. Iowa Southern Utility Co.)

Ultimately, Kiewit was delayed by the steel erection contractor, material delays, labor problems, equipment storage problems, heavy rain, and the frequent changes to the project schedule.

The court found that Iowa Southern and its agent, Black and Veatch, acted responsibly because there was nothing to indicate gross negligence or interference that would constitute bad faith or hindrance. Both parties acted to support the goal of prompt completion of the project. The schedule modifications were made within the contractual authority of the engineer. Accordingly, the no-damages-for-delay clause was enforced.

From a review of these and numerous other cases, it is recognized that delays that are contemplated by the parties can be termed ordinary and usual delays. Thus, contractors can recover damages only for extraordinary and unusual delays.

Ordinary and Usual Delays

The following examples illustrate ordinary and usual construction delays. In J. D. Hedin Construction Co. v. United States, the no-damages-for-delay clause was enforced where delays were caused by usual and long-established contractor means and methods used in the commercial world. In Broadway Maintenance Corp. v. Rutgers, the lack of elevators and stairs, lack of temporary light and power, increased wages, increased material costs, and lack of temporary heat were considered ordinary and usual, as were accidents, material shortages and delays in material deliveries, inclement weather, delayed subcontractor performance, and rework. Additionally, delays in the approval of shop drawings and labor inefficiencies are also likely to be defined as ordinary and usual.

Extraordinary and Unusual Delays

The following examples illustrate extraordinary and unusual construction delays. The no-damages-for-delay clause has not been enforced where delays were caused by delayed site access (Ace Stone, Inc., v. Township of Wayne), by deliberate delays to the delivery of five joints of 60-in. reinforced concrete pipe, an express contract obligation (Sandel and Lastripes v. City of Shreveport), when the owner ordered the contractor to proceed with the work when the owner knew that the contractor would not have access to the site for 14 weeks (Commonwealth of Pennsylvania Department of Transportation v. S. J. Groves and Sons Co.), by an unreasonable work schedule (Blake Construction Co. v. C. J. Coakley Co., Inc.), and by sporadic, unorthodox, and more expensive excavation at a medical outpatient facility (Chicago College of Osteopathic Medicine v. George A. Fuller Co.).

Did the Owner Act Responsibly?

If delays are within the scope of the clause and are ordinary and usual, the only remaining recourse for a contractor to avoid the harsh consequences of a no-damages-for-delay clause is to show that the owner demonstrated reprehensible behavior with regard to its contract obligations. This behavior can be in the form of fraud, malice, bad faith, hindrance, or gross negligence.

Various jurisdictions may apply different standards for deciding this issue. The more lenient standard is that if the owner showed gross negligence or hindered the contractor, then the owner acted irresponsibly, and the contractor will be allowed to recover delay damages. A few jurisdictions have applied a more stringent test by requiring a showing of malice or fraud. Thus, it is unwise to make too many generalizations regarding this issue. The various forms of irresponsibility are discussed below. The forms are listed generally in order of least to most stringent. For the more stringent criteria, it is extremely difficult for the contractor to prevail.

Simple Negligence

In general, a showing of more than simple negligence is necessary for the no-damages-for-delay clause to be set aside (Kalisch-Jarcho v. City of New York). In Anthony P. Miller, Inc., v. Wilmington Housing Authority, the court enforced the no-damages-for-delay clause where the delays were caused by simple negligence involving inaction, lack of diligence, and lack of effort. The Federal District Court felt that the actions of the housing authority fell short of bad faith.

Bad Faith

Every contract implies good faith and fair dealing between the parties. Bad faith includes gross negligence, arbitrary and capricious acts, and wrongful interference with the contractor's work. Wrongful, willful, or deliberate conduct by the owner is often the deciding factor in many judicial decisions, and where the standard test for owner responsibility is bad faith, the contractor may recover monetary damages despite a no-damages-for-delay clause (74 A.L.R.2d 215 (1967)).

A contractor may be entitled to monetary delay damages if the owner is grossly negligent (74 A.L.R.2d 216 (1967)). In Ozark Dam Constructors v. United States, the court set aside the no-damages-for-delay clause and allowed the contractor to recover monetary damages for delays and obstructions where the government acted in bad faith. Ozark contracted with the U.S. Army Corps of Engineers to build a concrete gravity dam and appurtenances work at the Bull Shoals dam site on the White River in Arkansas. Ozark was to furnish all labor, equipment, and materials except for the cement, which was to be ordered and furnished by the U.S. Army Corps of Engineers. The contract stated, "The Government will place an order with the mills for the Portland and natural cements which will be available approximately August 1, 1948." The contract contained a general no-damages-for-delay clause stating that if the work were delayed for causes beyond the control of the contractor, including strikes, the only remedy to the contractor would be a time extension.

For more than a year, employees of the Missouri Pacific Railroad had authorized a strike. The strike was called for July 11, 1948, but was postponed to allow the National Mediation Board to intervene. The board's efforts were unsuccessful, and the strike began on September 9 and ended on October 24. The events leading up to the strike were known to the U.S. government and to Ozark. The government, however, took no steps to investigate possible alternative ways of getting cement to the job until September 19, even though the cement could have been easily shipped by another railroad. On September 16, Ozark wrote to the contracting officer and requested that the project be suspended as allowed by the contact. The contracting officer refused the request. Because of the delayed deliveries, Ozark reduced its work force and slowed down the work. Ultimately, the untimely deliveries of cement delayed the job by 43 days and caused Ozark to perform much of its work during the winter months.

The court found that the Corps of Engineers' actions were grossly negligent. "The possible consequences were so serious and the action necessary to prevent those consequences so slight, that the neglect was almost willful. It showed a complete lack of consideration for the interests of (Ozark)."

In Peter Kiewit & Sons Co. v. Iowa Southern Utility Co., the contractor was delayed by the sanitary district's failure to obtain rights of way. The court held that a no-damages-for-delay clause was unenforceable because the cause of the delay was negligent, willful, and long-lasting. It felt that the district knowingly delayed the contractor and that their actions transcended "mere lethargy or bureaucratic bungling."

Arbitrary and capricious acts of the owner often reflect bad faith because such action is considered wrongful, willful, or deliberate conduct. Contractors have been awarded delay damages in many of these cases. In Hoel-Steffen Construction Co. v. United States, the court found that a contract officer's failure to determine if the contractor could substitute another subcontractor was capricious and arbitrary, and the contractor was awarded monetary delay damages.

In Housing Authority of Dallas v. J. T. Hubbell, the court found 11 separate acts or failures to act by the owner, which delayed the contractor's construction of a housing project. The court stated,

> The no-damage-for-delays provision was intended to protect Owner from damages for delays caused by others than Owner, and was intended also to protect Owner from damages for delays caused by Owner itself even if such delays were due to Owners negligence and mistakes in judgment. But the 'no-damage-for-delay' provision did not give Owner a license to cause delays 'willfully' by 'unreasoning action', 'without due consideration' and in 'disregard of the rights of other parties', nor did the provision grant Owner immunity from damages if delays were caused by Owner under such circumstances. (Housing Authority of Dallas v. J. T. Hubbell)

Thus, the housing authority was liable for delay damages despite the no-damages-for-delay provision.

Courts have applied the bad faith rule to resolve delay disputes involving the owner's interference in the contractor's work. Courts distinguish ordinary from bad faith or interference with the labels of "direct," "active," or "willful" interference with the contractor's performance. Most courts are in general agreement that a broad no-damages-for-delay clause will not protect an owner from liability for delay damages when the delay is caused by the owner's interference or bad faith (74 A.L.R.2d 219 (1967)). In Peter Kiewit & Sons Co. v. Iowa Southern Utility Co., the court held that to avoid the consequences of a no-damages-for-delay clause, active interference requires some affirmative, willful act, in bad faith, to unreasonably interfere with the contractor's compliance with the contract. Bad faith requires more than a simple mistake, error in judgment, lack of effort, or lack of diligence. In Cunningham Brothers, Inc., v. Waterloo, the court held that interference requires reprehensible conduct of the owner which is "in collision with or runs at cross purposes to the work." And in Northeast Clackamas County Electric Co-op., Inc., v. Continental Casualty Co., the no-damages-for-delay clause was not enforced when the owner breached its contract, without justification, by failing to clear a utility line right of way of heavy timber and trying to coerce the contractor to perform repair work that resulted because the co-op inadequately cleared the right of way. The court outlined the applicable contract interpretation by saying that the co-op's unjustified repudiation of the particular contract provision and subsequent wrongful termination of the entire contract made enforcement of the particular provision impossible in fact and inapplicable in law.

Bad faith may connote a dishonest purpose. In United States Steel Corp. v. Missouri Pacific Railroad Co., the court found that the owner acted in bad faith when it issued a notice to proceed to the contractor to build a bridge when the site was occupied by the substructure contractor, who was behind schedule. Thus, the United States Steel Corp. was delayed in completing the bridge. According to the court, the notice to proceed was dishonest (unworthy of trust or belief) because Missouri Pacific knew that the substructure contractor was behind schedule. In C. J. Langenfelder & Son v. Commonwealth of Pennsylvania, the Pennsylvania Department of Transportation acted in bad faith when it assured Langenfelder & Son that it could deposit sediment outside the right of way in wetlands, while simultaneously assuring environmental groups that construction in the marsh would be limited to the highway right of way. This was a dishonest act because the state compromised its integrity and truthfulness.

Generally, bad faith requires more than simple procrastination, and many cases involving allegations of bad faith have not been unsuccessful. In F. D. Rich Co. v. Wilmington Housing Authority, Rich argued that adverse soil conditions delayed a housing project, causing Rich additional expenses to provide heat for an additional winter season and to store electric refrigerators that were on hand but could not be installed. Rich's argument that the soil conditions constituted bad faith on the part of the housing authority was rejected.

Malice and Fraud

The New York Court of Appeals in Kalisch-Jarcho, Inc., v. City of New York outlined a strict approach to the enforcement of the no-damages-for-delay clause by applying a malice standard. The Court defined malice as "the state of mind intent on perpetrating a wrongful act to the injury of another without justification." The contractor, Kalisch-Jarcho, was delayed in the construction of the city police headquarters by the city's revisions of plans and the city's failure to coordinate the activities of the prime contractors. The court reasoned that these were all delays that were contemplated by the contract. The court of appeals determined that a stricter standard was required to void a no-damages-for-delay clause because the clause "would have little meaning if it were not read to extend acceptability to a range of unreasonable delay as well." The city's actions were not found to constitute malice, and the clause was enforced.

Fraud is the intentional deception of one person by another. The deception may consist of false statements or the partial concealment of information. In E. C. Nolan Co., Inc., v. State of Michigan, the court held that to prevail on a claim of fraud or misrepresentation, a contractor must establish that the owner made a material representation; that the representation was false; that the owner knew the representation to be false or made the representation recklessly, without any knowledge of the truth and as a positive assertion; that the representation was made with the intention that it should be acted on by the contractor; that the contractor acted in reliance on it; and that the contractor suffered an injury. Nolan was allowed to recover when the department of highways represented a progress schedule as showing firm starting dates when, in fact, the dates were merely projected, a fact known to the state.

In another example, a contractor falsely stated that it has access to special equipment required for the project (Sweet 1989, p. 504), and recovery was allowed.

Illustrative Examples

The application of the rules of application is illustrated with two appellate decisions addressing entitlement to delay damages that involved no-damages-for-delay clauses.

Case No. 1: Blake Construction v. Coakley

Statement of Facts

On July 9, 1974, C. J. Coakley Co. entered into a subcontract with Blake Construction Co. for the fireproofing of structural steel as part of the construction of the Walter Reed Army Hospital in the District of Columbia, a $100 million project. The subcontract amount was $570,000. Article 2(b) of the subcontract stated that,

> No such delay [caused by reasons beyond the Subcontractor's control] shall give rise to any right to the Subcontractor to claim damages therefor from the contractor. (Blake Construction Co. v. C. J. Coakley Co.)

Specification section 9K/6.6 provided

> Ducts, piping or conduit or other suspended equipment that could interfere with the uniform application of the fireproofing material are to be positioned *after* the application of the sprayed fireproofing. (Blake Construction Co. v. C. J. Coakley Co.)

Coakley's work was disrupted by Blake's delays in ordering and receiving structural steel; from stacking of trades, which caused overcrowding; and when other subcontractors damaged completed work of Coakley, necessitating rework. Blake failed to follow the project schedule because of the delays in ordering and receiving structural steel, which resulted in Coakley having to use scaffolding platforms. Coakley had to shift from floor to floor, rather than completing work on a systematic, scheduled, floor-by-floor basis. Furthermore, Blake did not provide a reasonably clear and convenient work area to Coakley. Blake failed to sequence the work reasonably to allow Coakley to perform on time, and failed to supervise other subcontractors, who ultimately disrupted Coakley's work.

Analysis

The no-damages-for-delay clause was broadly written; however, there were numerous references to scheduling and sequencing responsibilities of Blake. Specification provision 9K/6.6 is but one example of the kind of language that can be used to establish that the delays encountered by Coakley were clearly beyond the kind of delays contemplated by the parties. Coakley should be able to recover on this basis.

The prime contractor (Blake) had the obligation to provide suitable work areas and to coordinate the work of the various subcontractors. On this project, Blake hindered or prevented Coakley's performance; scheduled the on-site work in an unreasonable sequence; did not provide a job site in suitable condition for Coakley to perform its work; and did not cooperate with Coakley when necessary to ensure Coakley's performance. The court found no fraudulent deceit by Blake but presumed that Blake was more likely to have been disorganized and confused. However, the court of appeals determined that the delays resulted from conduct amounting to active interference, largely because of Blake's improper work sequencing. Thus, Coakley was allowed to recover damages despite the no-damages-for-delay clause.

Case No. 2: L. S. Hawley v. Orange County Flood Control District

Statement of Facts

In 1959, L. S. Hawley contracted with the Orange County Flood Control District to construct a sewer line and three manholes of the Huntington Beach Channel and a portion of the Talbert Channel in southern California. The specifications contained the following no-damages-for-delay provision:

> Furthermore, if the contractor suffers any delay caused by the failure of the District to furnish the necessary right-of-way or materials agreed to be furnished by it, or by failure to supply necessary plans or instructions concerning the work to be done after written request therefor has been made, the contractor shall be entitled to an extension of time equivalent to the time lost for any of the abovementioned reasons, but shall not be entitled to any damages for such delay. (L. S. Hawley v. Orange County Flood Control District)

Before the excavation, the flood control district notified Hawley that the plans contained the wrong type of manhole, and Hawley was instructed not to install any of the manholes until the plans were revised. Hawley began excavation on June 24, 1959, and completed the excavation on September 10, 1959. A particular section of trench contained one manhole that extended about 30–40 ft. The excavation depth was 6–8 ft. The water table in this area was above the bottom of the trench, and 12–14 in. of groundwater was present in the trench from July 15 through September 10. The revised manholes were not authorized until August 26, and Hawley began work on the manholes the next day.

On July 30, 1959, Hawley notified the flood control district that the open trench was a dangerous situation and requested permission to backfill the trench. The flood control district refused the request, opting instead to wait until testing was complete. On several other occasions, Hawley made similar requests, but the district would not allow Hawley to backfill any portion of the trench. As of September 10, Hawley had completed the excavation, laid all of the pipe for the sewer, completed the manholes, and left the excavation open until the system was tested. Before the sewer line could be tested, the banks of the trench caved in, knocking the sewer line out of alignment and causing it to separate at the joints. The cause of the trench cave-in was the gradual weakening of the banks of the trench that occurred as a natural result of allowing it to stand open from July 15 to September 10. The trenches had been open for up to 11 weeks, whereas the normal time for the district to revise and reissue the revised manhole drawings would have been about one week.

Analysis

Two arguments could be presented as causing the delay. The first argument is that the flood control district caused the delay by not providing correct plans. The second argument is that the delay could have been caused by the district by not allowing Hawley to backfill the trenches. Both arguments (failure to supply necessary plans or permission to backfill) are expressly contemplated by the contract. Thus, Hawley must resort to an examination of the district's actions.

The central question is whether the district was justified in denying Hawley's request to backfill the trench. If there were valid reasons for the denial, then Hawley cannot recover. However, if the district's actions are seen as arbitrary, then Hawley has valid arguments for recovery.

Expert testimony established that the only effective way to protect the trench was to cover it as requested by Hawley. The court determined that the district wrongfully interfered in the work by not allowing the contractor to backfill any portion of the trench until the sewer line had been fully tested. Their actions amounted to active hindrance. The court said that the district created an unreasonable delay in refusing to allow work progress of backfilling at a point where such progress was apparently the only way the contractor could effectively protect the work already done.

References

Ace Stone, Inc., v. Township of Wayne 221 A.2d 515 (1966).

A. Kaplen & Son, Ltd., v. Housing Authority of City of Passaic 126 A.2d 13 (1956).

American Institute of Architects (AIA). (1977). Engineers Joint Contract Documents Committee, Document 1910-8, American Institute of Architects, Washington, DC.

American Institute of Architects (AIA). (1987). *General Conditions of the Contract for Construction*, AIA A201, American Institute of Architects, Washington, DC.

American Institute of Architects (AIA). (1997). Standard Form 23A, Federal Contract, American Institute of Architects, Washington, D.C.

Anthony P. Miller, Inc., v. Wilmington Housing Authority 165 F.Supp. 275 (1958).

Blake Construction Co. v. C. J. Coakley Co., Inc. 431 A.2d 569 (1981).

Broadway Maintenance Corp. v. Rutgers 447 A.2d 906 (1982).

Chicago College of Osteopathic Medicine v. George A. Fuller Co. 776 F.2d 198 (1985).

Christhilf v. Mayor and City Council of Baltimore 136 A. 527 (1927).

C. J. Langenfelder & Son v. Commonwealth of Pennsylvania 404 A.2d 745 (1979).

Commonwealth of Pennsylvania, Department of Transportation v. S. J. Groves and Sons Co. 343 A.2d 72 (1975).

Cunningham Brothers, Inc., v. Waterloo 117 N.W.2d 46 (1962).

E. C. Nolan Co., Inc., v. State of Michigan 227 N.W.2d 323 (1975).

F. D. Rich Co. v. Wilmington Housing Authority 392 F.2d 841 (1968).

Hoel-Steffen Construction Co. v. United States 684 F.2d 843 (1982).

Housing Authority of Dallas v. J. T. Hubbell 325 S.W.2d 880 (1959).

J. D. Hedin Construction Co. v. United States 408 F.2d 424.

Kalisch-Jarcho, Inc., v. City of New York 448 N.E.2d 414 (1983).

L. S. Hawley v. Orange County Flood Control District 27 Cal.Rptr. 478 (1963).

McGuire and Hester v. City and County of San Francisco 247 P.2d 934 (1952).

Northeast Clackamas County Electric Co-op., Inc., v. Continental Casualty Co. 221 F.2d 329 (1955).

Ozark Dam Constructors v. United States 127 F.Supp. 187 (1955).

Peckham Road Co. v. State of New York (300 N.Y.S. 2d 174 (1969).

Peter Kiewit & Sons Co. v. Iowa Southern Utility Co. 355 F.Supp. 376 (1973).
Sandel and Lastripes v. City of Shreveport 129 So.2d 620 (1961).
Sheehan v. City of Pittsburgh 62 A. 642 (1905).
Sweet, J. (1989). *Legal Aspects of Architecture, Engineering, and the Construction Process,*
    5th Ed. West Publishing Co., St. Paul, MN, p. 504.
United States Steel Corp. v. Missouri Pacific Railroad Co. 668 F.2d 435 (1982).
Western Engineers, Inc., v. Utah State Road Commission 437 P.2d 216 (1968).

Additional Cases

The following are additional cases related to issues associated with no-damages-for-delay clauses. The reader is invited to review the facts of the case, apply the decision criteria, reach a decision, compare it with the judicial decision, and determine the rationale behind the judicial decision.

Clifford R. Gray, Inc., v. City School District of Albany (New York) 277 A.D.2d 843
    (2000).
J. A. Jones Construction Co. v. Lehrer McGovern Bovis, Inc., 89 P.3d 1009 (2004).
Mississippi Transportation Commission v. Ronald Adams Contractor, Inc., 753 So.2d
    1077 (2000).

Chapter 13

Substantial Completion

Timely completion of a construction project is one goal of the owner and contractor. When completion is delayed, each party is likely to incur additional costs and lose potential revenues. The liquidated damage assessment usually continues until there is substantial completion of the project. Thus, in determining amounts of liquidated damages, it becomes important to know when substantial completion has been reached.

Contract Language

The American Institute of Architects (AIA 1987) defines *substantial completion* as follows:

> Substantial Completion is the stage in the progress of the Work when the Work or designated portion thereof is sufficiently complete in accordance with the Contract Documents so the Owner can occupy or utilize the Work for its intended use. (AIA 1987)

The basic premise—when the project is complete in accordance with the contract documents and can be used for its intended purpose—is not always well defined and often requires the evaluation and certification of a design professional or professional engineer.

Background

Owners typically specify a completion date or the number of calendar days allowed for performance of work. Failure to complete a project on time can become a major source of dispute between the parties.

Usually, the contract specifies the procedures to be used to determine substantial completion. Normally, the contractor notifies the design professional that the project is substantially complete. If an inspection by the design professional confirms the declared project status, the design professional may prepare a certificate of substantial completion and delineate the punch-list items to be performed before final completion. The contractor is entitled to all payments except for the final payment, which is sufficient to cover the cost of the punch-list items.

Determining substantial completion can be a point of contention. As defined by one court, substantial completion is reached when the project is sufficiently complete so that the owner may occupy it or use it for its intended purpose, whether he does so or not (State of Louisiana v. Laconco). The contractor may contend that substantial completion has been reached, but a certificate may not be issued by the designer because of nonperformance of a minor item. The owner may express discontent toward the final product, its outcome, or nonconformity to specifications. The problem is further compounded if the contract does not identify procedures for determining substantial completion.

Significance of Substantial Completion

Substantial completion is an important project milestone for a number of reasons:

- entitlement to the final payment,
- the owner's right to use the facility,
- liquidated damages,
- default termination,
- responsibility for the facility, and
- statute of limitations.

Entitlement to the Final Payment

On reaching substantial completion, the contractor is entitled to the final or major payment (including retainage) minus deductions for work not yet completed. Deductions are generally limited to punch-list items or remedying defects in the work.

In the landmark case State of Louisiana v. Laconco, the contractor constructed a 60-person National Guard armory for the state of Louisiana. The architect certified that the building was substantially complete with the exception of punch-list items valued at 3% of the contract price. The state of Louisiana moved into the building and withheld 10% of the contract price until the punch-list items were complete. Laconco sued, and the court subsequently ruled that the state could withhold only a reasonable amount of money necessary to cover the completion of the punch-list items. The state was ordered to release 7% of the contract price.

Conversely, the final payment of retainage need not be paid if the contractor fails to substantially complete the project. In Keating v. Miller, the contractor, Keating, constructed 75% of a home for Miller before his contract was terminated. The electrical, plumbing, heating and air conditioning, and carpentry work had not been finished. In addition, the floors, kitchen appliances, and cabinets had not been installed. During construction of the home, a portion of a brick wall fell down. Miller wanted the entire wall replaced; Keating refused, opting only to fix the damaged portion of the wall. Miller terminated the contract, and Keating sued for the balance of the contract amount. Keating believed that he was due these payments because the home was substantially complete. The court ruled that because Keating completed only 75% of the home and Miller could not comfortably live in the unfinished structure, the work was not substantially complete and Keating was not entitled to the total contract amount less allowances for the uncompleted work.

Owner's Right to Use the Facility

Once substantial completion has been achieved, the owner has the right to occupy or operate the facility or structure for its intended purpose, whether he or she does so or not. As stated by one court,

> Substantial completion is a critical date, and it comes into exis-
> tence whether the owner needs the building or not. (Ricchini and
> O'Brien 1990)

Occupying or operating the facility by the owner is not a requirement for substantial completion; however, the owner must have the option to use the facility. Many landmark cases have considered this consequence as a criterion in determining if substantial completion was achieved.

In American Druggists Insurance Co. v. Henry Contracting, a dispute arose over the exact date of substantial completion. American had contracted with Henry for construction of an 8-in. water line from an existing pharmaceutical plant. Henry finished the project one day before the completion date, but failed to have the line sanitized and tested until 42 days later. The court ruled that the later date was the earliest time the owner could have used the new water line, and substantial completion was based on this date.

In another case, Rudy Brown Builders v. St. Bernard Linen Service, the dispute was similarly based on disagreements over the use for its intended purpose. St. Bernard contracted with Brown for the construction of a new building to house laundry and dry cleaning equipment. On completion of the project, St. Bernard withheld the final payments due Brown because the floor slab was not uniformly 6 in. thick, as required by the contract documents. St. Bernard hired expert witnesses who testified that the slab varied in thickness from 6.15 to 3.90 in., with an average thickness of 5.13 in. However, these witnesses also agreed that the slab could support the equipment loads, even though it was not quite to specifications. The court ruled that Brown had substantially completed the project because it could be used for its intended purpose. Brown's final payment was reduced because the slab was not constructed in accordance with the contract documents.

These two cases highlight the importance of use for its intended purpose as a measure of substantial completion. In the absence of the owner's opportunity of use, courts have frequently denied the existence of substantial completion.

Liquidated Damages

Liquidated damages are intended to compensate an owner for lost use or revenue if the contractor is late in completing the project. Liquidated damages cannot be assessed after the date of substantial completion. If the facility is suitable for occupancy and can be used for its intended purpose, the owner is no longer harmed by loss of its use.

Default Termination

Once substantial completion has been reached, the contractor cannot be considered in breach of contract for failing to complete the project, and the owner may not terminate the contract for default. By achieving substantial completion, the contractor has honored its contractual commitment and cannot be assessed damages because of default.

Responsibility for the Facility

There are owner responsibilities associated with the right to occupy or operate the facility. For example, operation and maintenance of the equipment (for example, heating, venting, and air conditioning) shifts to the owner. The owner is also responsible for security of the facility, utilities, and any necessary insurance costs. Warranty and guarantee periods for equipment in the facility (required by the contract documents) also begin on substantial completion, and the responsibility for their compliance falls on the owner.

Statute of Limitations

In many states, the statute of limitations for defective construction commences on substantial completion, not the completion date of component parts, subcontracts, or individual systems (Bramble 1987).

Evolution of Substantial Completion

In the mid-1800s, a contractor completing 99% of a project could not recover any withheld payments (and in some instances, the entire payment) because he or she breached the contract by not completing the remaining 1%. Gradually, a common-law principle known as *substantial performance* evolved. This principle ensures that contractors are compensated even though they have not performed the work covered by the contract to exact perfection. This principle also prevents owners from being unjustly enriched by receiving the benefits of an almost complete project without paying the full price (Black 1979).

In the 1890 decision of Leeds v. Little, the trial court discussed this principle:

> Substantial Performance of a contract to construct a building does not mean exact performance in every slight or unimportant detail. In many cases, such as building contracts, notwithstanding the most honest, diligent and intelligent effort to fully perform in every particular, yet owing to oversight, inadvertence, or some slight omissions or defects may be discovered. To hold that a builder could not in any such case recover on his contract would be too rigid a rule to apply to the practical affairs of life.

In a more recent decision, Bruner v. Hines, the court reinforced the application of this principle by stating,

> The doctrine of substantial performance is especially useful in building contracts because of the difficulty of reproducing on the construction site the precise specifications of blueprint drawings. Often comparable materials of different brands will have to be substituted for specified but unobtainable brands, and foundation specifications on drawings will bend somewhat to the realities of pouring concrete.

Primary Rules of Inquiry

Courts on numerous occasions have intervened to resolve disputes over substantial completion. Based on a review of multiple appellate decisions, the following are the primary inquiries that are made:

- How extensive is the claimed defect or incomplete work?
- To what degree was the purpose of the contract defeated?
- How easy is the defect to correct?
- Has the owner benefited from the work performed?

Obviously, the issue is not one simply of occupancy. All questions need to be addressed, and a harmonious outcome should be sought.

How Extensive Is the Claimed Defect or Incomplete Work?

Contractors and administrators need to consider the overall completeness of the facility by considering the extent of the defects or nonperformance of work and whether the building has met its essential purpose (Hadden v. Consolidated Edison Co. of New York). However, in one decision, substantial completion was achieved despite a cost of correction to the total contract price of 31% (Stevens Construction Corp. v. Carolina Corp.). The cost of correction versus the diminished value of the project may also be important (Jardin Estates v. Donna Brook Corp.; Plante v. Jacobs). Obviously, no steadfast rule specifies a ratio or percentage in determining substantial completion. To amplify, *Corbin on Contracts* states,

> In the case of a building contract, it is not easy to find the arithmetical ratio between the unperformed part and the full promised performance. The difference may be in quality of materials and workmanship, rather than in board feet or bags of cement. It is obvious that any rule stated in terms of extent of non-performance cannot be a rule of thumb.... The higher the ratio of this cost of curing defects to the total contract price, the less likely it is that the performance rendered will be held to be substantial performance. (Corbin 1963)

Several cases illustrate how the extent of defects may be considered.

In 1983, Minn-Dak Seeds contracted with Merrill Iron and Steel for the construction of mustard seed storage bins. Shortly after the project was completed, water leaked into the bins from gaps caused by defective welds and other holes in the storage pipes. In addition, a seed-level monitoring system did not operate properly. As determined by the court, the numerous defects, including the leveling system, were the result of the contractor's substandard construction methods. The court denied Merrill's claims because the contractor failed to substantially complete the project. The cost of correction was substantial, and the essence of the project had not been met (Merrill Iron and Steel v. Minn-Dak Seeds).

In 1961, David Fink contracted with Airco Refrigeration Service to move a 10-ton water-cooled air-conditioning unit from Fink's restaurant to an equipment room outside the building. On completion (during warm weather), the system did not adequately cool the restaurant. This failure occurred because the contractor did not insulate the metal ductwork. The court ruled in favor of Airco, citing the oversight in failing to insulate the ductwork as a minor defect that could be corrected at minimum cost to Airco. Airco had substantially completed the project, but the final payment was reduced by the cost of the insulation.

Although other factors may be relevant, generally, contractors must perform the contract in good faith, except for omissions or deviations caused by human error, skill level, or experience. Willful deviations are viewed unfavorably (Airco Refrigeration Service v. Fink).

To What Degree Was the Purpose of the Contract Defeated?

One must consider how the completed project meets the purpose of the original contract. Is the completed project or facility the thing for which the owner bargained? How much of a deviation exists? Has the essence of the contract been fulfilled?

Answering these questions requires studying the contract documents, plans, and specifications. It may also involve reviewing oral conversations, written correspondence, and the individual actions of the parties during construction.

In 1975, Gregory contracted with Wilson to construct a 30-ft by 60-ft swimming pool with a depth varying from 3 ft to 6 ft. Before the pool was finished, the parties orally agreed that Wilson would add a diving board and increase the depth of the pool to safely accommodate persons using the board. On completion of the pool, Gregory withheld the final payment, claiming that the pool size was 59.5 ft by 29.5 ft and the pool depth was not exactly 10 ft, in accordance with the specifications. The court ruled in favor of Wilson, stating,

> We find the deviations in dimensions, which could be discovered only by measuring the pool, in no way defeat the purpose of the contract...the method of constructing the pool made it impossible to achieve perfect compliance with the exact measurements called for by the contract. Therefore,...we find the slight devia-

tion was not a defect in construction and did not constitute a breach of contract. [In addition] the pool is deep enough to accommodate a diving board, and therefore, there is no defect in regard to depth of the pool. (Wilson v. Gregory)

B and B Cut Stone Co. v. Resneck is an example of how a project was not substantially complete because of the major degree to which the contract purpose was defeated. The Resnecks took exceptional pride in their home and were involved in an extensive remodeling effort. In 1982, they turned their attention to the master bedroom and bath. The Resnecks hired B and B to construct a marble fireplace with a hearth and large firewall, which would become the focal point for this phase of a redecoration effort. This desire was known to the contractor. Almost from the beginning of the project, problems arose. The Resnecks cited numerous defects, including misaligned marble panels and uneven seams. The finished product ultimately was unacceptable. The Resnecks withheld the final payment and sued for damages. B and B argued that the fireplace was entirely usable as a structure for burning wood and providing warmth and met all code requirements. The Resnecks claimed that the fireplace failed to serve its intended artistic purpose and did not provide "intellectual enjoyment and aesthetic appeasement" (B & B Cut Stone Co. v. Resneck). The court recognized that the Resnecks wanted a marble fireplace for its "impressive presence and beauty and elegance," not for its thermal value. The massive marble facing was to be used as a backdrop to hang their many modern paintings and "could not possibly serve any pertinent function to burning wood" (B & B Cut Stone Co. v. Resneck). The fact that the Resnecks matched their new furniture to the color of the marble reinforced this finding. The court ruled in favor of the Resnecks.

How Easy Is the Defect To Correct?

It is also important to consider the effort necessary to correct the defects of the facility. Can the defects be remedied with minimal effort, or is significant rework of the entire project required?

As demonstrated in Airco, the minor defects (lack of insulation on the ductwork) could be repaired relatively easily without removing the entire air-conditioning unit or adjusting the existing pipework. In Jerrie Ice Co. v. Col-Flake Corp., a similar situation occurred.

In 1956, Jerrie Ice Co. contracted with Col-Flake Corp. to construct an ice plant with a 200-ton ice storage capacity. The plant was the first one built by Col-Flake, which previously specialized in manufacturing ice machines. Once the project was completed, several deficiencies existed, including a shortage of ice storage capacity well below the 200 tons specified. Jerrie claimed that the reduced storage capacity would affect his ability to supply ice to a variety of customers, primarily to fishers and shrimp boats. He subsequently withheld the final payment and sued for damages. For two years before trial, Jerrie used the plant and successfully met customer demands on all but a few isolated occasions. The court considered various options to correct the storage bin deficiency. Evidence produced suggested that one way to increase the capacity would be to tear it down and rebuild the bin. The court could not, however, "in good conscience" have Col-Flake remove the bin and reconstruct it, given the usage record by Jerrie (Jerrie Ice Co. v. Col-Flake Corp.). The court found that even though the storage capacity was not in accordance with the specifications, it could be remedied by moving the compressors to another location in the plant, thus freeing some additional storage space near the bin. Therefore, substantial completion was achieved. Col-Flake's final payment was reduced by the cost of moving the compressors.

Conversely, Merrydale Glass Works v. Merriam is an illustration of a major effort or undertaking necessary to correct defects. In 1975, Merriam hired Merrydale Glass Works to install certain glass products in his home. This work entailed placing cathedral glass in the front of the house and installing mirrors on walls, ceilings, and doors throughout the remainder of the house. The cathedral glass was installed without any problems, but difficulties were encountered with the mirrors. The problems included uneven edges and the use of numerous smaller panels instead of the large panels requested. Merriam claimed that these problems were caused by Merrydale performing all cuts on the mirrors at the factory and not on site. He concluded that Merrydale was not performing satisfactorily and told them to discontinue the job. He contracted with another glass specialist to correct the problems and finish the project. Merriam withheld the remaining payments and sued for the cost of repairing and completing the work.

The court ruled that Merrydale had substantially completed the installation of the cathedral glass and was entitled to that portion of the withheld payment but that the mirror installation was not substantially complete because of the numerous defects present and the major rework required. "Correction of the defects could only be accomplished by replacing most of the panels" (Merrydale Glass Works v. Merriam).

Has the Owner Benefited from the Work Performed?

Administrators should evaluate if the owner has benefited from the work performed by the contractor. In the absence of express contract provisions governing substantial completion, the judicial attitude is that it is unfair for the owner to be unjustly enriched from the use of the facility without just compensation to the contractor.

In the landmark case of Neel v. O'Quinn, this rule was applied in determining if substantial completion was achieved. In July 1972, Neel contracted with O'Quinn to construct a home based on a rough, one-sheet floor plan and sketch. Work began immediately, and two months later, Neel moved into the home, even though he had identified to O'Quinn numerous defects in construction. There were 31 defects in the home, including a leaking roof and exterior substandard bricklaying. O'Quinn corrected some of the defects, but Neel retained the final payment and sued for damages.

The court ruled in favor of the contractor, O'Quinn, saying,

> Substantial performance (completion) is readily found, despite the existence of a large number of defects both in material and workmanship, unless the structure is totally unfit for the purpose for which it was originally intended. The evident purpose intended for construction was to provide living quarters for plaintiff and his family. The residence has been used as such since August 1972 (1+ years to date of trial), and under the facts of this case, we cannot say the contractor did not substantially comply with the contract. (Neel v. O'Quinn)

Ballou v. Basic Construction Co. is a case where the owner did not benefit from the incomplete or defective work of the project. In 1964, Basic, a general contractor, had subcontracted with Virginia Prestressed Concrete Corporation (VPCC) for the fabrication and delivery of precast concrete columns to be used in the construction of a hospital. The subcontract required that 200 concrete columns be made in strict compliance with the contract specifications, which called for a 2-in. cover of concrete over the reinforcing steel. During construction, the architects accepted only 45 of 139 columns made by VPCC because of improperly positioned rebar and failure to meet minimum cover specifications. Basic paid VPCC for the acceptable columns but withheld payments for the defective ones; subsequently, VPCC went bankrupt. Ballou was appointed the bankruptcy trustee. Ballou claimed that VPCC substantially completed the columns, was unfairly penalized, and was entitled to payment by Basic. The court ruled in favor of the general contractor, Basic, citing,

> Substantial performance (completion) is an equitable doctrine, intended to prevent unjust enrichment, which allows a contractor who has not complied with a contract in every detail to recover for work done which enriches the other contracting party...in this case, the non-complying columns were neither accepted nor used by the owner, there is no question of unjust enrichment. (Ballou v. Basic Construction Co.)

The court concluded that the doctrine (of substantial completion) was not available as a defense.

Certificate of Substantial Completion

The certificate of substantial completion is written confirmation that a contractor has substantially completed the project. It is recognized as the contractual document verifying release of the final payment to the contractor and confirming the owner's prerogative opportunity to use the facility.

The case review indicates that the courts place limited weight on the presence of this certificate as the sole factor in determining substantial completion. Not one of the cases studied was based solely on the presence of a certificate. Overall, the certificate is considered as just another piece of evidence.

Courts may review the certificate to ensure that it was not issued in collusion, as a result of a mutual mistake, or under fraudulent circumstances. The presence of a certificate may aid the contractor in proving that he or she achieved substantial completion. Also, it may be a formidable task to prove that the purpose of the contract was defeated if a certificate was issued by the design professional.

Illustrative Examples

To demonstrate the rules presented above, the disputes in Pioneer Enterprises v. Edens and Daspit Brothers Marine Divers v. Lionel Favret Construction Co. will be analyzed.

Pioneer Enterprises v. Edens

Statement of Facts

In 1981, Lee Edens contracted with Pioneer Enterprises to construct a grain storage facility. No certificate was required by the contract. Following completion, water leaked into the facility and caused extensive damage. After careful inspection, it was discovered that the facility had more than 57 leaks caused by unacceptable welds, missing washers, elongation of bolt holes, and improper caulking. In addition, the grain aeration system, an integral part of the storage facility, had not been installed properly and did not function. Edens abandoned the facility and withheld the final payment because the facility could not be used for the long-term storage of grain. Pioneer sought to recover the balance due on the contract. During the trial, experts stated that the facility was never fit for its intended purpose of storing grain and recommended that the facility be torn down and reconstructed.

Analysis

How Extensive Is the Claimed Defect or Incomplete Work?
There were more than 57 separate major deficiencies, which crippled the use of the facility for storing grain. Thus, the defects were extensive.

To What Degree Was the Purpose of the Contract Defeated?
The owner contracted for a grain storage facility and received a facility that was unfit for its intended purpose.

How Easy Is the Defect to Correct?
The facility would have to be entirely dismantled to correct the numerous problems.

Has the Owner Benefited from the Work Performed?
Edens did not operate the facility for grain storage or any other reason after the problems were noticed and subsequently did not benefit from its use.

Synopsis

There are no responses favorable to Pioneer Enterprises. Substantial completion was not achieved even though the Edens occupied the facility for a short time. This evaluation is consistent with the determination of the court (Pioneer Enterprises v. Edens).

Daspit Brothers Marine Divers v. Lionel Favret Construction Co.

Statement of Facts

In 1977, Daspit contracted with Favret, a franchise dealer, for the construction of a prefabricated steel building to store equipment incidental to marine diving. No certificate of substantial completion was required by the contract. Once completed, there were numerous defects noticed in the construction of the building. The steel had not been primed before erection; large cracks were prominent throughout the uneven floor slab; the roof leaked; and the sliding door and exterior lights were the wrong size. Daspit withheld the final payment and claimed that the building was unfit for its intended purpose. Daspit did, however, store equipment in the building. During trial, testimony was introduced that cracks in the slab were the normal result of expansion, contraction, and temperature. In addition, experts determined that the slab was capable of sustaining the load of the equipment Daspit intended to store in the building.

Analysis

How Extensive Is the Claimed Defect or Incomplete Work?

The deficiencies noted were minor compared to the completed work. The building was structurally sound and was constructed according to the plans, with some minor exceptions.

To What Degree Was the Purpose of the Contract Defeated?

Daspit contracted for a prefabricated steel storage facility and received one with minor defects. The facility could still be used as a storage facility for his equipment. The purpose of the contract was not defeated.

How Easy Is the Defect to Correct?

The steel could be primed; the slab was strong enough to support the load of the equipment and could be releveled with a topping mix; the roof could be repaired after closer examination; and the sliding door and exterior lights could be replaced. It would appear that the defects were not that difficult to correct.

Has the Owner Benefited from the Work Performed?

Daspit used the building to store equipment and, thus, benefited from the work.

Synopsis

There are no responses favorable to Daspit. Favret substantially completed the project and is entitled to the balance of the contract minus appropriate deductions to correct the deficiencies (Daspit Brothers Marine Divers v. Lionel Favret Construction Co.).

References

Airco Refrigeration Service v. Fink, 134 So.2d 880 (1961).

American Druggists Insurance Co. v. Henry Contracting, 505 So.2d 734 (1987).

American Institute of Architects (AIA). (1987). *General Conditions of the Construction Contract*, Document A201, American Institute of Architects, Washington, DC.

B & B Cut Stone Co. v. Resneck, 465 So.2d 851 (1985).

Ballou v. Basic Construction Co., 407 F.2d 1137 (1969).

Black, H. C. (1979). *Black's Law Dictionary*, 5th Ed., West Publishing Co., St. Paul, MN, 705.

Bramble, B. B. (1987). *Construction Delay Claims*, John Wiley and Sons, Inc., NY, 22.

Bruner v. Hines, 324 So.2d 265 (1975).

Corbin, A. L. (1963). *Corbin on Contracts, A Comprehensive Treatise on the Rules of Contract Law*, West Law Publishing Co., St. Paul, MN.

Daspit Brothers Marine Divers v. Lionel Favret Construction Co., 436 So.2d 1223 (1983).

Hadden v. Consolidated Edison Co. of New York, Inc., 34 N.Y.2d 88 (1974).

Jardin Estates, Inc., v. Donna Brook Corp., 126 A.2d 372 (1956).

Jerrie Ice Co. v. Col-Flake Corp., 174 F. Supp. 21 (1959).

Keating v. Miller, 292 So.2d 759 (1974).

Leeds v. Little, 44 N.W. 309 (1890).

Merrill Iron and Steel v. Minn-Dak Seeds, 334 N.W.2d 652 (1983).

Merrydale Glass Works v. Merriam, 349 So.2d 1315 (1977).

Neel v. O'Quinn, 313 So.2d 286 (1975).

Pioneer Enterprises v. Edens, 345 N.W.2d 16 (1984).

Plante v. Jacobs 103 N.W.2d 296 (1960).

Ricchini, J. A., and O'Brien, J. J. (1990). *Construction Documentation*, John Wiley and Sons, Inc., New York, 109.

Rudy Brown Builders v. St. Bernard Linen Service, 428 So.2d 534 (1983).

State of Louisiana v. Laconco, 430 So.2d 1376 (1983).

Stevens Construction Corp. v. Carolina Corp., 217 N.W.2d 291 (1974).

Wilson v. Gregory, 322 So.2d 369 (1975).

Additional Case

The following case deals with substantial completion. The reader is invited to review the facts of the case, apply the decision criteria, reach a decision, compare it with the judicial decision, and determine the rationale behind the judicial decision.

Sumrall Church of the Lord Jesus Christ v. Johnson 757 So.2d 311(2000).

Chapter 14

Liquidated Damages

When there is a delay in project completion, the owner can be harmed by added costs and lost revenue. Owners often attempt to avoid delay-related litigation by using a liquidated-damages clause in the construction contract. This provision imposes a daily dollar assessment for every day of delayed completion for which the contractor is responsible. This assessment continues until the date of substantial completion of the project.

Contract Language

Most public and many private contracts contain liquidated-damages clauses. Florida's Department of Transportation uses FLDOT Art. 8-10.3 Determination of Number of Days of Default and Art. 8-10.4 Conditions under Which Liquidated Damages Are Imposed, which are typical of the clauses found in most state department of transportation contracts. Amounts are specified as dollars per day for each day beyond the contract time that substantial or final completion has not been reached. Liquidated damages may also be assessed for delays in reaching intermediate project milestones. Thus, a careful reading of the contract is important.

In most contracts, the beginning of the contract time starts with the issuance of a notice to proceed. The contractor usually has 10 days after the date of this notice in which to begin work. In FLDOT (1996), Art. 8-7.2 Date of Beginning of Contract Time, the contract time beginning is defined as

> The date on which contract time will begin shall be either (1) the
> date on which the Contractor actually begins work or (2) the date

for beginning the charging of contract time as set forth in the proposal, whichever is earlier.

The time-extension clause is also frequently referred to when liquidated damages are discussed. The FLDOT standard specifications in
Art. 8-7.3.2 Contract Time Extensions states the following:

> The Department may grant an extension of contract time when a
> controlling item of work is delayed by factors not reasonably an-
> ticipated or foreseeable at the time of bid . . . . When failure by the
> Department to fulfill an obligation under the contract results in
> delays in the controlling construction operations, such delays
> will be considered as a basis for granting credit to the contract
> time. Extensions of contract time will not be granted for delays
> due to the fault or negligence of the Contractor.
>   Time extensions for delays caused by the effects of inclement
> weather will be handled differently from those resulting from
> other types of delay . . .

This clause can be contrasted with typical language found in commercial contracts. AIA A201 (1987), Art. 8.3.1, states,

> If the Contractor is delayed at any time in progress of the Work
> by any act or neglect of the Owner or Architect, or of an em-
> ployee of either, or of a separate contractor employed by the
> Owner, or by changes ordered in the Work, or by labor disputes,
> fire, unusual delay in deliveries, unavoidable casualties or other
> causes beyond the Contractor's control, or by delay authorized
> by the Owner pending arbitration, or by other causes which the
> Architect determines may justify delay, then the Contract Time
> shall be extended by Change Order for such reasonable time as
> the Architect may determine.

Background

Liquidated damages are generally deemed to be in lieu of actual delay
damages. When using a liquidated-damages clause, the owner is limited
to the delay damages stipulated and cannot seek actual damages result-

ing from the delay. However, the liquidated-damages clause does not prevent the owner from seeking compensation for damages resulting from negligence, poor workmanship, willful misconduct, or numerous other defaults by the contractor. Courts generally enforce the liquidated-damages clause so long as (1) the daily amount bears some resemblance to the actual damages, that is, the amount specified as liquidated damages is not perceived as a penalty, and (2) the damage amount is difficult or impossible to estimate (Restatement of Contracts 1932).

Basic Principles of the Liquidated-Damages Clause

The basic principles related to liquidated damages are explained here.

Delays Attributable to the Contractor

The contractor can only be assessed liquidated damages for nonexcusable delays. Because contractors are responsible for scheduling the work, managing subcontractors, and developing the means and methods of construction, shortcomings in any of these areas are considered nonexcusable.

Apportionment of Damages

If the owner is responsible for the delay, the owner forfeits all rights to recover under the liquidated-damages clause. The court reinforced this principle in the 1967 case of L. A. Reynolds Co. v. State Highway Commission of North Carolina. The state highway commission failed to have the right of way cleared so that Reynolds could work on a 4-mi stretch of road. The court would not allow Reynolds to be held accountable for the delay, and it could not be assessed liquidated damages.

In some instances, delays are attributable to both the owner and contractor (known as *concurrent* delays). Here, damages must be apportioned. An example illustrates the application of this principle.

In 1968, Butte-Meade Sanitary Water District (BMSD) contracted with Brunken and Sons Construction Co. to construct a water main, pump house, and reservoir foundation. Construction by Brunken on the water main and pump house took 1,100 more days than allowed by the contract. BMSD claimed that the contractor was responsible for the delays. After careful consideration, the court apportioned the delays between the owner and contractor (Aetna Casualty and Surety Co. v. Butte-Meade Sanitary Water District). Brunken was assessed liquidated damages only for the delay period for which they were solely responsible.

Liquidated Damage Amounts

The owner must evaluate the potential losses resulting from late completion. Factors to consider include lost revenue or rental value, user costs, engineering and administrative costs, additional wages, moving costs, interest, and extended management and overhead fees. Costs that cannot be calculated include the effect on follow-on contracts and loss of visibility.

Deciding the amount of liquidated damages is not a trivial matter. If the amount is too high, contractors may be reluctant to bid on the project or may include a high contingency amount in their bids to cover the possibility of paying these damages. Furthermore, the clause may be unenforceable should a dispute arise that is litigated in court. If the figure is too low, the owner may not be fully compensated for the delay.

Period of Assessment

Liquidated damages are usually assessed when the contractor fails to achieve substantial completion by the contracted date. Some contracts may not clearly define the assessment period or may specify another ending date, such as final completion.

In Ledbetter Brothers, Inc., v. North Carolina Department of Transportation (NCDOT), a dispute arose because the contract did not specify the date that damages would end, substantial or final completion. Ledbetter claimed the former; NCDOT argued for the latter date. The court resorted to the language of the contract, which mentioned "damages would take effect until successful completion of the work" (Ledbetter Brothers v. North Carolina Department of Transportation) The court determined that this date was equivalent to final completion, not substantial completion.

To avoid disputes, the language of the contract should be clear in specifying the assessment period, if the assessment is for workdays or calendar days, and if weekends and holidays are included.

How Are Liquidated Damages Withheld?

Generally, the owner will withhold liquidated damages from the contractor's final payment or retainage. If progress payments are made, the owner must be careful not to predict or anticipate liquidated damages for delays because the contractor may initiate a plan to bring the project back on schedule.

Relationship to Actual Damages

Generally, liquidated damages are the measure of the owner's recovery for delayed completion, and actual damages are not recoverable. In other words, liquidated damages are deemed to be in lieu of actual damages and are a reasonable estimate of foreseeable actual damages. The liquidated-damages clause does not prevent the owner from seeking compensation for damages resulting from negligence, poor workmanship, willful misconduct, termination, abandonment, or numerous other defaults by the contractor.

Liquidated Damages or Penalty?

Courts enforce liquidated-damages provisions when they are fair and reasonable estimates of anticipated losses caused by unexcused delays (Wise v. U.S.). Then, the clause is viewed as an effort by the owner and contractor to settle their differences without having to resort to costly legal remedies.

Courts do not enforce a liquidated-damages clause if it appears to be a penalty. A penalty is defined as a "sum that bears no apparent relationship to the injury, but is chosen by the party with the greater bargaining position to coerce performance by the other" (Rowland Construction Co. v. Beall Pipe and Tank Corp.). The distinction between damages and penalties was discussed in the decision of Westmount Country Club v. Kameny,

> Liquidated Damages is the sum a party to a contract agrees to pay if he breaks some promise, and which, having been arrived at by a good faith effort to estimate in advance the actual damage

that will probably ensue from the breach is legally recoverable as agreed damages if breach occurs.

A Penalty is the sum a party agrees to pay in the event of a breach, but which is fixed, not as a pre-estimate of probable actual damages, but as a punishment, the threat of which is designed to prevent the breach.

A liquidated-damages clause is not considered a "penalty" solely because it is labeled a penalty or forfeiture. Conversely, a contract provision is not automatically liquidated damages just because it is labeled a liquidated-damages clause.

Foreseeability of Damages

In evaluating the clause, courts generally only consider if the damage assessment was reasonable in light of the known circumstances at the time of contract execution (Southwest Engineering Co. v. U.S.; Priebe & Sons v. U.S.). If actual damages after completion are different from the liquidated amount, the parties are still bound by the liquidated-damages agreement. The judicial view is that proof of damages is not required for enforcement of liquidated damages unless proof is required by the contract. By entering into the contract, each party took a calculated risk and is bound by any reasonable contractual provisions pertaining to liquidated damages. Whether actual damages occurred does not prevent enforcement of the provision (Construction Briefings 1984).

Primary Rules of Application

Courts on numerous occasions have intervened to resolve disputes over liquidated damages. Based on a review of numerous appellate decisions, the following are the primary inquiries that are made:

- Is there a liquidated-damages clause?
- What were the intentions of the owner?
- Were actual losses difficult to predict?
- Is the stipulated sum reasonable?

All questions need to be addressed, and a harmonious outcome should be sought. Nevertheless, good judgment is required because the answers may be contradictory.

Is There a Liquidated-Damages Clause?

The first issue is to review the liquidated-damages clause in the contract. The essential facts to determine from the reading are the per diem figure or total sum to be liquidated; the assessment period (substantial or final completion); whether a day means a workday or a calendar day; and if weekends and holidays are included. If a clause is not present, the owner has no alternative but to seek actual damages.

What Were the Intentions of the Owner?

For the clause to be enforced, it must appear that the clause was a good-faith effort to preestimate actual damages suffered by the owner for delayed completion. If the purpose of the clause is to deter a breach (delayed completion) or to secure full performance by the contractor, the intent is to penalize or have an "in terrorem" effect (Black 1979). The specific language of the contract can be useful in determining intent.

The landmark case Bethlehem Steel Corp. v. City of Chicago reinforces this principle of intent to liquidate in determining enforceability of a clause. In 1962, Bethlehem signed a contract to complete the steel erection portion of a highway structure for the city of Chicago. Bethlehem's work on this project followed the construction of the foundation and piers by another contractor. Bethlehem, in turn, was followed by still another contractor, which constructed the deck and roadway. Bethlehem was 52 days late in finishing the steel erection, but the entire structure was completed on time. The city of Chicago, in accordance with the contract, withheld liquidated damages of $1,000 for each day of Bethlehem's delay. Bethlehem argued that these damages could not be assessed because the project overall was not delayed and the city suffered no damages. The court reviewed the contract to ascertain the city's intentions. The contract language specified that the damages were "to partially cover losses and expenses to the City of Chicago." The court stated,

> It was the parties intent that the [liquidated damages] be the sum recoverable for each day's delay in order to forestall legal proceedings for a determination of the precise amount of damages. (Bethlehem Steel Corp. v. City of Chicago)

The intent of the parties was to liquidate damages and compensate for delays, not to penalize Bethlehem. It was not foreseeable at the time the contract was executed that there would be no damages. The city did not have to prove actual damages for the clause to be upheld. The fact that the other contractors were able to make up the lost time and complete the structure on time was irrelevant.

Were Actual Losses Difficult to Predict?

Actual damages at the time the contract was drafted are often uncertain or difficult to predict. In other words, the harm that is caused by the delay may be difficult to estimate. Some projects have many intangible benefits, which are impossible to compensate for in the event of a delayed completion. The more certain the damages and the less costly they are to calculate, the less incentive the parties have to negotiate a stipulated-damages clause (Clarkson et al. 1978, p. 351).

Difficulty in preestimating damages was discussed in City of Fargo, North Dakota v. Case Development Co. In 1984, Case agreed to purchase and develop a city-owned building into an office complex for the city of Fargo. Later, Case abandoned the project for financial reasons. The city of Fargo, in accordance with the contract, assessed Case liquidated damages in the lump sum amount of $100,000 for delaying the project. Because this project was the first of its kind for Fargo, the court determined that the intangible benefits to the public and the monetary loss to the city were impossible to calculate at the time the parties entered into the contract. The court upheld the validity of the liquidated-damages clause.

Is the Stipulated Sum Reasonable?

This rule, sometimes known as the "reasonableness test," focuses on the time when the contract was formulated. Does the amount appear to be fair and reasonable in compensating the injured party? Is the figure disproportionate to any provable or conceivable damages? It must appear that the owner is not unjustly enriched as a result of a delayed completion. The recommended methodology for an owner is to base the determination on available information. If impossible, it can be arbitrarily determined, but the amount is closely scrutinized by a court if judicial proceedings ensue.

Both the Bethlehem and the City of Fargo cases are examples of how the courts have applied the "reasonableness" test. In both cases, the stipulated damages appeared to the courts to be fair and not extravagant, or disproportionate, so as to require the inference that the agreement must have been effected by fraud, oppression, or mistake (Bethlehem Steel Corp. v. City of Chicago). In Bethlehem, the liquidated damages were considered reasonable given the extent of damages the city could sustain in additional fees for the delayed completion of the highway structure. In Fargo, $100,000 was also considered just compensation for the damages suffered in lost tax revenues and future redevelopment projects for that area of the city.

Another example of how the courts have applied the reasonableness test is presented in Hicks Construction Co. v. Town of Hudson, Wyoming. In 1961, Hicks was contracted by the town of Hudson, Wyoming, to construct a sanitary sewer system. The project was completed 100 days late, and all of the delays were attributable to Hicks's inefficient construction methods and scheduling. These nonexcusable delays included failure to backfill and properly install culverts and to clean out irrigation ditches. Hudson, in accordance with the contract, withheld $50/day or $5,000 liquidated damages. Hicks sued for the withheld payment. The court ruled that the $50/day was reasonable as a preestimate for the delayed completion of a new town sewer system. The court said,

> ... the evidence clearly demonstrates that the liquidated damages were in lieu of actual damages suffered by the town and its inhabitants because of the default of the contractor. We do not consider this as a penalty. (Hicks Construction Co. v. Town of Hudson, Wyoming)

Strategy

All questions must be considered in determining enforceability of the clause. Although intention of the parties is not always the governing factor, it does assist in establishing the "environment" in which the contract was drafted. In some cases, the question about intentions helps solve the labeling problem (penalty versus liquidated damages).

Liquidated-damages provisions have seldom been voided solely because the damages are easy to estimate. Sometimes owners included liquidated-damages clauses despite having good estimates of potential damages. The clause represents a less expensive alternative than proving actual damages in court.

Courts consider the intention of the parties and the degree of uncertainty as influential factors in determining reasonableness of the estimate. *Reasonableness* is the major criterion for determining enforceability of the liquidated-damages clause. This test ensures that at the time the contract was formulated, the liquidated damages specified were a reasonable estimate of potential delay damages. The damages prescribed must not be so disproportionate as to overcompensate, profit, or unjustly enrich the injured party.

If the clause successfully addresses the four questions, it is probably enforceable, regardless of the actual damages. Courts have traditionally decided on the enforceability of the clause by looking at the facts at the time of contract execution and not after the breach.

Illustrative Examples

Two recent appellate court cases illustrate the application of the rules described above.

Osceola County, Florida, v. Bumble Bee Construction Co.

Statement of Facts

In 1981, the County of Osceola, Florida, and Bumble Bee Construction Co. (BCC) entered into a contract for the construction of a tourist information center. The contract called for liquidated damages of $250 per day. Each day of delay was defined as a workday, and the assessment period ended on substantial completion. The $250 per day was an arbitrary figure selected by the county.

The project was completed 114 days late. The delay causes were nonexcusable. The county assessed liquidated damages for the delay;

BCC sued for this withheld payment, claiming the clause was a unenforceable penalty.

Analysis

Is There a Liquidated-Damages Clause?
Yes, workdays were defined, and the period of assessment was specified as substantial completion.

What Were the Intentions of the Owner?
This question is difficult to answer.

Were Actual Losses Difficult to Predict?
The damages sustained by the county for a late tourist center were difficult to ascertain because the facility would not directly generate revenue (there were no admission fees). At the tourist center, tourists could make hotel reservations or obtain information about sightseeing locations within the county. The county would ultimately benefit from the taxes on tourist expenditures.

Is the Stipulated Sum Reasonable?
The sum of $250 per day was considered a reasonable preestimate of actual damages. It was not so excessive or disproportionate as to unjustly enrich the county.

Synopsis

The arbitrary selection of $250 per day was justified because the damages were impossible to calculate. The clause was not considered a penalty and was therefore enforceable.

Rohlin Construction Co. v. City of Hinton

Statement of Facts

In 1988, Rohlin was awarded a contract for resurfacing selected roads for the city of Hinton, Iowa. The contract called for the work to be finished within a 40-day period. The contract also included a clause establishing $400 per day as the amount of liquidated damages. The measure specified was working days. Rohlin completed the project 26 days late because of nonexcusable reasons. The city withheld $10,400 as liquidated damages. Rohlin sued, claiming that the clause was an unenforceable penalty. Hinton could not present evidence or data to justify the $400-per-day damage figure. The city adamantly indicated that they wanted the roads finished before the increased traffic associated with the school year and annual grain hauling season.

Analysis

Is There a Liquidated-Damages Clause?
Yes, workdays were defined, and the period of assessment was specified.

What Were the Intentions of the Owner?
The city's intentions are difficult to determine. The city's strong desire to complete the project before the heavy traffic season may imply an "in terrorem" effect.

Were Actual Losses Difficult to Predict?
It is difficult to calculate the monetary loss to the local community as a result of delayed completion of the road resurfacing project.

Is the Stipulated Sum Reasonable?
The $400 per day was not considered a reasonable preestimate of actual damages. The liquidated damages were considered disproportionate and overcompensated the city. The high figure supports an "in terrorem" view by providing financial security to the city.

Synopsis

The liquidated damages of $400 per day failed the reasonableness test because the amount bore no relation to any conceivable loss that the city could substantiate. Rohlin relied on an Iowa Department of Transportation construction manual that contained a schedule of suggested rates for liquidated damages based on the engineer's estimate of the contract price. That manual suggested $100 per day. The clause was considered an unenforceable penalty because the liquidated damages were unrealistic and excessive. It strongly appeared that the city was using the liquidated-damages clause to coerce the contractor into timely completion.

Exercise 14-1: S. J. Otinger Construction Co. and Montgomery, Ala., Water Works & Sanitary Sewer Board

S. J. Otinger Construction Co. entered into an agreement for $125,000 with the Water Works and Sanitary Sewer Board of the city of Montgomery, Alabama, to construct trunk sewers, force mains, and a lift station in the area of Catoma Creek. Otinger commenced work on October 17, 1961. The work was to be completed within 120 calendar days, but Otinger did not complete the work until 250 calendar days later, exceeding the contract time by 130 days.

The contract provided for $50 per day as liquidated damages. Under the contract, the board was entitled to withhold $6,500, but the board actually withheld $3,872.45, which represented the additional engineering fees paid by its consulting engineers.

The relevant contract language stated,

> It is mutually agreed between the parties hereto that time is the essence of this contract, and in the event the construction of the work is not completed within the time herein specified, it is agreed that from the compensation otherwise to be paid to the Contractor, the second party may retain the sum of $50.00 per day for each day thereafter, *Sundays and holidays included*, that the work remains uncompleted, which sum shall represent actual damages which the owner will have sustained per day by failure of the Contractor to complete the work within the time stipulated, and this sum is not a penalty, being the stipulated damage the second party will have sustained in the event of such default by the first party. (emphasis added)

Otinger now argues that by including Sundays and holidays, the provision is a penalty, not liquidated damages.

What are the damages that may occur? Is the liquidated-damages clause enforceable?

References

Aetna Casualty and Surety Co. v. Butte-Meade Sanitary Water District, 500 F.Supp. 193 (1980).

American Institute of Architects (AIA). (1987). *General Conditions of the Construction Contract*, Document A 201, American Institute of Architects, Washington, DC.

Bethlehem Steel Corp. v. City of Chicago, 234 F.Supp. 726 (1965).

Black, H. C. (1979). *Black's Law Dictionary*, 5th Ed., West Publishing Co., St. Paul, MN, 705.

City of Fargo, North Dakota, v. Case Development Co., 401 N.W.2d 529 (1987).

Clarkson, K., Miller, R. L., and Muris, T. (1978). "Liquidated Damages v. Penalties: Sense or Nonsense." *Wisconsin Law Review*, 78, 351–390.

Florida Department of Transportation (FLDOT). Art. 8-10.3 "Determination of Number of Days of Default" and Art. 8-10.4 "Conditions under Which Liquidated Damages Are Imposed." Restatement of Contracts. (1932). Para. 339(1).

Hicks Construction Co. v. Town of Hudson, Wyoming, 390 F.2d 84 (1976).

L. A. Reynolds Co. v. State Highway Commission of North Carolina, 155 S.E.2d 473 (1967).

Ledbetter Brothers, Inc., v. North Carolina Department of Transportation, 314 S.E.2d 761 (1984).

"Liquidated Damages." (1984). *Construction Briefings*, No. 84-4, Federal Publications, Inc., Washington, DC.

Osceola County, Florida, v. Bumble Bee Construction Co., 479 So.2d 310 (1985).

Priebe & Sons, Inc., v. United States, 332 U.S. 407 (1947).

Restatement of Contracts. (1932). Para. 339(1).

Rohlin Construction Co. v. City of Hinton, Iowa, 476 N.W.2d 78 (1991).

Rowland Construction Co. v. Beall Pipe and Tank Corp., 540 P.2d 912 (1975).

Southwest Engineering Co. v. United States, 341 F.2d 998 (1965).

Westmont Country Club v. Kameny, 197 A.2d 379 (1964).

Wise v. United States, 249 U.S. 361 (1919).

Appendix A

Additional Exercises

The problems in this appendix are more complex than the problems at the ends of the chapters. They afford the opportunity to practice complex disputes and partitioning the dispute into specific issues that can be resolved using multiple flowcharts. There may be a hierarchy to how these disputes are resolved. For instance, for some of the problems, notice is an issue. The notice issue must be successfully resolved in the contractor's favor before an entitlement issue can be addressed. Multiple issues are common among the problems in this appendix, and no solutions are offered.

Exercise 1

In 1976, the West Plains Bridge and Grading Company, Inc., submitted a successful bid and thereafter contracted with the Missouri Highway Department to build a bridge. This construction necessitated excavation of earth and rock. West Plains and N. B. Harty General Contractors, Inc., entered into a written subcontract by which Harty was to do the drilling and blasting of the "Class C excavation" on the project. West Plains was to remove the material after the blasting. The subcontract stated,

> Now, therefore, the said WEST PLAINS does hereby sublet to said HARTY the following items of the contract between the State and WEST PLAINS at the unit prices stated in the following schedule;

343

NO.	DESCRIPTION	QUANTITY	UNIT PRICE	AMOUNT
203 20.00	Class C. Excavation (Drilling) O.C.	20,014 C.Y.	@ $1.63	$32,622.82

The word "Drilling" was inserted in handwriting by Oakley Carte, the West Plains president, before the agreement was sent to Harty. His initials appear immediately under that word. The contract between Harty and West Plains makes no other reference to "Class C Excavation."

The agreement between West Plains and the highway department contained two classifications of excavation work. "Class A" excavation is earth removal, and "Class C" excavation is removal of rocks larger than 6 in. In West Plains' bid documents and in the contract, the highway department estimated that there would be 20,014 yard^3 of Class C excavation on the project. The contract between Harty and West Plains also incorporated this figure, but both knew that this figure was an estimate of the volume of rock and there could be "overruns and underruns."

West Plains excavated with earthmoving equipment without drilling and blasting until it reached solid rock, at which time Harty commenced drilling and blasting. Harty drilled and blasted 10,194 yard^3 of solid rock. While West Plains was removing earth, "percentage rock" was discovered. It is a mixture of earth and rock that is more difficult to remove than earth. Percentage rock can be drilled and blasted, but here it was not necessary to do so. The presence of percentage rock had not been contemplated. Harty's president stated that he had been in the drilling and blasting business for 20 years and this was the second time he had ever encountered percentage rock. The district construction engineer for the highway department stated that its estimate of 20,014 yard^3 of Class C excavation did not include percentage rock, but that once it is "percentaged," it becomes a part of Class C excavation.

For the purposes of payment to West Plains, the highway department applied the rock portion of the percentage rock to Class C excavation and the earth portion to Class A. There were three different areas of percentage rock. One was classified as 30% rock and 70% earth, one as 50% rock and 50% earth, and one as 60% rock and 40% earth. West Plains was paid for 28,388 yard^3 of Class C excavation. That consisted of 10,194 yard^3 of solid rock that was drilled and blasted by Harty, and for which Harty was paid, and 18,194 yard^3 that was the portion of percentage rock determined by the highway department to be rock.

Harty contended that it should be paid for the 18,194 yard^3 of rock. During the job, no protest was made by Harty that it should have been allowed to remove the percentage rock. The dispute as to the payment arose after the job was completed.

Exercise 2

Tri-Ad Constructors contracted in 1987 with the U.S. Department of the Navy for construction work at San Nicolas Island Naval Air Station in California. The contract in section 605(a) provided,

> All claims by a contractor... relating to a contract shall be in writing and shall be submitted to the contracting officer for a decision.

Paragraph 66 of the contract stated that the officer in charge of construction (OICC) was the authorized representative of the contracting officer with respect to supervising the contract work, but not with respect to the handling of disputes. Section 605(c)(5) provided,

> Any failure by the contracting officer to issue a decision on a contract claim within the period required will be deemed to be a decision by the contracting officer denying the claim and will authorize the commencement of the appeal or suit on the claim...

For claims over $50,000, Section 605(c) established the time period requirement, which commenced on the contracting officer's receipt of a submitted certified claim. Section 605(c) provided, in pertinent part,

> (2) A contracting officer shall, within 60 days of the receipt of a submitted certified claim over $50,000—
> > (a) issue a decision; or
> > (b) notify the contractor of the time within which a decision will be issued.
>
> (3) The decision of a contracting officer on submitted claims shall be issued within a reasonable time, in accordance with regulations promulgated by the agency, taking into account such factors as the size and complexity of the claim and the adequacy of the information in support of the claim provided by the contractor.

Tri-Ad submitted to the Navy a letter dated April 14, 1988, which specifically requested, in two places, a final decision by the contracting officer. The first paragraph of the letter stated,

> We hereby submit the request for a contracting officer's final decision on behalf of Tri-Ad in accordance with the Disputes Clause, FAR 52.233-1, of the general paragraphs of the subject contract.

The last paragraph, repeating the same basic request, stated,

> By this claim, we request a contracting officer's final written decision.

The letter, which was addressed to the OICC, detailed instances of allegedly improper delay by the government and sought payments totaling more than $118,000. The letter was 12 pages long and described in detail the factual and legal basis of Tri-Ad's request. By letter dated June 6, 1988, the resident office in charge of construction (ROICC) forwarded Tri-Ad's April 14 letter to the contracting officer.

The contracting officer never sent Tri-Ad a response that could be characterized as a final decision on the claim. However, Tri-Ad did receive a response from the assistant resident office in charge of construction (AR-OICC), dated June 28, 1989, which specifically addressed each item raised in Tri-Ad's April 14, 1988, claim letter. The AR-OICC found some items meritorious and rejected all others. The response included a proposed contract modification that would compensate Tri-Ad for the claim items found meritorious. The response noted that if Tri-Ad failed to sign the proposed contract modification, the Navy would issue it unilaterally. On August 1, 1989, a change order was issued to compensate Tri-Ad for the items presented in the letter of April 14, 1988, which the government concluded were valid.

Tri-Ad contended that the pertinent jurisdictional requirements for a direct action suit in the court were present. Tri-Ad contended that the April 14, 1988, letter constituted a proper written claim under Section 605(a). As to the requirement of a contracting officer's decision on the claim, Tri-Ad contended that pursuant to Section 605(c)(5), a contracting officer's decision denying the claim should be "deemed" to exist because the contracting officer received the claim on or about June 6, 1988, and failed thereafter to issue a final decision on the claim within the Section 605(c) time period.

In response, the government disputed that the April 14, 1988, letter qualified as a claim under Section 605(a). The government did not dispute that the letter provided adequate notice as to the basis and amount of the claim or that the letter unequivocally requested a final decision by the contracting officer. Indeed, the government found no fault with the body of the letter. Rather, the government's sole criticism was that the letter was addressed to the OICC and not to the contracting officer.

Exercise 3

Background

Applied Construction Co., Inc. (Applied), is a Pennsylvania corporation with its principal office in Uniontown, Pennsylvania. It is a general contractor, prequalified by the Pennsylvania Department of Transportation (PennDOT) to bid for contracts for the construction of highways throughout Pennsylvania. On an annual basis, Applied derives substantial revenues from contracts entered into with PennDOT, and its very existence depends on its status as a PennDOT prequalified contractor.

PennDOT is an administrative agency of the commonwealth of Pennsylvania. It is responsible for letting competitive bidding contracts for the construction of highways within the commonwealth of Pennsylvania. These contracts are awarded to the lowest responsive and responsible bidder. PennDOT is responsible for prequalifying contractors for PennDOT work. In all other respects, PennDOT represents the commonwealth as the owner of state and interstate highways.

Obtaining and maintaining its status as a prequalified contractor is an express condition of being a responsible bidder on contracts for the construction of highways for PennDOT and other state and local agencies, including the Pennsylvania Turnpike Commission.

Bid Solicitation

Before February 19, 1990, PennDOT solicited bids for the construction of a section of Interstate 5 (I-5), in Green County, Pennsylvania. On February 19, 1990, PennDOT opened the bids for the contract and determined that Applied was the lowest responsible bidder.

The contract incorporated plans, drawings, and PennDOT's Form 408 Specifications (1987) as amended. For reasons not attributable to Applied, the contract was not formally awarded by PennDOT until April 2, 1990, even though the anticipated notice-to-proceed date used by PennDOT for the calculation of contract time and as set forth in the bid documents was March 9, 1990. The total contract duration was 617 calendar days. Applied's bid was $60 million.

Project Description

The project can be generally described as the reconstruction of a segment of old, four-lane state highway (S.R. 1) to interstate highway standards (expanding the lane width and adding a full-width shoulder) and the construction of a new interchange and a segment of new interstate highway east of the interchange on virgin ground. The reconstructive work on S.R. 1 included the widening of old S.R. 1.

S.R. 1 is cut through a ridge of one of the Blue Mountains. The widening of S.R. 1 required the construction of two retaining walls (No.1 and No. 2) at the toes of the slopes of the cut into the mountain. The retaining walls were each approximately 2,400 ft long and varied in height from 5 ft to 30 ft above the roadway.

Traffic Control

Because S.R. 1 is an important access highway to the area, strict traffic control requirements were incorporated into the contract traffic control plan. Throughout the course of the construction, PennDOT required that traffic along S.R. 1 be maintained at all times with a minimum of two lanes of traffic open during the summer months and all four lanes open between November 23 and March 30. As a result, the traffic control plan contained compulsory construction phases. The compulsory construction phases were developed before the bid by PennDOT to accommodate the various traffic patterns to be maintained throughout the course of the work. The construction phases were a contractual requirement.

The traffic control plans defined areas where work was "required" or "allowable" during each phase of construction. In addition, the compulsory construction phases specified the work that was required by PennDOT to be commenced and completed both before and after the

1990–1991 winter shutdown period. During the winter shutdown, no traffic restrictions were permitted by the contract.

Scheduling Requirements

The contract documents required Applied to incorporate the compulsory construction phases into a CPM schedule and to schedule its construction operations "as indicated for the various phases of construction." Nothing in the contract documents gave Applied the right to alter or change unilaterally the compulsory construction phases or any other part of the traffic control plans.

The traffic control plans provided specific milestones when traffic was to be changed so that the concrete paving of the lanes of traffic could proceed. These milestones were also to be incorporated into Applied's CPM schedule.

Liquidated Damages

Liquidated damages of $7,500 per day were specified for every day after the dates specified for the road to be open to traffic (i.e., November 24, 1990, for the winter shutdown and November 6, 1991, the date specified for the road to be opened to unlimited traffic). Additionally, engineering liquidated damages of $2,400 per day assessed to the contractor were also specified for every day the road remained incomplete after November 30, 1991.

Project Execution

The two retaining walls were critical to the completion of the project. All work on retaining wall No. 2 was required by the compulsory construction phases to be completed before the winter shutdown of 1990–1991. The completion of retaining wall No. 1 was both a physical and planning constraint to paving any of the southbound lanes of S.R. 1, which work was required by the contract's compulsory construction phases to be completed before the winter shutdown of 1990–1991. For the work on retaining wall No 1. to be commenced and completed in the 1990 construction season, the work on the wall had to begin no later than July 15, 1990. Work on each wall was anticipated to require 103 calendar days.

The project was defined by PennDOT as a "fast track" project. At the April 13, 1990, preconstruction conference, PennDOT stated that it would review and return all submittals within 3 days and shop drawings and other plans within 10 days of the date of the receipt by PennDOT, its consultants, or PennDOT's construction management consultant.

Also at the April 13, 1990, preconstruction conference, Applied submitted to PennDOT plans and drawings for an alternate design of the retaining walls. At PennDOT's request, these plans and drawings formed a part of a value-engineering proposal from Applied to PennDOT.

PennDOT returned the value-engineering proposal to the contractor stamped "Approved as to Design" on October 23, 1990. However, PennDOT did not issue a written work order accepting Applied's value-engineering proposal or authorizing the work. Pending the receipt of approved drawings, Applied was unable to commence any substantial work in the area of retaining wall No. 1. As a result, the excavation adjacent to the retaining wall No. 1 on existing S.R. 1 southbound and the paving of these lanes before the compulsory winter shutdown period 1990–1991 could not be accomplished.

After the compulsory winter shutdown period expired and no formal acceptance of the proposal was issued by PennDOT, Applied commenced its work in the area of retaining wall No. 1 on April 14, 1990. Applied started in the areas that were common to both the contract's original design and the value-engineering design, with the expectation that the work order accepting its value-engineering proposal would be issued.

When Applied began the layout work on retaining wall No. 1, it noticed that the ground lines and ground slope described in the cross sections in the contract differed from those actually existing in the field. The cross sections described the existing slopes and showed the final slopes adjacent to wall No. 1 to be at 2:1. Applied discovered that the existing slopes were steeper than indicated and that if the slopes were cut to 2:1, the cuts would extend beyond the legal right of way. To stay within the right of way, a slope steeper than 2:1 would be required. Applied brought these conditions to PennDOT's attention on April 30, 1990.

Additionally, the traffic pattern immediately in front of retaining wall No. 1 was changed by PennDOT because an adjacent contractor's work required a shift of traffic from the southbound lanes. To accommodate the other contractor's work, PennDOT directed Applied to install crossovers, which restricted the work area in front of wall No. 1 to one lane instead of two.

One of the more significant effects of the lane restriction was that trucks removing the excavated materials were required to back down the single lane to be loaded because Applied's excavator occupied the single lane. This changed Applied's planned "loop" of using both lanes of traffic to remove the excavated materials (trucks running in lane 1, excavator situated in lane 2).

The retaining wall systems specified in the contract and the value-engineering proposal by Applied both used a tieback system. After completing some of its work and after discovering field conditions different from those anticipated, Applied notified PennDOT on May 5, 1990, that it was withdrawing its value-engineering proposal and requested further direction. It pointed out the following conditions:

- While cutting into the mountain, Applied discovered that during the original construction of S.R. 1, areas of the slope had been drilled and shot (blasted). Apparently the presplitting operation had not been successful, and there had been considerable overbreak. Importantly, the rock in the existing slope was significantly fractured. Applied had not considered the possible compromised integrity of the rock slope in completing its proposed tieback design.

- While completing cuts into the slope, Applied discovered seams of sand and decomposed rock throughout the mountain. The design computations Applied had furnished with its value-engineering proposal had assumed friction angles greater than zero.

- The legal right of way for the highway extended approximately 50 ft beyond the proposed face of the wall. Based on Applied's recomputation of the friction angle and anticipated need for extended tiebacks or rock anchors, the tiebacks would extend well beyond the legal right of way.

- Cutting the slopes behind wall No. 1 to the planned 2:1 slope would require a cut beyond the legal right of way. No soil analysis was available to Applied to determine the angle of repose of the existing and varied soil types discovered to propose a steeper slope.

- Importantly, the blasting that occurred during the original construction caused the slopes to be unstable. A nationally known soil consultant retained by Applied confirmed in a report to Applied that the slopes were unstable and unsafe. On further examination, movement in the slope over a period of time was visible. Further work on the slope was judged by Applied to be a violation of OSHA regulations.

On May 7, 1990, Applied suspended all operations at wall No. 1 pending further investigation. On May 17, 1990, PennDOT directed Applied in writing to proceed with the construction of the wall according to Applied's value-engineering proposal, but no work order was issued accepting the value-engineering proposal. PennDOT informed Applied on June 7, 1990, that Applied was in breach of its contract for improperly suspending work on retaining wall No.1 and threatened to impose liquidated damages and reevaluate Applied's prequalification status unfavorably if Applied did not immediately resume work on the retaining wall. At this time, no testing had been done by PennDOT or Applied to determine the actual conditions of the mountain.

Exercise 4

The Olshan Demolishing Co. entered into a contract with the Angleton Independent School District for the demolition of a two-story building on a campus of the school district. The contract called for the demolition and removal of the building, foundation, paving, and sidewalks. After the removal of these items, the site was to be regraded. Work on the project started on May 29, 1982, and was completed by August 10, 1982.

Paragraph 4.2.2 of the Supplemental Conditions to the contract provides,

> After reporting to the Architect any error, inconsistency, or omission he may discover in the Contract Documents, the Contractor shall not proceed with any Work affected without obtaining specific written instructions from the Architect.

Paragraph 4.2.1 of the General Conditions provides,

> The Contractor shall carefully study and compare the Contract Documents and shall at once report to the Architect any error, in-

consistency or omission he may discover. The Contractor shall not be liable to the Owner or the Architect for any damage resulting from any such errors, inconsistencies or omissions in the Contract Documents. The Contractor shall perform no portion of the Work at any time without Contract Documents or, where required, approved Shop Drawings, Product Data or Samples for such portion of the Work.

The contract also contained in paragraph 12.2.3 the following notice provision:

[The Contractor shall] give written notice to the Architect before executing the work if Olshan wished to make a claim for an increase in the Contract Sum, he shall give the Architect written notice hereof within twenty days after the occurrence of the event giving rise to such claim. This notice shall be given by the Contractor before proceeding to execute the Work, except in an emergency endangering life or property in which case the Contractor shall proceed in accordance with Paragraph 10.3. No such claim shall be valid unless so made. If the Owner and the Contractor cannot agree on the amount of the adjustment in the Contract Sum, it shall be determined by the Architect. Any change in the Contract Sum resulting from such claim shall be authorized by Change Order.

A concealed conditions clause was also in the contract. It stated in paragraph 12.2.1,

Should concealed conditions encountered in the performance of the Work below the surface of the ground or should concealed or unknown conditions in an existing structure be at variance with the conditions indicated by the Contract Documents, or should unknown physical conditions below the surface of the ground or should concealed or unknown conditions in an existing structure of an unusual nature, differing materially from those ordinarily encountered and generally recognized as inherent in work of the character provided for in the Contract, be encountered the Contract Sum shall be equitably adjusted by Change Order upon claim by either party made within twenty days after the first observance of the conditions.

On August 3, 1982, Olshan discovered three additional slabs under-neath at least part of the building. These additional slabs were not shown on the architectural drawings on which Olshan had based its bid. The school district knew that there was more than one slab under the building but believed that it had informed all of the contractors bidding on the project. However, Olshan did not know of the existence of the three additional slabs. Before a written request for additional compensation could be filed, the three slabs were removed. Thereafter, Olshan notified the school district's architect about the additional slabs and requested additional payment for the work and materials. The claim was made within the required 20 days.

Exercise 5

On August 29, 1967, Inland Bridge Co., Inc., and the North Carolina State Highway Commission entered into an agreement for the relocation of U.S. 21 in Charlotte, North Carolina, from the south city limits north to a point approximately 0.4 mi south of Shuman Road. Part of the relocation consisted of the building of certain embankments, which required excavation, hauling, drying, backfilling, and compaction.

The contract contained the following language related to soil classifications:

> 22-3.8 The classification of all roadway and drainage excavation shall be made by the Engineer as the work progresses and the classification as determined by the Engineer for the work completed each month will be included in the current monthly estimate. If classification thus allowed is protested by the contractor, claim must be made within 30 days after the current estimate is mailed to him.

> 22-1.1 Description of Roadway and Drainage Excavation. This item shall consist of the removal and satisfactory disposal of all ... unsuitable subgrade material and the replacement of such unsuitable material with satisfactory material.

> 22-1.2 The classification of all materials excavated shall be as follows:

. . .

(b) Unclassified Excavation shall include all excavation within the limits of the original slope stakes. . .

Unsuitable material shall be classified as any material which is unsatisfactory for use under a base course or pavement. It shall not include any rock undercut in the roadbed.

The proposal also contained the following information related to soil classification:

UNCLASSIFIED EXCAVATION

This item shall include the removal of all existing flexible pavement, walls, steps and other masonry items inside or outside the limits of the right of way which in the opinion of the Engineer is rendered useless for highway purposes by the construction of this project. These items that are removed shall be used in embankments or disposed of in waste areas furnished by the Contractor.

These items that are removed will be measured and paid for at the contract unit price per cubic yard, "Unclassified Excavation." The cost of disposal shall be included in the unit price bid per cubic yard "Unclassified Excavation."

UNSUITABLE MATERIAL EXCAVATION

The Contractor will be required to remove unsuitable material at locations as shown in the plans and other locations as the Engineer may direct. . .

NOTE TO CONTRACTOR

Note to the contractor: the contractor's attention is directed to the fact that the natural moisture of the material to be placed in the embankment is approximately 40%. This material shall be dried to optimum moisture as determined by the engineer.

Also contained in the contract were the following clauses pertaining to changes in the project:

4.3A Should the Contractor encounter or the Commission discover during the progress of the work conditions at the site differing materially from those indicated in the contract, which conditions could not have been discovered by reasonable examination of the site, the Engineer shall be promptly notified of such conditions and if he finds they do so materially differ and cause a material increase or decrease in the cost of performance of the contract, an equitable adjustment will be made and a supplemental agreement entered into accordingly.

4.4B Request for Authorized Modification:

Whenever the Contractor is required to perform work which is, in his opinion, extra work, and an authorized modification therefor has not been issued by the Engineer, then the Contractor may make a written request for an authorized modification at any time before beginning any of the alleged extra work.

4.4C If the Contractor's request for an authorized modification has been denied by the Engineer and the Contractor intends to file a claim for payment for performing such alleged extra work, he shall notify the engineer in writing of his intention to file a claim for such payment and shall receive written acknowledgment from the Engineer that such notification has been received before he begins any of the alleged extra work. In such case the Contractor will be required to keep an accurate and detailed cost record which will indicate the cost of performing the extra work. Such cost record will be kept with same particularity as force account records and the Commission shall be given the same opportunity to supervise and check the keeping of such records as is done in force account work.

In preparing its bid, Inland relied on the note to contractors that the natural moisture content of the soil was approximately 40%. Additionally, Inland obtained a copy of the subsurface report from the commission, which showed the results of 10 moisture content tests to be 29.4%, 31.6%, 33.5%, 34.0%, 35.3%, 36.8%, 39.1%, 39.2%, 40.5%, and 43.3%. Inland bid $342,650 for excavation of unclassified material based on $0.77 per cubic yard. The unit price for excavation of unsuitable material was $2.50 per cubic yard and amounted to $36,250.

Inland began work in August 1967 and almost immediately, it became apparent that it was going to be difficult to dry the soil to the optimum moisture content or to compact it to the required 95% density as required by the contract. The work progressed slowly that fall, and on November 20, 1967, Inland and the commission met to discuss the problems that had arisen. At that time, the commission's engineers made certain suggestions and also began on site soil moisture testing. There was little improvement in progress, and on March 26, 1968, the parties met again to discuss the soil drying operation. Inland stated that every possible means of drying the soil was being used, but that it was not possible to dry the material "within practical methods" and still complete the project on time.

In April 1968, Inland began its own moisture content testing. A total of 16 tests were made, with the highest value recorded as 66.6%. Only a few test results were below 40%. It was Inland's position that more than half of the "unclassified material" was in fact "unsuitable material."

On May 6, Inland's president, Fred Triplett, wrote the following letter to J. F. Warren, resident engineer for the commission:

Re: Construction Conference of March 26, 1968
N.C. Project 8.1654707—etc.

Dear Sir:
Subsequent to the above conference, we have redoubled our efforts toward drying this soil. We have, as your records will bear out, exhausted every practical resource and are yet not even close to drying this unusual material anywhere near rapidly enough to allow us to prosecute this job to a practical conclusion. We would like to reiterate that this material has been shown to successfully resist even extreme practical methods to dry to optimum moisture; therefore, we will, under the present circumstances, be forced to present a claim in this connection in the future based on an engineering impracticability.

In the interest of a workable solution we have investigated the use of hydrated lime to dry and possibly improve the soil. The use of this material in an appropriate quantity and manner throughout the work could give us a workable situation. If after your investigation, this special treatment is indicated, we offer to place this material in the fill according to standard practices listed for lime modification of subgrade material in the Lime Stabilization Construction Manual at a price of $30.00 per ton in place. In this way, you will be in complete control of the amount of application as well as the scope of the entire operation.

Please advise us as soon as possible if you are in favor of a trial of this method on this basis. Awaiting your valued reply we are

> Yours truly,
> INLAND BRIDGE COMPANY, INC.
> *Fred Triplett*

On May 8 and May 14, 1968, several conferences were held where the use of lime was discussed. On May 27, Mr. Triplett wrote John Davis, assistant chief engineer of the state highway commission in Raleigh,

Re: N.C. Project 8.1654707

Dear Sir:

We doubled and redoubled our efforts to dry this soil on something akin to a production basis to no apparent avail. In fact, we have not been able to dry one spot anywhere to optimum moisture. We believe that the inherent moisture of this soil (up to 50%) coupled with its extreme capillarity and affinity for water makes it impractical from an engineering standpoint to dry to anywhere near optimum moisture or compact to 95% density. We respectfully request, therefore, that this material be classified as unsuitable material and we be allowed to waste it.

If this remedy does not seem entirely practical to you, we are, of course, amenable to any alternatives you may suggest which will, in fact, afford us to complete this work satisfactorily to us both.

Please advise when we may meet with you in order to pursue this matter toward a sensible solution.

> Yours truly,
> *Fred A. Triplett, Jr.*

On June 6, 1968, the results of tests conducted by the commission indicated that it would be beneficial to use lime to stabilize the soil. Subsequently, an extra work order was executed. In addition to agreeing to payment for the cost of liming the soil, the commission agreed to grant a time extension and to reduce the compaction requirement to 90%. The work was completed some 10 months later in April 1969. On August 25, 1969, Inland submitted a claim for $169,821. In the claim, Inland argued that the conditions at the site were substantially different from those indicated in the contract and that the material should have been classified as "unsuitable" instead of "unclassified." Because the material was classified incorrectly, Inland was induced to submit a bid that was lower than it would have otherwise made.

What were the intentions of the parties regarding the classification of soil? What are the issues in this dispute? What type of claim is this, and what chapters in this book should be used to resolve this situation? Should Inland be entitled to additional compensation? Why or why not?

Exercise 6

On or about April 20, 1966, Blankenship Construction Co. received an invitation to bid on Projects 8.1657505 and 8.1657507 in Mecklenburg County, advertised by the North Carolina State Highway Commission. The projects were a segment of Interstate 85 northeast of Charlotte, extending to the Mecklenburg–Cabarrus County line and the Interstate 29 Connector, which is roughly 3 mi long and connects Interstate 85 and U.S. 29, 2 mi west of the Mecklenburg–Cabarrus County line. The bidding period was three weeks. Blankenship spent one full day inspecting the site and encountered no evidence of rock or unusual conditions.

The commission had conducted a subsurface investigation. That report, which Blankenship requested and received before the bid, showed approximately 27,000 yard³ of rock throughout the project. An amended earthwork summary mailed to Blankenship a few days before the date bids were due reflected 135,000 yard³ or rock. This information was consistent with Blankenship's analysis of the profile sheets, which indicated approximately 129,000 to 130,000 yard³. However, the subsurface report did not indicate that the commission had discovered rock at several other drilling locations.

Section 22 of the Standard Specifications contained the following description related to "Roadway and Drainage Excavation":

> 22-1.1 DESCRIPTION. This item shall consist of the removal and satisfactory disposal of all materials excavated within the limits of the right of way, including such intersecting roads, driveways, streets, outlooks, parking areas, unsuitable subgrade material and the replacement of such unsuitable material with satisfactory material, and shall include such excavation as is necessary for berm, inlet and outlet and lateral drainage ditches; for stripping

material pits, and for the formation, compacting and shaping of all embankments, subgrade, shoulders, slopes, intersections, approaches, and private entrances to conform to the typical cross section shown on the plans and to the lines and grades set by the Engineer. . .

Section 22-1.2 defines four classes of excavation: (a) Solid Rock Excavation; (b) Unclassified Excavation; (c) Drainage Ditch Excavation; and (d) Stripping Excavation. Unclassified excavation encompasses any and all of the other classes of excavation within the original slope stakes: "Unclassified Excavation shall include all excavation within the limits of the original slope stakes."

The contract contained the following language from the Standard Specifications under the heading of "Changed Conditions":

> The Commission reserves the right to make, at any time during the progress of the work, such increases or decreases in quantities and such alterations in the details of Construction, including alterations in grade or alignment of the road or structure or both, as may be found to be necessary or desirable. Such increases or decreases and alterations shall not invalidate the contract nor release the Surety, and the Contractor agrees to accept the work as altered, the same as if it had been a part of the original contract.
>
> Under no circumstances shall alterations of plans or of the nature of the work involve work beyond the termini of the proposed construction except as may be necessary to satisfactorily complete the project.
>
> Unless such alterations and increases or decreases materially change the character of the work to be performed or the cost thereof, the altered work shall be paid for at the same unit prices as other parts of the work. If, however, the character of the work or the unit costs thereof are materially changed, an allowance shall be made on such basis as may have been reached, then the altered work shall be paid for by force account in accordance with Article 9.4.
>
> Should the Contractor encounter or the Commission discover during the progress of the work conditions at the site differing materially from those indicated in the contract, which conditions could not have been discovered by reasonable examination of the site, the Engineer shall be promptly notified in writing of such

conditions before they are disturbed. The Engineer will thereupon promptly investigate the conditions and if he finds they so
materially differ and cause a material increase or decrease in the
cost of performance of the contract, an equitable adjustment will
be made and a supplemental agreement entered into accordingly.

In the event that the Commission and the Contractor are unable to reach an agreement concerning the alleged changed conditions, the Contractor will be required to keep an accurate and
detailed cost record which will indicate not only the cost of the
work done under the alleged changed conditions, but the cost of
any remaining unaffected quantity of any bid item which has
had some of the quantities affected by the alleged changed conditions, and failure to keep such a record shall be a bar to any recovery by reason of such alleged changed conditions. Such cost
records will be kept with the same particularity as force account
records and the Commission shall be given the same opportunity to supervise and check the keeping of such records as is
done in force account work.

Related to extra work, the standard specifications also contained the
following:

... (The Contractor) shall notify the Engineer in writing of his intention to file a claim for such payment and shall receive written
acknowledgment from the Engineer that such notification has
been received before he begins any of the alleged extra work. In
such case the Contractor will be required to keep an accurate and
detailed cost record which will indicate the cost of performing
the extra work. Such cost records will be kept with the same particularity as force account records and the Commission shall be
given the same opportunity to supervise and check the keeping
of such records as is done in force account work.

The proposal called for approximately 2,076,000 yard^3 of unclassified
excavation. Blankenship bid $612,420 for this item. The bid was based
on two major cost considerations: first, the estimated quantity of solid
rock (130,000 yard^3) was estimated at $1.50 per cubic yard; and the re-
maining amount of excavation was calculated at $0.22 per cubic yard.
Blankenship's total bid was $1,570,369. Blankenship was awarded the
project on May 11, 1966, and work began in late June 1966.

As the work progressed, Blankenship encountered more rock than originally anticipated. The first cut was made at Station 180 in July 1966, and it was mostly rock. Likewise, the next cut consisted of a large quantity of rock. None of this rock appeared in the subsurface information supplied by the commission. In the fall, the first major cut also contained large quantities of rock and at this time, Blankenship realized that its original estimate of solid rock excavation was grossly inaccurate. Rather than the original of 130,000 yard³, Blankenship now estimated the amount of rock to be between 750,000 and 800,000 yard³.

Because of the large quantities of rock, Blankenship fell behind schedule. The construction engineer wrote in his report dated January 18, 1967,

> Unclassified material has high percentage rock making it ideal for working this time of year. Contractor has placed another culvert crew in the project in an attempt to increase culvert construction which is falling behind.

In April 1967, Malcolm Blankenship called John Davis, the commission's chief engineer. Blankenship told Davis that he would like to come to Davis's office and discuss the amount of rock they were encountering on the job. There was some discussion about the subsurface report, and Blankenship told Davis that what was being found differed substantially from the report. That was the essence of the conversation.

The contract completion date was originally December 1, 1967. This date was extended to April 21, 1968. Blankenship did not complete the project until March 9, 1969. Liquidated damages in the amount of $64,000 were withheld by the commission. Approximately one year after the project was completed, Blankenship filed a claim for $4,167,276.

Exercise 7

In June 1962, the Department of Public Works (DPW) of the Commonwealth of Massachusetts requested proposals for the construction of a section of Interstate Route 495 (including two bridges) in the city of Haverhill. State Line Contractors, Inc., submitted a bid and was awarded the contract on August 21, 1962.

The contract called for the construction of 1.231 mi of the highway between stations 255+00 and 320+00, including two bridges, one over

Newton Road; the other bridge with approach ramps over Amesbury Road. In State Line's bid, the following line items and unit prices are of particular importance:

LINE ITEM	QUANTITY (YARD^3)	UNIT PRICE
A2-1, Roadway earth excavation	885,600	$0.56
A2-2, Class A rock excavation	166,500	$0.56
A6-1, Ordinary borrow	Not available	$0.60

The contract was originally to be completed by July 15, 1964. This date was extended to September 17, 1964, but the work was not finished until December 2, 1964.

The Standard Specifications contained the following relevant provisions:

> Article 22. Alteration of work.... In case of any alterations, so much of the contract as is not necessarily affected by such alterations shall remain in force upon the parties thereto, and such alterations shall be made under the terms of and as a part of the contract...
>
> Article 23. Extra Work. The Contractor shall do any work not herein otherwise provided for when and as ordered in writing by the Engineer, such written order to contain particular reference to this article.
>
> If the Engineer directs, the Contractor shall submit promptly in writing to the Engineer an offer to do the required work on a lump sun or unit price basis, as specified by the Engineer.
>
> If the Contractor claims compensation for the extra work not ordered as aforesaid, or for any damage sustained, he shall, within one week after beginning of such work or of the sustaining of any such damage, make a written statement to the Engineer of the nature of the work performed or damage sustained and shall on or before the fifteenth day of the month succeeding that in which any such extra work shall have been sustained, file with the Engineer an itemized statement of details and amount of such work or damage; and unless such statement shall be made as so required, his claim for compensation shall

be forfeited and invalid, and he shall not be entitled to payment on account of any such work or damage.

Article 58. Claims of Contractor .... All claims of the Contractor for compensation other than as provided in the contract on account of any act of omission or commission by the ... D.P.W. must be made in writing to the Engineer within one week after the beginning of any work or the sustaining of any damage .... Unless such statement shall be made as so required, the Contractor's claim for compensation shall be forfeited and invalid, and he shall not be entitled to payment on account of any such work or damage.

There was no concealed or differing site conditions clause.

The DPW's original cross-sectional plans for the proposed highway indicated the existence of solid ledge "in the main highway between Stations 295+00 and 308+00 and also in some areas of the North West ramp." In places where ledge was to be encountered, the DPW plans called for the construction of a four-to-one (4:1) slope. Where no ledge was anticipated, the plans called for construction of a two-to-one (2:1) earthen slope. From the original quantity estimates and cross sections and by implication from the inclusion of an item of ordinary borrow, the project was planned as a borrow job with no waste.

Before commencing work, State Line reviewed the subsurface report made available by the DPW and concluded that the indications of solid ledge on the plans were inaccurate. The DPW resident engineer was so informed. New borings were made by the DPW resident and bridge engineers, and no ledge was encountered in the areas shown on the DPW original plans. Additional borings were then made, and this information was adopted by the engineer, who drew up new cross-sectional plans indicating no ledge in the line of the proposed excavations. These new details required a basic change in the original plans. All slopes were now required to be two-to-one (2:1) earthen; the four-to-one (4:1) slopes were eliminated. The result was to change the project from a borrow job to a waste job.

During the later part of 1962 and through the spring of 1963, various discussions were held between State Line and the DPW regarding the changes in the slopes. Additionally, the DPW wanted to modify the contract to build a rest area adjacent to the highway between stations 281+00 and 298+00 at the same unit prices quoted by State Line in the original contract. State Line informed the DPW that it would build the rest area at those prices only if it could be constructed simultaneously with the main highway embankment between these stations. State Line was advised by the DPW personnel that it could be done and on March 18, 1963, the district engineer in writing outlined to State Line a proposed alteration with estimated additional cost covering in part the use of surplus excavated materials to construct the rest area and to flatten the median slopes. On March 27, 1963, State Line advised in writing that it would perform the proposed alteration No. 1 at the specified prices provided it built the rest area simultaneously with the roadway embankment. In the interval between the March 18 and March 27 letters, State Line spoke with DPW personnel regarding their understanding that the rest area and the main road were to be built simultaneously.

Alternative No. 1 was finally approved by the DPW on April 23, 1963. It stated, "Reason for Alteration: Additional borings taken by contractor give no indication of ledge, also a reappraisal of available preliminary information shows that the presence of solid ledge is doubtful." The alteration revised the slopes to 2:1 and called for the construction of a rest area for which there was no provision in the original design. The transmittal letter stated, "Alteration #1 is considered to be in accordance with the applicable provisions stipulated under Article XXII of the . . . Standard Specifications." The schedule annexed to the alteration showed that 35 items in the original contract were affected: increases in 27 and decreases in 8. The item principally affected was item A2-1, roadway earth excavation, which was increased by 128,252 yard^3 at the same unit price as originally bid. There was no reference in Alteration No. 1 to item A2-2, class A rock excavation. Alteration No. 1 also said nothing about simultaneous performance of the rest area work and adjacent highway work.

The claim presented by State Line covered five major issues, which are detailed here.

The Rest Area Claim

On April 23, 1963, the DPW directed State Line to begin work on the rest area. However, on April 29, 1963, State Line received a "right of entry" notice informing it for the first time that it could not proceed with the rest area until an easement was obtained beyond the layout for the slope and drainage. It was ordered to stop work on the rest area. Work within the layout at that point was impractical because the existing drainage ditch would be blocked off. The DPW had done nothing about an easement and did not work out a plan for securing the easement until July 2, 1963. The easement was not secured until September 3, 1963. At that time, State Line was told to resume work on the rest area. Unfortunately, at this time, State Line's operation was distant from the rest area site, and because the work area provided no direct access to the rest area, the job was completed by hauling fill by truck over public roads. On September 9, State Line filed notice of claim with the DPW.

Ledge Removal Claim

After execution of Alteration No. 1, and as the work progressed, State Line encountered solid ledge in excavating the northwest ramp in an area where neither the original nor the revised DPW plans and boring information indicated the existence of ledge. (The original DPW plans had indicated ledge in *some* areas of the northwest ramp.) State Line removed 82,000 yard^3 of ledge, for which it was paid $0.56 per cubic yard. State Line now requested an equitable adjustment based on $1.25 per cubic yard. The basis of State Line's position was that Alteration No. 1 removed all ledge from the contract at the unit price of $0.56 per cubic yard. The work now constituted extra work, for which an equitable price was $1.25 per cubic yard. There was no record of written communication between State Line and the DPW.

Newton Road Claim

As part of the contract, State Line built a bridge over Newton Road, which intersected Interstate 495. During the relocation of Newton Road, while the bridge was under construction, it was necessary for State Line to construct a detour. After State Line completed the relocation of Newton Road, it notified the DPW resident engineer that the new roadway would be ready to use at noon on June 21, 1963. When the traffic division of the DPW failed to put in signs, paint lines, and install directional signals, the resident engineer ordered State Line to reopen the detour. State Line now claims the costs of reopening the detour as an item of damages. There is no record of written communication between State Line and the DPW.

Waste Disposal Claim

As a result of the redesign of the slopes because of the unexpected absence of ledge, and also as the result of other changes in the design made during the project, State Line was never required to provide any ordinary borrow as called for in the original contract. Instead, State Line had to dispose of some 204,000 yard^3 of excavated materials outside the job site. State Line was paid for this work at the unit price for roadway earth excavation of $0.56 per cubic yard but now argued that $0.56 per cubic yard is not adequate compensation for disposing of the material off-site. There is no record of written communication between State Line and the DPW.

Earth Excavation Claim

State Line's superintendent took independent measurements of the elevations throughout the entire project. His survey, the accuracy of which is not disputed, revealed discrepancies in the measurements taken by the DPW. The measurements were used to compute the amount of earth excavation between stations 300+00 and 307+00. The earthwork measurements by the DPW were underestimated by 18,893 yard^3. The commonwealth contends that State Line is precluded from recovering these charges because of Article 58.

Appendix B

Exercise Solutions

Solution: Exercise 3-1

The facts of this exercise are based on the case of Kreisler Borg Florman General Construction Co., Inc., v. Rosen and Morelli Masons et al. (181 A.D.2d 813), presented before the Supreme Court, Appellate Division, New York in 1992.

Competent Parties

There is no reason to doubt, and no arguments were made to suggest, that the parties were anything but competent. The answer is YES.

Offer and Acceptance

The key issue is whether attending job meetings, submitting a "trade payment breakdown," and providing certificates of insurance constituted acceptance of the offer made by Kreisler Borg. Because acceptance must be unconditional and absolute, it seems highly unlikely that these actions could be construed as acceptance of any technical provisions of the contract offer. The answer to this question is NO.

Reasonable Certainty of Terms

This was not an issue in this dispute. The answer is YES.

Proper Subject Matter

This was not an issue in this dispute. The answer is YES.

Considerations

This was not an issue in this dispute. The answer is YES.

Synopsis

The court felt that there was no "meeting of the minds" and thus there was no contract. They wrote,

> ... there was testimony from the defendant's (Rosen) president that many of the actions relied on by the plaintiff (Borg) as evidencing the existence of a contract were undertaken at the plaintiff's (Borg) request and represented the normal accommodations extended to a general contractor as part of the negotiation process. Moreover, the record contains evidence that there were several areas of disagreement with respect to various components of the masonry subcontract which were never completely resolved prior to the defendant's decision to leave the job.

Because there was no "meeting of the minds," there was no contract.

Solution: Exercise 6-1

The facts of this exercise are based on the case of Philip Chiappisi and others v. Granger Contracting Co, Inc., and another case (223 N.E.2d 924), presented before the Supreme Judicial Court of Massachusetts in 1967.

Does the Notice Clause Apply?

YES. Article 16 specifically states that notice is required as a prerequisite for any claim for extra cost. Article 37 also says that the time for making a claim for extra cost is one week.

Did the Owner (Contractor) Have Knowledge?

Chiappisi was told what to do to comply with his contract: take it up with the office. This he did not do. This can be argued as the only way Granger could limit their liabilities. Without knowledge, "there is no evidence that...Granger ordered Chiappisi to do any extra work or authorized any extra work." As stated by the court,

> We think that the decisive issue is whether Chiappisi followed the procedures governing claims for extra work.

With the answer to this question as NO, the flowchart leads to the conclusion that notice was not given, and there can be no recovery. All other questions in the flowchart are now irrelevant. There is no evidence of sloppy contract administration that would have waived the requirement.

Synopsis

The court denied recovery to Chiappisi, citing two reasons:

> ...arts. 16 and 37 reasonably require (a) that prompt inquiry by someone in behalf of Chiappisi at "the office" as Gillette had suggested, and (b) Chiappisi give immediate written notice to the architect.... Because Chiappisi proceeded without such notice and postponed until after the work was completed all written mention of any claim for extra work, there can be no recovery.

Solution: Exercise 6-2

The facts of this exercise are based on the case of Linneman Construction, Inc., v. Montana-Dakota Utilities Co., Inc. (504 F.2d. 1365), which was brought before the Eighth Circuit, U.S. Court of Appeals in 1974.

Does the Notice Clause Apply?

At first glance, the clause would not seem to apply because it states, "Contractor shall be allowed no additional compensation for any extras." The movement of gas mains is not an extra. However, the clause goes on to say "or any work performed by the Contractor not contemplated by the agreement..." This catchall phrase now covers all additional cost for any reason, so the clause applies.

Did the Owner Have Knowledge?

Of course, the owner knew because the owner initiated the action.

Did the Owner Know the Contractor Was Expecting Additional Compensation?

The owner did not know until some 10 months after completion of the work. This was too late for the owner to control its liabilities, so the contractor's action to recover additional monies should not prevail because of this inquiry.

Synopsis

The court ruled against Linneman. Much time was spent in discussing extras and Linneman's compliance with the contract. The court wrote,

> It is undisputed that no written orders were prepared, nor any claim made until some ten months after completion of the job. Non-compliance with this "extras" clause bars recovery for alleged extra work performed under the contract. Although waiver of the contractual requirement is possible, a definite agreement to pay is required to establish a waiver. Here there is no evidence of a definite agreement to pay. The conduct of Linneman during performance of the contract will not support a conclusion of waiver. No demand for extra compensation was made during the performance of the contract, nor until 10 months after completion of the job. This is a strong indication that the laying of mains behind the curbs, while more costly, was not regarded by Linneman as an extra under the contract.

The court further defined extra work:

> "Extra work" as used in connection with building contracts means work which arises outside of and independent of the original contract; that is, something not required in the performance

of the original contract, not contemplated by the parties, and not controlled by the terms of such contract.

Solution: Exercise 6-3

The facts of this exercise are based on the case of Acchione and Canuso, Inc., v. Commonwealth of Pennsylvania, Department of Transportation (461 A.2d. 765), which was heard before the Supreme Court of Pennsylvania in 1983. The flowchart analysis examines only the issue of notice.

Does the Notice Clause Apply?

YES, the clause applies.

Did the Owner Have Knowledge?

YES, PennDOT knew that much of the conduit was not reusable. They paid the unit price for the new trenching, so they knew.

Did the Owner Know the Contractor Was Expecting Additional Compensation?

NO. It was not until the work was almost complete that Acchione made its intentions clear. So the notice requirements were not satisfied, and recovery is unlikely.

Synopsis

The Pennsylvania Board of Claims, an administrative forum, awarded Acchione damages. This decision was overturned on appeal to the Commonwealth Court. The Pennsylvania Supreme Court overturned the Commonwealth Court decision and sided with the Board of Claims. The Supreme Court did not address the issue of notice but instead only spoke to the issue of constructive fraud on the basis of a misrepresentation (see Chapter 10). Apparently, they were persuaded by the affirmative statement that 50% of the conduit was reusable. The flowchart analysis in this instance is inconsistent with the court's decision, but different issues were involved.

Solution: Exercise 6-4

The facts of this exercise are based on the case of McKeny Construction Co., Inc., v. Town of Rowlesburg, West Virginia (420 SE2d 281), which was presented before the Supreme Court of Appeals of West Virginia in 1992.

The issue is about extra work that McKeny claimed he was directed to do, but the town claims was volunteer work.

Does the Notice Clause Apply?

YES, the contract says, "No claim for an addition to the contract sum . . .," which covers everything.

Did the Owner Have Knowledge?

There is no indication that the town or its engineers knew of the extras for which McKeny would later make claims.

Did the Owner Know the Contractor Was Expecting Additional Compensation?

This is probably an irrelevant point if the additional work was not known. Nevertheless, there is no indication that anyone knew that McKeny would seek additional monies until after the job was completed.

Was the Requirement Waived?

There is no indication of sloppy contract administration that would constitute a waiver of the written notice requirement. The court noted that eight changes were given in accordance with the contract, and these were entered into by agreement and approved by the town.

Synopsis

McKeny was not allowed to recover the cost for the extras, which is consistent with the flowchart analysis.

Solution: Exercise 6-5

The facts of this exercise are based on the case of Northern Improvement Co., Inc., v. South Dakota State Highway Commission (267 N.W.2d 208), which was presented before the Supreme Court of South Dakota in 1976.

The flowchart performs a contract analysis. However, the Supreme Court of South Dakota took a different approach, instead finding that the South Dakota State Highway Commission had breached the contract. Thus, the two approaches are not comparable.

Does the Notice Clause Apply?

YES, the contract covers all types of changes.

Did the Owner Have Knowledge?

Of course the owner knew, especially of the detour problems.

Did the Owner Know the Contractor Was Expecting Additional Compensation?

YES, the contactor repeatedly requested supplemental agreements.

Was the Notice Timely?

The contractor made requests for supplemental agreements before doing the work.

Was the Requirement Waived?

There is no indication of sloppy contract administration, so the contractor should be able to present its claim because constructive notice was given.

Synopsis

The Court was of the opinion that the actions of the state's project engineer constituted a waiver of the notice requirement and that Northern should be able to recover. In discussing the requirements of the contract, the court was inclined to apply the Plumley doctrine, that is, that only a written notice will suffice.

Solution: Exercise 6-6

The facts of this exercise are based on the case of D. Federico Co., Inc., v. Commonwealth of Massachusetts (415 N.E.2d 855) and were presented before the Appeals Court of Massachusetts in 1979.

Does the Notice Clause Apply?

It is assumed the answer to this inquiry is YES.

Did the Owner Have Knowledge?

Of course the owner knew of the excess excavation and borrow requirements because the contractor was paid at the contract unit prices.

Did the Owner Know the Contractor Was Expecting Additional Compensation?

NO, there is no indication that the owner knew this information until the end of the job.

Was the Requirement Waived?

There is no indication of sloppy contract administration, so the contractor should not be able to present its claim because constructive notice was not given.

Synopsis

There was an enormous overrun in excavation quantities. Because no borings were taken, the quantities were given as an estimate only. The borrow material alone went from an estimated 30,150 yard^3 to an actual amount of 278,493 yard^3. In discussing the mistake, the court said,

The mistake here is of a different type: a guess or assumption made from insufficient factual data. Nothing in the master's report suggests that the basis for the engineer's estimates was not available to bidders for the asking; but all parties apparently proceeded content to rely on the estimates rather than go to the trouble of making test borings. The contract made clear that the excavation unit price was to control regardless of actual quantities and covered peat-removal, boulder-removal, and ledge blasting; it also made clear that the estimate shown for unit price items were for purposes of bid-comparison only and were not guaranteed.

Federico lost its bid for compensation partly because "no timely claim for extra compensation was made."

Solution: Exercise 7-1

The facts of this exercise are based on the case of Lazer Construction Co., Inc., v. James and Kathy Long (370 SE2d 900), and the case was heard before the Court of Appeals of South Carolina in 1988.

The contract contained a guaranteed maximum price of $99,500. Because of oral changes made by the Longs, the cost of the home rose to $140,771.50. When the Longs stopped making payments, they had paid $116,713.57. Lazer filed a lien; the Longs brought action to vacate the lien. A central issue was the payment for orally directed changes.

Is There a Statutory Requirement?

NO, because this is a private dealing.

Does the Changes Clause Apply?

YES, the clause can be read to cover all types of changes.

Did the Owner Have Knowledge?

YES, the owner knew.

Did the Owner Promise to Pay?

YES, by authorizing the changes, the owner promised to pay, even if those exact words were never spoken.

Did the Agent Have Authority?

There was no agent in this instance.

Synopsis

Based on the flowchart analysis, the Longs should pay for all the authorized changes. In addressing the changes issue, the court noted that all the changes were orally authorized by the Longs. To this point, they said,

> Moreover, a written contract may be modified by oral agreement even when it explicitly states all changes must be in writing . . . . Under this rule, Lazer's failure to execute change orders for additional work did not necessarily preclude recovery if it could show that Long approved the changes.

On this basis, the court refused to vacate the lien. The court did not address the question of why the Longs had already paid Lazer an amount exceeding the guaranteed maximum price.

Solution: Exercise 7-2

The facts of this exercise are based on the case of Henry's Electric Co. v. Continental Casualty Co. (the surety) and Marilyn Apartments, Inc., (366 F.Supp. 954), which was heard before the U.S. District Court, W.D. Oklahoma in 1973. The owner was Atrium Corp., which was bankrupt at this time.

On this appeal, most of the attention was focused on whether the contract clause was a "pay when paid" clause or a "pay if paid" clause. The court found it to be a "paid when paid" clause. The general contractor, Marilyn, did not receive full payment from the owner, apparently because the owner was not satisfied with the work. Marilyn argues that its contract with Henry is a "pay if paid" clause.

Is There a Statutory Requirement?

NO, because this is a private dealing.

Does the Changes Clause Apply?

YES, the clause can be read to cover all types of changes.

Did the Owner Have Knowledge?

YES, the contractor knew.

Did the Contractor Promise To Pay?

YES, by verbally directing Henry to obtain and install certain light fixtures, it promised to pay.

Did the Agent Have Authority?

There was no agent in this instance.

Synopsis

Based on the flowchart analysis, Marilyn should pay Henry for all the fixtures requested, provided the contract clause was determined to be a "pay when paid" clause. The court found this to be the case.

Solution: Exercise 7-3

The facts of this exercise are based on the case of Care Systems, Inc., v. Edward Laramee (166 A.D.2d 770), which was heard before the Supreme Court, Appellate Division, New York.

Care Systems refused to pay for extra work required arising from a faulty foundation, so Laramee filed a lien. Care Systems now brings action to vacate the lien. The key issue is about Care Systems's obligation to pay for orally directed changes.

Is There a Statutory Requirement?

NO, because this is a private dealing.

Does the Changes Clause Apply?

YES, the clause can be read to cover all types of changes.

Did the Owner Have Knowledge?

YES, it is difficult for Care Systems to argue that they did not know.

Did the Owner Promise to Pay?

This is the central issue. If directing Laramee to proceed with corrections that were knowingly outside the scope of the contract promises constitutes a promise to pay, then the answer would be YES. Because the problems with the foundation were known in advance and were not caused by Laramee, it seems the answer should be YES.

Did the Agent Have Authority?

There was no agent in this instance.

Synopsis

The court addressed the issue of obligation to pay by saying,

> When an owner knowingly receives and accepts extra work orally directed by himself and his agents, that owner is equitably bound to pay the reasonable value thereof, notwithstanding the provisions of the contract that any extra work must be supported by a written authorization signed by the owner; such conduct constitutes a waiver of that requirement.

The court refused to vacate the lien.

Solution: Exercise 7-4

The facts of this exercise are based on the case of Owens Plumbing and Heating v. City of Bartlett, Kansas (528 P.2d 1235), which was heard before the Supreme Court of Kansas in 1974.

The issue involved in this action is whether rock removal was part of the work or an extra to the contract. Apparently, it could not be determined from a reading of the contract. The city's defense was that there was no written directive, as required by the contract.

Is There a Statutory Requirement?

NO, there is no indication of a statute or regulation.

Does the Changes Clause Apply?

YES, the clause can be read to cover all types of changes, including extra work.

Did the Owner Have Knowledge?

YES, the mayor was fully aware, and he kept the council advised of his actions.

Did the Owner Promise to Pay?

This is the central issue. If directing Laramee to proceed with corrections that were knowingly outside the scope of the contract promises constitutes a promise to pay, then the answer would be YES. Because the problems with the foundation were known in advance and were not caused by Laramee, it seems the answer should be YES.

Did the Agent Have Authority?

The mayor was the agent. He probably did not have authority, but the city council knowingly let him negotiate with Owens and endorsed his actions, so the answer is YES.

Synopsis

Like the flowchart analysis, the court ruled in favor of Owens. The court was persuaded by the city's rental of the compressor and jackhammer and the payment by the city for some rock removal as shown in one of Owens's itemized statements. The rationale behind their decision was quite lengthy and stated in part,

> However, if the original provisions of the contract are mutually rescinded by agreement of the parties the contractor is free of any obligations to obtain an order in writing for extra work or materials and is no longer obligated to perform the work for the amount specified in the original contract. Rescission depends upon the intention of the parties as shown by their words, acts, or agreement. Parties to a contractual transaction may mutually rescind the transaction although neither party has a right to compel rescission. In such a case the terms of written contract may be varied, modified, waived, or wholly set aside by any subsequent executed contract whether such subsequent executed contract be in writing or parol (oral).

Solution: Exercise 7-5

The facts of this exercise are based on the case of Thorn Construction Co., Inc., v. Utah Department of Transportation (598 P.2d 365), which was heard before the Supreme Court of Utah in 1979. The issue involved statements made by a low-level state employee that a borrow pit called the Utelite property was available and could be used for borrow material.

Is There a Statutory Requirement?

NO, there is no indication of a statute or regulation.

Does the Changes Clause Apply?

YES, the clause can be read to cover all types of changes, including modifications.

Did the Owner Have Knowledge?

UNKNOWN, it is not known if Virgil Mitchell was told to advise all prospective bidders that the Utelite property was available.

Did the Owner Promise to Pay?

YES, relative to Thorn this occurred when Mitchell said the Utelite property was available.

Did the Agent Have Authority?

NO, Mitchell was a low-level employee of the state, and would ordinarily not be empowered to convey such information, unless specifically told to do so.

Synopsis

On the basis of the flowchart analysis, Thorn would not likely recover the cost of using another borrow pit. But, Thorn successfully sued the state on the theory of a misrepresentation, which is a breach of contract action. Thorn argued that Mitchell made a positive material statement of fact that the Utelite property was available, which turned out not to be true. Generally, Thorn would not be responsible for checking up on Mitchell's authority. This case along with others in Chapter 10 illustrate the extent to which the courts will protect a contractor from inducements to lower their bids.

Solution: Exercise 7-6

The facts of this exercise are based on the case of Security Painting Co. v. Commonwealth of Pennsylvania (357 A.t.2d 251), which was heard before the Commonwealth Court of Pennsylvania in 1975.

The work was for the sandblasting and painting of 12 highway bridges. Security alleges it was ordered to do extra work.

Is There a Statutory Requirement?

NO, there is no indication of a statute or regulation.

Does the Changes Clause Apply?

YES, the clause can be read to cover all types of changes, including modifications.

Did the Owner Have Knowledge?

YES, the additional sandblasting was ordered by the PennDOT inspectors.

Did the Owner Promise to Pay?

NO, simply ordering the contractor to follow the specifications is not a promise to pay.

Synopsis

In reconciling the language of the contract with the footnotes, the court said,

> Surely the discretionary and broad language does not alter the clear provisions found in form 409. It is well established that when a written contract is clear and unequivocal, its meaning must be determined by its contents alone. It speaks for itself and a meaning cannot be given to it other than that expressed. Where the intention of the parties is clear, there is no need to resort to extrinsic aids or evidence.

The mere act of requiring Security to follow the specifications did not constitute extra work.

Solution: Exercise 8-1

The facts of this exercise are based on the case of Barash v. State of New York (154 N.Y.S.2d 317), which was heard before the Court of Claims of New York. It involved the rehabilitation of the ceiling in one of the rooms of the state Capitol building.

Do the Terms Have Plain Meaning?

This dispute involves more than an interpretation of a word or phrase. However, for whatever it is worth, cabinetry work would seem different from ceiling rehabilitation.

Is the Ambiguity Patent?

NO. Although the ambiguity is obvious, it does not raise a duty to inquire because Addendum 2 is clear on its face.

Was There Mutual Understanding?

NO. Barash did the work under protest.

What Does the Contract Say When Read as a Whole?

When read as a whole, Addendum 2 deleted the entire Section 19A, which contained the ceiling work. It seems unnecessary and perhaps impractical to issue a new plan sheet showing the ceiling work deleted, a point on which Addendum 2 left little doubt.

Synopsis

The court of claims awarded damages to Barash, stating that the ceiling work was clearly deleted. They further stated that when the architect ordered the work to be done, he overstepped the authority granted to him by paragraph 50 of Article 12 and thereby breached the contract.

Solution: Exercise 8-2

The facts of this exercise are based on the case of Lazer Construction Co., Inc., v. James and Kathy Long (370 SE2d 900), which was heard before the Court of Appeals of South Carolina in 1988.

The contract contained a guaranteed maximum price of $99,500. Because of oral changes made by the Longs, the cost of the home rose to $140,771.50. When the Longs stopped making payments, they had paid $116,713.57. Lazer filed a lien; the Longs brought action to vacate the lien. A central issue is whether Lazer is entitled to the total amount owed. Long argues no because the guaranteed maximum price has already been reached.

Do the Terms Have Plain Meaning?

This dispute involves more than an interpretation of a word or phrase.

Is the Ambiguity Patent?

NO. This is not an issue in this matter.

Was There Mutual Understanding?

YES, Long has already paid $116,713.57. They both agreed that payment was due. Lazer billed; Long paid.

Synopsis

The court agreed with Lazer; Long had to pay.

Solution: Exercise 8-3

The facts of this exercise are based on the case of W. H. Armstrong and Co. v. United States (98 Ct.Cl. 519), which was heard before the U.S. Court of Claims. The work began in 1932. Some of the law may be interpreted differently today. The work involved a masonry contract. The government supplied the bricks.

Do the Terms Have Plain Meaning?

This dispute involves more than an interpretation of a word or phrase. However, there is ample reference in the contract to the term "common brick." Additionally, Armstrong was shown a pile of common brick and was told that he was to use this pile. This appears to be information on which he could rely. "Common brick" is a technical term in the brick industry that conveys certain absorption properties and uniformity in size. It would seem that the contracting officer overstepped his authority by ordering Armstrong to use brick from another pile that were not common bricks.

Is the Ambiguity Patent?

NO. There is no ambiguity. The issue here is over the authority of the contracting officer.

Was There Mutual Understanding?

NO.

What Does the Contract Say When Read as a Whole?

There are numerous references in the contract to "common brick," so that it seems clear that when read as a whole, the government intended Armstrong to use common brick from a certain stockpile. The government pointed to a specific stockpile during Armstrong's site visit. There are similarities between this case and the Thorn case in Chapter 10.

Synopsis

The court of claims ruled in favor of Armstrong. However, they said that the government had the authority to change stockpiles, but there was an obligation to pay the difference in costs.

In another part of the decision, the court addressed the lack of a written change order and the language of Article 5, specifically the part which read,

> ...unless the same has been ordered in writing and the price stated in such order.

In the opinion of the court, the price could not be determined until after the work was finished. It was at this time that the order would be reduced to writing. Therefore, Armstrong could not be barred from recovering because there was no written order.

Based on the definition of extra work, it may be successfully argued that the change was not actually extra work. If the argument was successful, then Article 5 would not apply, and no written order would be required.

Solution: Exercise 8-4

The facts of this exercise are based on the case of R. B. Wright Construction Co. v. United States (919 F.2d 1569), which was heard before the U.S. Court of Appeals in 1990. The dispute involved an interpretation of a painting specification that was in conflict with industry custom.

Do the Terms Have Plain Meaning?

The dispute does not involve the meaning of a word or phrase. The subcontractor, Rembrandt, Inc., argues that the work he is required to do is beyond the norm in the painting business. However, the argument is inconclusive because there is other contract language.

Is the Ambiguity Patent?

NO. There is no ambiguity.

Was There Mutual Understanding?

NO.

What Does the Contract Say When Read as a Whole?

The contract expressly requires three coats of paint on certain surfaces. Paragraph 14 states that all surfaces "listed in the PAINTING SCHEDULE ... will receive the ... number of coats prescribed in the schedule."

Is There One Logical Conclusion?

YES, if Rembrandt's interpretation is adopted, then the entire painting schedule is meaningless. This is an unreasonable interpretation.

Synopsis

Rembrandt and Wright lost in court because they could not overcome the express language of the painting schedule.

Solution: Exercise 8-5

The facts of this exercise are based on the case of Kaiser Construction Co. v. Lyon Metal Products, Inc. (461 S.W.2d 893), which was heard before the Kansas City Court of Appeals, in Kansas City, Missouri, in 1970. The issue at hand was whether the work performed by Kaiser was extra work (not called for in the contract) or additional work (work incidental to the contract).

Do the Terms Have Plain Meaning?

NO.

Is the Ambiguity Patent?

NO. Although it could be argued that because Kaiser was on the project helping the Holm Co., they knew that this work was required. Maybe Kaiser thought that Lyon was going to prepare the site so that all Kaiser had to do was installation. This we will never know.

Was There Mutual Understanding?

NO.

What Does the Contract Say When Read as a Whole?

It is clear from the contract language that Kaiser was hired to

> ... unload, distribute, erect, install locks, adjust doors and equipment in place; touch up all mars and scratches and dispose of all crates, boxes and packing ...

They had to do much more, and should be compensated.

Synopsis

The court ruled in favor of Kaiser. In their decision, they defined extra and additional work as follows:

Extra work, as used in connection with building contracts, means work of a nature not contemplated by the parties and not controlled by the contract 17A C.J.S. Contracts... (p. 411); Extra work is entirely independent of the contract, something not required in its performance.... Additional work is something necessarily required in the performance of the contract, and without it the work could not be carried out...

Because the finding was that it was extra work, the parties probably had some discussion about these matters before signing the contract. Clearly, this work was not called for in Kaiser's contract.

Solution: Exercise 8-6

The facts of this exercise are based on the case of Environmental Utilities Corp. v. Lancaster Area Sewer Authority (453 F.Supp, 1260), which was heard before the U.S. District Court in 1978.

Environmental was constructing a sewer line. At issue is the difference between pay widths and excavation widths.

Do the Terms Have Plain Meaning?

One could argue YES on this point, but we will continue the flowchart analysis.

Is the Ambiguity Patent?

NO, this is not an issue in this dispute.

Was There Mutual Understanding?

NO, although one could argue YES because Environmental did not invoice the sewer authority for any extras and waited more than two years before filing a claim.

What Does the Contract Say When Read as a Whole?

It is clear from the contract language that Environmental was to be paid based on pay widths.

Synopsis

The court ruled in favor of the Lancaster Area Sewer Authority.

Solution: Exercise 8-7

The facts of this exercise are based on the case of Metro Insulation Corp. v. Leventhal and others (294 N.E.2d 508), which was heard before the Appeals Court of Massachusetts in 1973.

The issue was whether the sprinkler piping was required to be insulated. The Boston Housing Authority argued that it was. Metro's argument was that it is customary in the insulation trade not to insulate sprinkler piping unless specifically directed to do so.

Do the Terms Have Plain Meaning?

This was not an issue in this dispute, although Metro argued that custom in the trade precluded insulation.

Is the Ambiguity Patent?

NO, this is not an issue in this dispute.

Was There Mutual Understanding?

NO.

What Does the Contract Say When Read as a Whole?

The contract was silent on this issue.

Is There One Logical Conclusion?

YES, the housing authority's conclusion is unreasonable because it requires the acceptance of a secret or undisclosed intention and it also requires one to rewrite the contract. The contractor's conclusion is the only logical conclusion, so Metro should prevail.

Synopsis

The court ruled in favor of Metro but offered little rationale for its decision.

Solution: Exercise 8-8

The facts of this exercise are based on the case of Jasper Construction Co., Inc., v. Foothills Junior College District of Santa Clara County, California (91 Cal. App.3d 1), which was heard before the First District Court of Appeals, California, in 1979.

The issue was over the location joints for reinforced concrete walls.

Do the Terms Have Plain Meaning?

This was not an issue in this dispute.

Is the Ambiguity Patent?

NO, this is not an issue in this dispute.

Was There Mutual Understanding?

NO.

What Does the Contract Say When Read as a Whole?

The contract addresses the joint issue with a clause that said,

11. CONSTRUCTION JOINTS

(A) Location and details of construction joints shall be as indicated on the structural drawings, or *as approved by the Architect*. Relate required vertical joints in walls to joints in finish. In general, approved joints shall be located to least impair the strength of the structure.

The construction joints were not shown on the structural drawings, but Jasper did not consult the architect. (You cannot ignore the clear and unambiguous language of the contract.) In this dispute, there was no ambiguity.

Is There One Logical Conclusion?

YES, Jasper's interpretation is unreasonable because he did not follow the contract, and to adopt his interpretation (that the contract was silent) means that one would have to make useless part of the contract language.

Synopsis

The dispute was argued as a defective specification case, which made it difficult to reconcile or understand. However, the court did say,

> A contractor who submits a bid for public work which proves unprofitable because of his negligence in failing to ascertain all the facts concerning it from sources readily available, cannot thereafter throw the burden of his negligence on the shoulders of the state by asserting that the latter was guilty of fraudulent concealment in not furnishing him with information which he made no effort to secure for himself.

Jasper lost.

Solution: Exercise 8-9

The facts of this exercise are based on the case of Western Contracting Corp. v. State Highway Department of Georgia (187 SE2d 690), heard before the Court of Appeals of Georgia in 1971.

The issue was over the access to designated borrow pits during the construction of a portion of Interstate 95. The plans showed numerous borrow pits, and the specifications said, "The Department will obtain all necessary options from the owners..."

Do the Terms Have Plain Meaning?

This was not an issue in this dispute.

Is the Ambiguity Patent?

NO, this is not an issue in this dispute.

Was There Mutual Understanding?

NO.

What Does the Contract Say When Read as a Whole?

The critical information is contained on Sheet 22.

Is There One Logical Conclusion?

YES, it is clear from Sheet 22 and other contract language that the intent of the parties was that the contractor was to obtain material from the designated borrow pits and deliver this material to specified locations on the project.

Synopsis

The court spent much time discussing Sheet 22. They said,

> ... this sheet depicts five clearly defined and numbered borrow pits. Furthermore, this sheet of the plans is drawn to scale of 1" = 1,000" and therefore shows with a certain amount of precision the location and boundaries of the borrow pit in question. It further shows, within each pit location, the type of borrow available and whether it is usable with or without "muck removal." It shows as to each pit the volume (in cubic yards) of "material available, ..." It further shows where in the project in delineated lengths and widths and in what volume the material taken from each respective pit is to be placed to create the embankment. It shows the names of the owners of the property though which the embankment and various waterways run.
>
> The Defendant (DOT) places heavy reliance on an admittedly conspicuous notation appearing on Sheet 22, to wit: "Note: These Pits Are Shown As Possible Sources of Material."
>
> In view of *all* the information contained on Sheet 22, we do not find that the above notation *must* imply to one studying the

plans and preparing a bid or proposal and preparing to enter into a contract thereon that the parties did not contemplate at least the *availability* of the borrow pits *shown* on the plans as sources of embankment material. It may or may not have so implied. The notation is susceptible to construction as a caution to one studying the plans that some of the material in the pits, when exhumed, may or may not meet other contract requirements as to *quantity* and *quality* . . .

Western prevailed.

Solution: Exercise 8-10

The facts of this exercise are based on the case of Liberty Mutual Insurance Co. v. Bob Roberts Co., Inc. (357 S,O,2d 968), heard before the Supreme Court of Alabama in 1978.

The issue in this dispute is a narrow one as to which contractor, Dawson, the prime contractor, or Roberts, the subcontractor, is responsible for fixing defective exposed aggregate panels.

Do the Terms Have Plain Meaning?

This was not an issue in this dispute.

Is the Ambiguity Patent?

NO, this is not an issue in this dispute.

Was There Mutual Understanding?

NO.

What Does the Contract Say When Read as a Whole?

Dawson argues that as per paragraph 16, one subcontractor is to do this work. Jacksonville counters that paragraph 34 makes the contractor responsible for everything, so Dawson should fix the panels.

Is There One Logical Conclusion?

YES, if Dawson's interpretation is accepted, there is no meaning to paragraph 34. If Jacksonville's interpretation is accepted, there is meaning to all parts. It is clear that Jacksonville wanted one contractor to do the work (Roberts), but for one contractor to have overall charge of the work (Dawson). Dawson should be responsible for fixing the panels.

Synopsis

In an earlier decision, Dawson was required to fix the panels.

Solution: Exercise 8-11

The facts of this exercise are based on the case of Security Painting Co. v. Pennsylvania Department of Transportation (PennDOT) (357 A.2d 251), heard before the Commonwealth Court of Pennsylvania in 1975.

The issue in this dispute was allegations by Security that PennDOT inspectors required them to sandblast the bridges beyond what the specifications required.

Do the Terms Have Plain Meaning?

This was not an issue in this dispute.

Is the Ambiguity Patent?

NO, this is not an issue in this dispute.

Was There Mutual Understanding?

NO.

What Does the Contract Say When Read as a Whole?

The specifications spoke with clarity about what was required. Security points to several footnotes that allegedly give Security wider latitude than the inspectors allowed.

Is There One Logical Conclusion?

YES, PennDOT's conclusion is most logical. The specifications spoke with clarity about what Security was required to do. Security's position is that the footnotes should control. However, to adopt this interpretation would mean that substantial parts of the specifications would be inoperable or useless. Clearly, this interpretation violates one of the principal rules of interpretation, so Security's interpretation is unreasonable.

Synopsis

The commonwealth court sided with PennDOT, saying,

> It is well established that when a written contract is clear and un-equivocal, its meaning must be determined by its contents alone. It speaks for itself and a meaning cannot be given to it other than that expressed.

Solution: Exercise 8-12

The facts of this exercise are based on the case of D'Annunzio Brothers, Inc., v. New Jersey Transit Corp. (586 A.2d 301), heard before the Superior Court of New Jersey in 1990.

Do the Terms Have Plain Meaning?

This was not an issue in this dispute.

Is the Ambiguity Patent?

YES, D'Annunzio had a duty to inquire.

Synopsis

The court rejected D'Annunzio's claim on the basis of a patent ambiguity.

Solution: Exercise 8-13

The facts of this exercise are based on the case of Blake Construction Co., Inc., v. Lawrence Garrett, Secretary of the Navy (956 F.2d 1174), heard before the U.S. Court of Appeals in 1992.

The issue in this dispute was over a claim for an equitable adjustment for the cost of encasing electrical control system wiring.

Do the Terms Have Plain Meaning?

This was not an issue in this dispute.

Is the Ambiguity Patent?

NO, this is not an issue in this dispute.

Was There Mutual Understanding?

NO.

What Does the Contract Say When Read as a Whole?

Blake relies on a clause in the contract that requires each contractor to avoid interference with one another, so Blake wants to install certain electrical systems in an underground duct bank.

The government points to paragraph 3.1.1.8, which requires certain electrical systems to include outlets and terminals interconnected with empty conduit. This requirement can only be met if conduit is installed in the ceiling as depicted in a diagrammatic drawing.

Is There One Logical Conclusion?

YES, if Blake's interpretation is adopted, the diagrammatic drawing and paragraph 3.1.1.8 have no meaning, so Blake's interpretation is unreasonsble.

Synopsis

Blake lost in court on the basis of an unreasonable interpretation.

Solution: Exercise 8-14

The facts of this exercise are based on the case of Forest Construction Co., Inc., v. Farrell-Cheek Steel Co. (484 So.2d 40), heard before the District Court of Appeals, Florida, in 1986.

At issue is the unit price to be paid for asphalt paving.

Do the Terms Have Plain Meaning?

This was not an issue in this dispute.

Is the Ambiguity Patent?

NO, this is not an issue in this dispute.

Was There Mutual Understanding?

Arguably, YES. When Farrell-Cheek failed to discuss price until some later time, they essentially agreed with Forest that the price was $180/ton.

Synopsis

The court agreed that the correct price was $180/ton. Their rationale was stated thus:

> Generally, the law implies an obligation to pay reasonable cost for the extras not provided for in a contract, and the price of extras should be computed at a reasonable rate, unless the price is otherwise agreed upon. In this case, a written change order provided that the owner agreed to pay the sum of $180 per ton for the 130 tons of leveling asphalt. Because the subsequent 381.15 tons delivered was of the same nature and character as provided for in the first change order, the work should be chargeable at the same rate.

Solution: Exercise 8-15

The facts of this exercise are based on the case of Philip Chiappisi v. Granger Contracting Co., Inc. (223 N.E.2d 924), heard before the Supreme Judicial Court of Massachusetts in 1967.

The issue in this dispute was over insulation for roof decking. The decking was shown upside down in the plans.

Do the Terms Have Plain Meaning?

This was not an issue in this dispute.

Is the Ambiguity Patent?

NO, this is not an issue in this dispute.

Was There Mutual Understanding?

NO.

What Does the Contract Say When Read as a Whole?

The specifications required the decking to be installed as per the manufacturer's instruction. Chiappisi erroneously relied on a diagrammatic sketch to do his quantity takeoff.

Is There One Logical Conclusion?

YES, it is unreasonable to use a diagrammatic sketch to do a quantity takeoff.

Synopsis

The Court denied Chiappisi any equitable adjustment because he failed to give notice. Had notice been given, Chiappisi would have likely lost anyway.

Solution: Exercise 8-16

The facts of this exercise are based on the case of J. A. Jones Construction Co. v. United States (395 F2d 783), heard before the U.S. Court of Claims in 1968.

The facts in this case were both detailed and confusing. At issue was a conflict in the zinc coating requirements between two paragraphs. The contract did not specify the class of material.

Do the Terms Have Plain Meaning?

This was not an issue in this dispute.

Is the Ambiguity Patent?

NO, this is not an issue in this dispute.

Was There Mutual Understanding?

NO.

What Does the Contract Say When Read as a Whole?

The government did not want Class C material, as Jones assumed they did. The contract would have been clearer on this issue if paragraph 14-02 had not been included. They wanted ¾ oz of zinc on both sides as specified in paragraph 14-04.

Is There One Logical Conclusion?

YES, by applying the canon of specific over general, paragraph 14-04 controls, and the government's position is logical. It is assumed that QQ-I-716 covers more than gages, classes, and coatings, so it is not rendered useless. If Jones's position that paragraph 14-02 controls, then parts of paragraph 14-04 are useless; therefore their position is unreasonable.

Synopsis

The court ruled in favor of the government but on the basis of a patent ambiguity. They felt that by not specifying a class of material,

Jones should have known something was amiss. They went on to say that,

> Plaintiff (Jones) argued that its interpretation of the specifications, i.e., that QQ-I-716 controlled, that not less than 0.75 oz of zinc coating meant 1.50 oz on both sides was a reasonable construction of Government drafted specifications and therefore should control.... While the "reasonableness" of the contractor's interpretation may sometimes be controlling in resolving a controversy of this nature, ... that argument overlooks Article III of the contract and the legal significance the inclusion of that clause carries.

Solution: Exercise 9-1

The facts of this exercise are based on the case of Blauner Construction Co. v. United States (94 Ct.Cl. 503), heard before the U.S. Court of Claims in 1941.

The issue in this case was over the depth of rock for a post office foundation.

Is the Contract Silent?

NO.

What Does the Contract Indicate?

The rock ledge slopes from east to west and south to north.

Were the Conditions Different?

From the information given in the contract, the contractor could expect the rock ledge near or at the surface at the southeast corner of the building and probably no rock ledge at the northwest corner (assuming that these were shallow footings). It is not known where the contractor began excavation.

Was Reliance on the Information Justified?

NO, the contractor did not visit the site.

Synopsis

Blauner lost. The court of claims did not discuss if there were material differences, but instead did not favor Blauner because it did not perform a site visit.

Solution: Exercise 9-2

The facts of this exercise are based on the case of Stuyvesant Dredging Co. v. United States (834 F.2d 1578), heard before the U.S. Court of Appeals in 1987.

The issue in this dispute was over maintenance dredging of the Corpus Christi Entrance Channel and the character of the material.

Was the Contract Silent?

According to the court, "Stuyvesant had not encountered a site condition different from that described in the contract because the contract did not indicate or describe the materials or their characteristics,..." Their position was that density alone does not indicate the character of the materials; it is but one factor. Therefore, this is a type II differing site condition (DSC).

Were the Conditions Unknown and Unusual?

NO.

Was Reliance on the Information Justified?

The court spent considerable effort in describing how Stuyvesant did not visit the site and did not examine the materials made available by the Corps of Engineers. So the answer is NO.

Synopsis

The court stated what Stuyvesant had to prove to prevail. They said,

> To prevail on a claim for differing site conditions, the contractor must prove, by the preponderance of the evidence, "that the conditions 'indicated' in the contract differ materially from those it encounters during performance." The conditions must be reasonably unforeseeable based on all the information available to the contractor at the time of bidding. The contractor must show that it reasonably relied upon its interpretation of the contract and the contract-related documents and that it was damaged as a result of the material variation between the expected and encountered conditions.

Stuyvesant could not prove its demand for an equitable adjustment.

Solution: Exercise 9-3

The facts of this exercise are based on the case of Morrison and Lamping v. Oregon State Highway Commission (357 P.2d 389), heard before the Supreme Court of Oregon in 1960.

This DSC case was argued as a type II, although it could also have been argued as a type I. Only a type II flowchart analysis was performed.

Were the Conditions Unknown and Unusual?

The contract allows relief if there are "unknown conditions of an unusual..." The question here is were the conditions unknown? NO, it is highly unlikely that this is the case because Morrison and Lamping added 5 cents to their unit price to cover delays from irrigation. The conditions for this type of irrigation do not seem unusual. Therefore, there should be no recovery.

Synopsis

The court denied any recovery to Morrison and Lamping, saying,

Mr. Morrison...testified that he was familiar with irrigation problems, knew the irrigation season was July and August, and that he had made a careful investigation of the area before submitting his bid. He also stated that in bidding the job he took into consideration delays and shutdowns because of the irrigation and allowed an additional 5 cents per yard for the possible interruptions caused by irrigation. He also stated that he knew it would be difficult to operate his road equipment if the right of way was wet and muddy. Mr. Lamping...testified that he had been advised by Estes (the farmer) at the approximate time he started construction that the irrigation would start the following Monday.

Solution: Exercise 9-4

The facts of this exercise are based on the case of Umpqua River Navigation Co. v. Crescent City Harbor District (618 F.2d 588), heard before the U.S. Court of Appeals in 1980.

Is the Contract Silent?

YES, so this is a type II DSC. Technically, the contract said nothing because the borings were not taken where the dredging basin was located. If the answer is YES, then the borings indicated a large quantity of sand and silt, some cobbles, and one boulder.

Were the Conditions Unusual?

NO, it seems the conditions were not unusual.

Synopsis

Without addressing what the contract indicated or whether there were material differences, the court said Umpqua's actions were unreasonable because Umpqua failed to seek or review the backup information to sheet 002 even though their own backhoe investigation showed inconsistent results with sheet 002 and should have put them on notice that something was amiss.

Solution: Exercise 9-5

The facts of this exercise are based on the case of Moorhead Construction Co. v. City of Grand Forks, North Dakota (508 F.2d 1008), heard before U.S. Court of Appeals in 1974.

Moorhead was constructing a portion of a sewage treatment plant (phase II). Their work depended on the phase I contractor complying with the city's specifications, which Moorhead alleges they did not. He brought action as a type I DSC.

Is the Contract Silent?

NO.

What Does the Contract Indicate?

Because Moorhead was given a notice to proceed just as the phase I contractor was beginning work, the actual conditions could not be observed. All Moorhead could rely on was what was said in the phase I contract, which was 90% compaction.

Were the Conditions Different?

YES, apparently significantly so.

Synopsis

The court awarded Moorhead damages because conditions were materially different and Moorhead was justified in relying on the language of the phase I contract.

Solution: Exercise 9-6

The facts of this exercise are based on the case of Western Contracting Corp. v. California State Board of Equalization (39 Cal. App.3d 341), heard before the Court of Appeal of California in 1974.

This dispute evolved when the legislature increased the sales and use tax by 1%. Western had signed a contract to construct the Castaic Dam in Los Angeles. The tax increase was enacted after the contract was signed.

Western brought the claim based on a type I DSC, but this dispute is not a DSC because it is not a hidden physical condition at the site.

Synopsis

Western did not prevail in its claim for a variety of reasons.

Solution: Exercise 9-7

The facts of this exercise are based on the case of P. J. Maffei Building and Wrecking Corp. v. United States (732 F.2d 913), heard before the U.S. Court of Appeals in 1984.

The issue was over the amount of salvageable steel in a building that Maffei was demolishing and whether the amount was represented in the contract. Maffei brought action under the DSC clause, which read in part "Subsurface or latent physical conditions at the site differing materially from those *indicated in this contract...*" It is questionable if the DSC applies because the amount of steel was subsurface or latent. This issue was not argued before the court, so we will proceed with the flow-chart analysis.

Is the Contract Silent?

YES, the as-built drawings on which Maffei relied formed no part of the contract; they were specifically excluded. Therefore, this can only be a type II claim.

Were the Conditions Unusual?

NO, so Maffei should not be compensated.

Synopsis

The court rejected Maffei's claim, saying,

> ...we agree with the Board that the contract documents did not "indicate" the amount of steel recoverable from the pavilion.

Solution: Exercise 9-8

The facts of this exercise are based on the case of Carlos Teodori v. Penn Hills School District Authority (196 A.2d 306), heard before the Supreme Court of Pennsylvania in 1964.

The issue is over a 6-in. high-pressure gasoline pipeline buried beneath an athletic field that Teodori was excavating. It was not shown on the plans.

Is the Contract Silent?

NO, this is one of those instances where not showing something is an indication that nothing is there. It is similar to not showing a water table, so it is a type I DSC.

What Does the Contract Indicate?

The contract indicates no pipeline.

Were the Conditions Different?

YES, there was a pipeline present.

Was Reliance on the Information Justified?

YES, Teodori had no way of knowing the pipeline was there.

Synopsis

The contractor prevailed in court because the conditions found were materially different from those indicated in the contract documents.

Solution: Exercise 10-1

The facts of this exercise are based on the case of J. A. Thompson and Son, Inc., v. State of Hawaii (465 P.2d 148), heard before the Supreme Court of Hawaii in 1970.

On a highway project, Thompson alleged that the state misrepresented the boring information in hole No. 6, showing a shrinkage factor when there should have been a swell factor, and withheld vital information. In the analysis, we will ignore the withholding of information.

Was There a Positive Representation?

YES and NO. The state did not misrepresent what was found in hole No. 6. The state did show a shrinkage factor where there should have been a swell factor. This problem will be dealt with later.

Was There an Intent to Deceive?

NO.

Did the Conditions Differ from Those Represented?

NO.

Was the Representation Complete?

YES. Thompson alleges that the state withheld information, but it is not evident what that information was.

Was the Contractor Misled?

NO.

Was Reliance on the Information Justified?

NO. Rock outcroppings were evident during the site visit, so Thompson should have expected rock, even though there was a shrinkage factor given.

Synopsis

According to the court,

> Plaintiff (Thompson) also contends "the State misrepresented on the plans that the excavated material would shrink by 23 percent. In fact, the material swelled by 25.4 percent... this representation involved a 48.4 percent error involving 118,000 cubic yards of excess material."

In resolving this matter, the court turned to the site visit and the disclaimer. In discussing the disclaimer, the court said,

> Here the State by a provision in the contract, Section 2.4n2 of the Standard Specifications, explicitly informed the plaintiff and other bidders that "the swell of the excavated material and the direction and quantities of overhaul... are for the purpose of design only," that it "assumes no responsibility whatever in the interpretation or exactness of any of the information... and does not, either expressly or implicitly, make any guarantee of the same" and that it "reserves the right to change the direction and quantities of overhaul and the swell and shrinkage factors shown on the mass diagram and no additional compensation will be allowed by reason of such changes..."

Solution: Exercise 10-2

The facts of this exercise are based on the case of Michigan Wisconsin Pipeline Co. v. Williams-McWilliams Co. v. United States (551 F.2d 945), heard before the U.S. Court of Appeals in 1977.

In a dredging project, Williams struck and damaged a 30-in. gas pipeline owned by Michigan Wisconsin. The issue was who was responsible.

Was There a Positive Representation?

YES, by not showing the pipeline, the government was saying it was not there. This is because government plans always show pipelines.

Was There an Intent to Deceive?

NO.

Did the Conditions Differ from Those Represented?

YES.

Was the Representation Complete?

NO, but the government did not withhold information, so the representation was complete as far as they knew.

Was the Contractor Misled?

YES.

Was Reliance on the Information Justified?

YES.

Synopsis

In stating the court's logic for finding the government responsible, they said,

> These circumstances lead us to the conclusion that the Engineers regularly depict pipelines on dredging specification drawings, and that they failed to do so in this case through mistake. Williams-McWilliams was justified in relying on the Engineers' prior practice of depicting pipelines. Thus the failure of the drawings in this case to depict pipelines across Atchafalaya Bay in the area being dredged amounted to a representation of "positive assertion" of their absence .... We conclude that a representation was made.

The government was found to be responsible.

Solution: Exercise 10-3

The facts of this exercise are based on the case of P. T. and L. Construction Co., Inc., v. State of New Jersey (531 A.2d 1330), heard before the Supreme Court of New Jersey in 1987.

The work involved construction of a small portion of Interstate 78. P. T. and L. alleged that the state misrepresented the work would be in the dry and also withheld vital information from the bidders. This case illustrates the obligation of owners to share information, especially when there is no DSC clause.

Was There a Positive Representation?

An argument can be made for YES because the term "stripping" implies dry conditions and certain types of equipment, the culvert had no sheeting, and the project design was based on less-porous zone 3 material. Nevertheless, a stronger case can be made for withholding of information, so the analysis proceeds along those lines.

Was There Intent to Deceive?

NO.

Did the Conditions Differ from Those Represented?

YES.

Was the Representation Complete?

NO. Vital information about the Corps of Engineers' removal of the obstructions and the letter from the local engineer were withheld.

Was the Contractor Misled?

YES.

Was Reliance on the Information Justified?

YES, the withheld information was known only to the state, so P. T. and L. should be compensated.

Synopsis

The court examined the facts for positive material representations and found none. A new state law may have been a factor. Next, the court looked at withholding information. Two issues were relevant.

Here there are two aspects of the case in which the trial court found that facts were not known to or reasonably discoverable by the plaintiff (P. T. and L.). First, there was the so-called "Madigan-Hyland letter" dated December 30, 1964. This letter clearly disclosed to the state that working conditions at the site would impose unusual difficulties for a construction contractor. In the fill area between the proposed Springfield Avenue bridge and the cut and fill line, i.e., the area that is the focus of this litigation, Madigan-Hyland stated that the "extent of the removal of wet excavation . . . will depend on the climatic conditions and the time of year in which the construction will be accomplished." The trial court was most aggrieved by this feature of the case, saying that "[t]he withholding of the Madigan-Hyland information represented a *misrepresentation, if not fraud*, for the withholding of material information." Second, the design of the project itself was predicated on the completion of a project to drop a branch of the East Rahway River in accordance with the plan of the U.S. Army Corps of Engineers, a factor relevant to the court's conclusion concerning culvert construction. These factual omissions and assumptions were held by the trial court to constitute, along with other design features, "design defects, and those design defects [did] constitute misrepresentations to this contractor at the time [it] bid."

P. T. and L. recovered their damages.

Solution: Exercise 10-4

The facts of this exercise are based on the case of Ruby-Collins, Inc., v. City of Charlotte, North Carolina (930 F.2d 23), heard before the U.S. Court of Appeals in 1991.

In constructing a water main, Ruby-Collins encountered unusually wet soil. The contract did not contain a DSC clause.

Was There a Positive Representation?

NO, a positive statement about the soil moisture content was not made, so Ruby-Collins cannot prevail.

Synopsis

Ruby-Collins did not prevail in court.

Solution: Exercise 10-5

The facts of this exercise are based on the case of Pennsylvania Turnpike Commission v. Smith et al. (39 A.2d 139), heard before the Pennsylvania Supreme Court in 1944.

York Engineering and Construction Co. alleged that the turnpike commission misrepresented the character of materials in the construction of a portion of the Pennsylvania Turnpike. The contract did not contain a DSC clause.

Was There a Positive Representation?

YES, the contract showed a shrinkage factor indicating mostly earth. The work was mostly limestone.

Was There Intent to Deceive?

NO.

Did the Conditions Differ from Those Represented?

YES, the work was mostly rock.

Was the Representation Complete?

MAYBE, engineers for the turnpike knew that there was a substantial amount of rock present, but this may have been a weak argument for York.

Was the Contractor Misled?

YES.

Was Reliance on the Information Justified?

YES, York only had 11 days in which to submit its bid. This was insufficient time for an independent investigation, so York had to rely on the bid documents, even though rock outcroppings were visible in numerous locations. York should recover damages.

Synopsis

York prevailed because they were only given 11 days to bid. They had no choice but to rely on the information provided by the turnpike commission.

Solution: Exercise 10-6

The facts of this exercise are based on the case of Thorn Construction Co., Inc., v. Utah Department of Transportation (598 P.2d 365), heard before the Supreme Court of Utah in 1979.

The issue arose when a low-level state employee informed Thorn that a certain borrow pit was available and contained suitable material.

Was There a Positive Representation?

YES, the employee made an affirmative statement that the Utelite pit was available and that it contained suitable material. It is not Thorn's responsibility to check everyone's credentials or the validity of the positive statements they make. Obviously, Virgil Mitchell had some authority or responsibility because he was in charge of conducting the site tour.

Was There Intent to Deceive?

NO.

Did the Conditions Differ from Those Represented?

YES, Thorn could not use the Utelite pit.

Was the Representation Complete?

YES.

Was the Contractor Misled?

YES.

Was Reliance on the Information Justified?

YES, so Thorn should prevail.

Synopsis

In siding with Thorn, the court said,

> Here, according to the facts as the District Court found them to
> be, the representative of the Department stated the material at
> the Utelite pit could be used in the project. He did not simply re-
> port the findings of any test made, but made a positive repre-
> sentation that was untrue, on which the plaintiff's (Thorn)
> representatives relied in computing their bid. Their reliance was,
> in our opinion, justified . . .

Solution: Exercise 10-7

The facts of this exercise are based on the case of City of Indianapolis, Indiana, v. Twin Lakes Enterprises, Inc. (568 N.E.2d 1073), heard before the
Court of Appeals of Indiana in 1991.

The work involved a lake-dredging contract in which the contractor
was told to use a suction hose. The misrepresentation analysis could be
done based on a positive representation or withholding of information.
Only the positive representation analysis is shown below.

Was There a Positive Representation?

YES, the contract provided that Twin Lakes was to "dredge the site
or the rowing course of silt and sand by means of a suction hose." A
suction hose will only work where there is sand and silt. By specifying
the suction hose method, the city was representing this as a sand and
silt job.

Was There Intent to Deceive?

NO.

Did the Conditions Differ from Those Represented?

YES, Twin Lakes found large obstructions on the lake bottom. The suction hose method would not work.

Was the Representation Complete?

NO, information about the obstructions was withheld from Twin Lakes.

Was the Contractor Misled?

YES.

Was Reliance on the Information Justified?

YES, so Twin Lakes should recover damages.

Synopsis

In court, the city was found to have breached its contract with Twin Lakes.

Solution: Exercise 10-8

The facts of this exercise are based on the case of Post and Front Properties, Ltd., v. Roanoke Construction Co., Inc. (449 SE2d 765), heard before the Court of Appeals of North Carolina in 1994.

The issue here was that of an owner conveying to a contractor that it had sufficient funds to renovate a building.

Was There a Positive Representation?

YES, P&F said affirmatively that they had access to bank loan funds of $180,000.

Was There Intent to Deceive?

YES, so Roanoke should prevail.

Did the Conditions Differ from Those Represented?

YES, P&F had no money.

Synopsis

At trial, P&F was found to have committed a fraudulent act.

Solution: Exercise 10-9

The facts of this exercise are based on the case of Ideker Corp., Inc., v. Missouri State Highway Commission (654 S.W.2d 617), heard before the Missouri Court of Appeals in 1983.

The issue centered around whether the project was represented as a balanced job.

Ideker prevailed in court. In their decision, the court set forth the elements that Ideker had to prove:

(1) a positive representation by a government entity,

(2) a positive representation of a material fact,

(3) a positive representation that is false or incorrect,

(4) lack of knowledge by a contractor that the positive representation of material fact is false or incorrect,

(5) reliance by a contractor on the positive representation of material fact made by the governmental entity, and

(6) damages sustained by a contractor as a direct result of the positive representation of material fact made by the government entity.

Idecker was able to prove each of these points.

Solution: Exercise 10-10

The facts of this exercise are based on the case of the Daniel Hamm Drayage Co. v. Waldinger Corp. v. W. E. O'Neil Construction Co. (508 F.Supp. 390), heard before the U.S. District Court in 1981.

The issues centered around an equipment delivery schedule.

Was There a Positive Representation?

YES, the wrong equipment delivery dates were conveyed.

Was There Intent to Deceive?

Probably, YES.

Did the Conditions Differ from Those Represented?

YES, the equipment arrived later than stated.

Was the Representation Complete?

NO, Roth knew the delivery dates he was citing were wrong.

Did the Owner (Contractor) Have Knowledge of the Withheld Information?

YES.

Was the Contractor Misled?

YES.

Should the Contractor Have Known of the Condition?

NO.

Synopsis

Hamm prevailed. In examining Waldinger's actions, the court said,

To recover for fraud, plaintiff (Hamm) must establish that defendant (Waldinger) knowingly made a false representation of a material fact intending that plaintiff (Hamm) act upon the representation. Plaintiff (Hamm) must further establish a right to rely on defendant's (Waldinger's) representations, actual reliance, that such reliance was reasonable, and that by reason thereof, plaintiff (Hamm) was damaged.

Hamm was able to prove each of these points, and the court found Waldinger's actions fraudulent.

Solution: Exercise 10-11

The facts of this exercise are based on the case of the Pinkerton and Laws Co. v. Roadway Express, Inc., v. Owens-Corning Fiberglas Corp. (659 F.Supp. 1138), heard before the U.S. District Court in 1986.

The issues centered around extremely wet soil conditions at a freight terminal that Pinkerton was constructing for Roadway.

Was There a Positive Representation?

NO, in no way did the contract documents say anything about the wet soil conditions.

Was the Representation Complete?

NO, Roadway waited some seven months after the project started before giving Pinkerton a geotechnical report that described in detail the wet conditions at the site.

Did the Owner (Contractor) Have Knowledge of the Withheld Information?

YES.

Was the Contractor Misled?

YES.

Should the Contractor Have Known of the Condition?

NO.

Synopsis

Pinkerton failed to show that the conditions were misrepresented, but successfully showed that Roadway withheld vital information. The court said that Roadway's actions were fraudulent.

Solution: Exercise 10-12

The facts of this exercise are based on the case of Flippin Materials Co. v. United States (312 F.2d 408), heard before the U.S. Court of Claims in 1963.

Flippin alleged that the government misrepresented what was in limestone cavities in rock that it was crushing for the production of concrete.

Was There a Positive Representation?

NO, the contract used a nondescript symbol to highlight void areas. They never defined the material.

Was There an Intent to Deceive?

NO.

Was the Representation Complete?

YES, the government made all their information available. Even the cores were available, which showed many of the cavities to be filled with clay. Flippin apparently did not consult these borings.

Synopsis

Flippin did not recover any additional compensation.

Solution: Exercise 10-13

The facts of this exercise are based on the case of Public Constructors, Inc., v. State of New York (55 A.D.2d 368), heard before the Supreme Court of New York in 1977.

Public alleged that on a highway project, the state misrepresented the character of material Public encountered.

Was There a Positive Representation?

YES, the information given to Public showed primarily coarse materials that would shed water relatively easily.

Was There an Intent to Deceive?

NO.

Did the Conditions Differ from Those Represented?

YES, the materials encountered were fine-grained materials that were difficult to compact because of excessive moisture.

Was the Representation Complete?

NO.

Was Relevant Information Withheld?

YES, There were multiple subsurface investigations done on the site beginning in the early 1950s. Two in particular, the Albany report and the Binghamton report, gave conflicting views of the problems likely to be encountered. Only the most favorable report was provided to Public.

Was the Contractor Misled?

YES.

Should the Contractor Have Known of the Condition?

NO.

Was Reliance on the Information Justified?

YES.

Synopsis

Public showed that there was a misrepresentation and prevailed in court.

Solution: Exercise 10-14

The facts of this exercise are based on the case of E. H. Morrill Co. v. State of California (423 P.2d 551), heard before the Supreme Court of California in 1967.

The state stated the size and frequency of boulders, which Morrill alleged was incorrect.

Was There a Positive Representation?

YES, the specifications affirmatively stated that boulders varied "in size from one foot to four feet in diameter. The dispersion of boulders varies from approximately six feet to twelve feet in all directions, including the vertical."

Was There an Intent to Deceive?

NO.

Did the Conditions Differ from Those Represented?

YES, the boulders were larger and more numerous than represented.

Was the Representation Complete?

YES.

Was the Contractor Misled?

YES.

Was Reliance on the Information Justified?

YES.

Synopsis

The state tried to rely on the site visitation clause, but that notion was rejected as a way to overcome a positive representation. The court said,

> In its positive assertion of the nature of this much of the work (the Government) made a representation upon which the claimants had a right to rely without an investigation to prove its falsity....
>
> The responsibility of a governmental agency for positive representations it is deemed to have made through defective plans and specifications is not overcome by the general clauses requiring the contractor to examine the site, to check up the plans, and to assume responsibility for the work.... (United States v. Spearin 248 U.S. 132)

Morrill prevailed.

Solution: Exercise 11-1

The facts of this exercise are based on the case of Blount Brothers Corp. v. United States (872 F.2d 1003), heard before the U.S. Court of Appeals in 1989.

The dispute arose because Blount Brothers could not satisfy the color of aggregate specified by the contract.

What Caused the Failure?

Blount Brothers was unable to procure satisfactory tan and brown aggregate.

Who Had Control?

The contract allowed the contractor to procure aggregate from any source. Because the contractor had control, this was a performance directive.

Was Performance Impossible?

YES, Blount investigated quarries in Ohio, Indiana, Kentucky, Alabama, and Georgia, but all were unsatisfactory. They asked the government for a reference, but that quarry was also unsatisfactory. Because this effort was far beyond what was contemplated in the contract, finding satisfactory aggregate was commercially impractical.

Did the Contractor Assume the Risk?

The warranty clause was a "garden variety" materials and workmanship clause for which the contractor bore any unusual risk.

Synopsis

The court ruled in favor of Blount Brothers, saying,

> We hold, however, that Blount Bros. has carried the burden of establishing legal impossibility by proving the contract specifications as written were defective.

Solution: Exercise 11-2

The facts of this exercise are based on the case of Marine Colloids, Inc., v. M. D. Hardy, Inc. (433 A.2d 402), heard before the Supreme Judicial Court of Maine in 1981.

The dispute arose over a firewall that collapsed during a storm.

What Caused the Failure?

According to a report prepared by a referee, the wall collapsed because

> Marine Colloids' decision not to expand the Pilot House left the firewall "to stand exposed and unsupported on its north side and serving the purpose of an exterior end wall rather than the purpose of a firewall for which it was designed and built.

Who Had Control?

The decision to not expand the pilot house was Marine Colloids' alone, so the collapse was caused by a method directive.

Was It a Preconstruction Loss?

The collapse was not related to protecting a partially completed structure.

Was the Defect Patent?

When the defect became obvious, Hardy advised Marine Colloids, who chose to do nothing. The burden for the failure thus shifted to Marine Colloids.

Did the Contractor Deviate?

Hardy did not deviate.

Did the Contractor Assume the Risk?

The warranty clause said that Hardy would "guarantee soundness of construction..." But that guarantee was for an interior fire (curtain) wall, not as an end wall. Therefore, Hardy assumed no unusual risk.

Synopsis

The court found that Hardy was not responsible for fixing the wall.

Solution: Exercise 11-3

The facts of this exercise are based on the case of J. D. Hedin Construction Co. v. United States (347 F.2d 235), heard before the Court of Claims in 1965.

This case is often cited in delay claims, but there were three defective specification issues involved: pile driving, spread footers, and stormwater sewers. This particular exercise deals with the pile driving.

What Caused the Failure?

When the subcontractor drove piles adjacent to ones previously driven, the ones previously driven collapsed and failed. The soil was a glacial moraine consisting of earth, stone, sand, silt, clay, and gravel, with occasional boulders. It was very dense material. The cause of the failure was the fact that the thin-shelled piles could not withstand the pressures exerted from the driving of subsequent piles.

Who Had Control?

The owner (government) had control because they specified the wall thickness (0.05 in.). The contractor had no choice. Thus, it was a method directive.

Was It a Preconstruction Loss?

NO.

Was the Defect Patent?

NO.

Did the Contractor Deviate?

NO.

Did the Contractor Assume the Risk?

NO, there was a "garden variety" warranty clause. The subcontractor should be paid for the added costs.

Synopsis

Hedin prevailed.

_____

Solution: Exercise 11-4

The facts of this exercise are based on the case of J. D. Hedin Construction Co. v. United States (347 F.2d 235), heard before the Court of Claims in 1965.

This case is often cited in delay claims, but there were three defective specification issues involved: pile driving, spread footers, and storm sewers. This particular exercise deals with the spread footings.

What Caused the Failure?

The gravel stratum was not at the elevations thought.

Who Had Control?

The government, because they told the contractor where the bottom elevation of the footing was supposed to be. Thus, it was a method directive.

Was It a Preconstruction Loss?

NO.

Was the Defect Patent?

NO.

Did the Contractor Deviate?

NO.

Did the Contractor Assume the Risk?

NO, so the contractor should be paid for the additional work.

Synopsis

Hedin prevailed.

Solution: Exercise 11-5

The facts of this exercise are based on the case of J. D. Hedin Construction Co. v. United States (347 F.2d 235), heard before the Court of Claims in 1965.

This case is often cited in delay claims, but there were three defective specification issues involved: pile driving, spread footers, and storm sewers. This particular exercise deals with the storm sewer pipe.

What Caused the Failure?

The ground was too soft to support the pipe and fill.

Who Had Control?

The government, because they knew of the conditions and specified the pipe size and fill. The contractor had no choice. Thus, it was a method directive.

Was It a Preconstruction Loss?

NO.

Was the Defect Patent?

NO.

Did the Contractor Deviate?

NO.

Did the Contractor Assume the Risk?

NO, so Hedin should prevail.

Synopsis

Hedin prevailed.

Solution: Exercise 11-6

The facts of this exercise are based on the case of Republic Floors of New England v. Weston Racquet Club, Inc. (520 N.E.26 160), heard before the Appeals Court of Massachusetts in 1987.

The dispute arose over delamination, separation, and bubbles in an asphalt tennis court.

What Caused the Failure?

The bubbling and separation were due to vapor pressure, the fact that such pressure was endemic to a concrete floor, and the problem that adhesive flooring such as ChemTurf should not be used on concrete. There is also evidence that there was soap in the concrete before Republic's application and that the soap prevented bonding.

Who Had Control?

If the cause of the failure was the vapor pressure, then this is a method directive because the contract required ChemTurf. The owner had control. If the cause was the soap, then it is a performance directive because the contractor was likely not told how to clean the surface. Because the cause of the failure cannot be clearly established, both types of directives are investigated.

Was It a Preconstruction Loss?

NO, protection of incomplete work was not an issue.

Was the Defect Patent?

NO, if examined more in depth, it may have become known, but it does not appear readily obvious.

Did the Contractor Deviate?

NO.

Did the Contractor Assume the Risk?

YES, Condition A.2 in the contract says, "Republic will guarantee the ChemTurf installation..." The magic word is "guarantee," which amounts to a promise. Republic should be responsible. Thus, it is a performance directive.

Was Performance Impossible?

There is nothing to suggest that this work was impossible or commercially impractical.

Did the Contractor Assume the Risk?

The same logic as before is followed. Republic is responsible for the damages.

Synopsis

There was no court decision because the case was remanded to the lower court for further proceedings. However, the Appeals Court spent considerable time discussing the warranty clause.

Solution: Exercise 11-7

The facts of this exercise are based on the case of Rhone-Poulenc Rorer Pharmaceuticals, Inc., v. Newman Glass Works v. Spectrum Glass Products, Inc. (112 F.3d 695), heard before the U.S. Court of Appeals in 1996.

The issue was over delaminated opaque glass.

What Caused the Failure?

It appears that the failure was caused by defective glue used in the lamination process.

Who Had Control?

Because the contract specified the glass to be used, this is a method directive. The owner had control.

Was It a Preconstruction Loss?

NO, there are no indications that this was an issue.

Was the Defect Patent?

NO.

Did the Contractor Deviate?

NO.

Did the Contractor Assume the Risk?

NO. This is a manufacturer's defect, and Spectrum must supply new glass. However, under the warranty clause, Newman must install the glass. Any action Newman has against Spectrum is another matter.

Synopsis

In the court's decision, the subcontractor was required to correct the defects.

Solution: Exercise 11-8

The facts of this exercise are based on the case of American and Foreign Insurance Co. v. Bolt Construction Co. (106 F.3d 155), heard before the U.S. Court of Appeals in 1996.

The issue involved the negligence for a roof collapse caused by excessive snow and ice.

What Caused the Failure?

The failure was caused by the improper installation of purlins on a roof to strengthen it against excessive snow loads.

Who Had Control?

Because it was the manner of installation, and Bolt had allegedly been told how to proceed, this dispute is analyzed as a method directive.

Was It a Preconstruction Loss?

NO.

Was the Defect Patent?

YES, Bolt knew the manner of installation was wrong, but he proceeded anyway. Therefore, Bolt is responsible.

Synopsis

The court examined this issue on the basis of negligence rather than defective specifications. Nevertheless, Bolt was found to have been negligent. The court said,

> Thus, the liability imposed on the defendant's (Bolt) failure to exercise skill and care in the performance of his work as required, independent of the contractual requirements undertaken by this additional duty of care.
>
> ... In this case, Bolt knew the manner in which he installed the additional purlins was wrong. He also knew the purpose of the additional purlins was to strengthen the roof to prevent it from collapsing under the weight of ice and snow.

Case Table

A. Kaplen & Son, Ltd., v. Housing Authority of City of Passaic 126 A.2d 13 (1956).

Ace Stone, Inc., v. Township of Wayne 221 A.2d 515 (1966).

Aetna Casualty and Surety Co. v. Butte-Meade Sanitary Water District, 500 F.Supp. 193 (1980).

Airco Refrigeration Service v. Fink, 134 So.2d 880 (1961).

Al Johnson Construction Co. v. Missouri Pacific Railroad Co., 426 F.Supp. 639, 647 (8th Cir. 1976), aff'd 553 F.2d 103 (8th Cir 1976).

Allied Contractors, Inc., v. United States, 381 F.2d 995, 999, 180 Ct.Cl. 1057 (1967).

Alpert v. Commonwealth, 357 Mass. 306, 258 N.E.2d 755 (1970).

American Druggists Insurance Co. v. Henry Contracting, 505 So.2d 734 (1987).

Anthony P. Miller, Inc., v. Wilmington Housing Authority 165 F.Supp. 275 (1958).

Arundel Corp. v. United States, 103 Ct.Cl. 688, 712 (1945), cert. denied, 326 U.S. 752, rehearing denied, 326 U.S. 808 (1945).

Ashton Co. v. State, 9 Ariz. App. 564, 454 P.2d 1004 (1969).

B & B Cut Stone Co. v. Resneck, 465 So.2d 851 (1985).

B's Co., Inc., v. B. P. Barber and Associates, Inc., 391 F.2d 130, 137 (1968).

Ballou v. Basic Construction Co., 407 F.2d 1137 (1969).

Bartlett v. Stanchfield, 148 Mass. 394, 19 N.E. 549, 2 L.R.A. 625 (1889).

Baton Rouge Contracting Co. v. West Hatchie Drainage District, 304 F.Supp. 580, 587 (1969).

Beacon Construction Co. of Massachusetts v. United States, 314 F.2d 501 (1963).

Berg v. Kucharo Construction Co., 21 N.W.2d 561, 567 (1946).

Bergman Construction Corp., ASBCA 9000, 1964 BCA 4426.

Bethlehem Corp. v. United States, 462 f.2d 1400, 1404, 19 Ct.Cl. 247 (1972).

Bethlehem Steel Corp. v. City of Chicago, 234 F.Supp. 726 (1965).

Bilt-Rite Contractors, Inc., v. The Architectural Studio (Pennsylvania).

Blair v. United States, 66 F.Supp. 405, 405 (1946).

Blake Construction Co. v. C. J. Coakley Co., Inc. 431 A.2d 569 (1981).

Blake Construction v. Upper Occoquan Sewage Authority, Supreme Court of Virginia, 10/31/03.

Blankenship Construction Co. v. North Carolina State Highway Commission, 222 S.E.2d 452, 462, 28 N.C.App. 593 (1976).

Blauner Construction Co. v. United States, 94 Ct. Cl. 503, 511 (1941).

Blount Bros. Construction Co., ASBCA 4780, 1 G.C. 686, 59–2 BCA 2316.

Blount Brothers Construction Co. v. United States, 346 F.2d 962 (1965).

Dugan & Meyers Construction Co., Inc. v. State of Ohio. Ohio Court of Appeals.

E. C. Ernst, Inc. v. Koppers Co., Inc., 476 F.Supp. 729, 626 F.2d 324 (3rd Cir., 1980).

E. C. Ernst, Inc., v. General Motors Corp., 482 F.2d 1047 (5th Cir., 1973).

E. C. Nolan Co., Inc., v. State of Michigan 227 N.W.2d 323 (1975).

E. H. Morrill Co. v. State, 423 P.2d 551, 554, 56 Cal.Rptr. 479, 65 Cal.2d 787 (1967).

Earl T. Browder, Inc., v. County Court, 143 W.Va. 406, 102 S.E., 2d., 425 (1958).

Edsall Construction Co., Inc., ASBCA.

Elkan v. Sebastian Bridge District, 291 F. 532, 538 (8 Cir. 1923).

Elter, S.A., ASBCA 52327.

Emerald Forest Utility District v. Simonsen Construction, 679 SW.2d 51, 53, (Tex. App. 14 Dist. 1984).

F. D. Rich Co. v. Wilmington Housing Authority 392 F.2d 841 (1968).

Farnsworth & Chambers Co. v. United States, 171 Ct.Cl. 30, 376 F.2d 577 (1965).

Flippin Materials v. United States, 312 F.2d 408, 414, 160 Ct.Cl. 357 (1963).

Flour Mills of America, Inc., v. American Steel Building Co., 449 P.2d 861, 878 (1969).

Foster Construction C.A. & Williams Bros. Co. v. United States, 193 Ct. Cl 587, 602, 603, 604, 624, 435 F.2d 873 (1970).

Foundation Co. v. State of New York, 135 N.E. 236, 238/239, 233 N.Y. 177 (1922).

Foundation Intern v. E. T. Ige Construction 81 P.3d 1216, Supreme Court of Hawaii, 2003.

Fox v. Mountain West Electric, Inc.

Framlau v. United States, 568 F.2d 687 (1977).

Franchi Construction Co. v. United States, 609 F.2d. 984 (1979).

Frank Sullivan v. Midwest Sheetmetal Works, 335 F.2d 33 (1964).

Fru-Con v. U.S., 43 Fed. Cl. 306 (1999).

G. R. Osterland Co. v. Cleveland, Court of Appeals of Ohio, 11/20/00.

Gill Construction, Inc., v. 18th Vine Authority of Kansas City, Missouri.

Guy F. Atkinson Co., IBCA 385, 65–1 BCA 4642.

Hadden v. Consolidated Edison Co. of New York, Inc., 34 N.Y.2d 88 (1974).

Haehn Management Co. v. United States, 15 Cl.Ct. 50, 56 (1988).

Haggard Construction Co. v. Montana State Highway Commission, 149 Mont. 422, 427 P.2d 686 (1967).

Havens Steel Co. v. Randolph Engineering Co., 613 F.Supp. 514, 527 (D.C. Mo. 1985).

Hensel Phelps Construction Co. v. United States, 314 F.2d 501 (1963).

Hicks Construction Co. v. Town of Hudson, Wyoming, 390 F.2d 84 (1976).

Hoel-Steffen Construction Co. v. United States, 456 F.2d 760 (Ct.Cl., 1972).

Hoel-Steffen Construction Co. v. United States, 684 F.2d 843 (1982).

Hoffman v. United States, 166 Ct.Cl. 39, 340 F.2d 645 (1964).

Hollerbach v. United States, 233 U.S. 165, 169, 43 S.Ct. 533, 58 L.Ed. 898 (1914).

Housing Authority of Dallas v. J. T. Hubbell 325 S.W.2d 880 (1959).

Illinois Central Railroad Co. v. Manion, 113 Ky. 7, 67 S.W. 40, 101 Am. St. Rep. 345 (1902).

Interstate Contracting, Inc.. v. City of Dallas, Texas, U.S. Court of Appeals for the Fifth Circuit, 4/22/05.

J. A. Johnson & Son, Inc., v. State, 51 Hawaii 529, 465 P.2d 148 (1970).

J. A. Jones Construction Co. v. Lehrer McGovern Bovis, Inc., 89 P.3d 1009 (2004).

J. D. Hedin Construction Co. v. United States 408 F.2d 424.

Weeshoff Construction Co. v. Los Angeles County Flood Control District, 88 C.A.3d 579, 152 Cal.Rptr. 19 (1979).

Western Contracting Corp. v. Sooner Construction Co. 256 F.Supp. 163 (1966).

Western Engineers, Inc., v. Utah State Road Commission 437 P.2d 216 (1968).

Western Well Drilling Co. v. United States, 96 F.Supp 377, 379 (9 Cir 1951).

Westmont Country Club v. Kameny, 197 A.2d 379 (1964).

Wiebner v. Peoples, 142 P 1036 (1914).

Wiechmann Engineers v. California State Department of Public Works, 107 Cal.Rptr. 529, 535, 31 Cal.App.3d 741 (1973).

William F. Wilke, Inc., ASBCA Nos. 33,233, 33,748, 88–3 BCA 21,134.

Wilson v. Gregory, 322 So.2d 369 (1975).

Wise v. United States, 249 U.S. 361 (1919).

Woodcrest Construction Co. v. United States, 408 F.2d 406, 410, 187 Ct.Cl. 249 (1969), cert. denied, 398 U.S. 958, 90 S.Ct. 2164, 26 L.Ed.2d 542 (1970).

WPC Enterprises v. United States, 323 F.2d 874 (1964).

Wunderlich v. State of California, 423 P.2d 545, 548, 56 Cal.Rptr. 473, 65 Cal.2d 777 (1967).

Index

Page numbers in italics indicate illustrative material (figures, charts, and tables). Please check the table of cases for individual case law.

About the Authors

H. Randolph Thomas, Ph.D., P.E., is a professor of civil engineering at Pennsylvania State University. He received his Ph.D. in civil engineering from Vanderbilt University in 1976. For many years, Dr. Thomas has been teaching and conducting research in the areas of construction labor productivity and construction contracts. He has written extensively on both topics in various ASCE journals. In 1998, he was awarded a Fulbright scholarship to study construction practices in Eastern Europe and, in 2000, he received ASCE's Peurifoy Construction Research Award.

Ralph D. Ellis Jr., Ph.D., P.E., is a professor of civil engineering at the University of Florida, where he heads the Construction Engineering Management program. He has 15 years of industry experience as a construction projects manager and business manager. He received his MBA from Nova University in 1987 and his Ph.D. in civil engineering from the University of Florida in 1989. Dr. Ellis teaches and performs research in the legal aspects of construction and engineering project management.